INFERENTIAL STATISTICS and NUMERICAL ANALYSIS

Dr. Vandana Bagla (March, 2018)

Price: $38.50

Publisher: CreateSpace Independent Publishing Platform (A unit of Amazon)

Marketed by

Preface

It is my pleasure to present the first edition of the book titled 'Inferential Statistics and Numerical Analysis'. The book provides an insight into elementary inferential statistical methodologies including probability and sampling techniques; with basic numerical procedures of equation solving and interpolation. The content elaborates statistical and numerical methods that are significantly useful in science and engineering undergraduate courses; engrossing basic concepts and applications of statistical techniques and numerical methods in scientific, research and industrial applications. With this first edition, I have strived to meet the expectations of the students by explaining the text and related theory right from the elementary level. With a substantial emphasis on conceptual knowledge, the book provides working methodologies with sufficient number of illustrative examples. Practicing students and budding scientists will find it pragmatic in research related application problems. Level of the book has been kept moderately rudimentary and plain sailed to provide its' readers with lucidity and perceptibility. An exuberant attempt has been made for the students to comprehend and develop an envision for the concepts exemplified. It is hoped that this book would be determinedly advantageous to the tutees and prove to be serviceable.

Dr. Vandana Bagla

B.Sc. (Computers), M.Sc. (Mathematics), MBA (HR), Ph.D. (Operations Research)

Assistant Professor at Maharaja Agrasen Institute of Technology,

Guru Gobind Singh Indraprastha University, Delhi, India

DEDICATED TO MY LOVING FATHER
MR. JAGAN NATH DHINGRA

Contents

Chapter 1: Introduction to Inferential Statistics Numerical Methods 9

1.1 Inferential Statistics 9

    1.1.1 Measures of central tendencies 9

        1.1.1.1 Mean 9

        1.1.1.2 Median 10

        1.1.1.3 Mode 10

    1.1.2 Measures of Variation (Dispersion) 10

        1.1.2.1 Range 10

        1.1.2.2 Quartile Deviation 11

        1.1.2.3 Standard Deviation 11

        1.1.2.4 Variance 11

1.2 Numerical Methods 11

    1.2.1 Approximate Numerals 12

    1.2.2 Types of Errors 13

    1.2.3 Calculation of Errors 13

Chapter 2: Theory of Probability 14

2.1 Introduction 14

2.2 Mathematical Definition of Probability 14

2.3 Addition Law of Probability 18

2.4 Conditional Probability 20

    2.4.1 Multiplicative Law of Probability 20

2.5 Bay's Theorem 22

2.6 Random Variable and Probability Distributions 25

    2.6.1 Discrete Probability Distributions 25

    2.6.2 Continuous Probability Distributions 28

2.7 Moments 30

2.8 Moment Generating Function (MGF) 31

2.9 Skewness 32

2.10 Kurtosis 33

Chapter 3: Theoretical Probability Distributions 36

3.1 Introduction 36

3.2 Binomial Distribution 36

3.2.1 Mean of Binomial Distribution...36

3.2.2 Variance of Binomial Distribution..36

3.2.3 Moment Generating Function of Binomial Distribution 37

3.3 Poisson Distribution 42

3.3.1 Mean of Poisson Distribution...43

3.3.2 Variance of Poisson Distribution..43

3.3.3 Moments of Poisson Distribution 43

3.4 Normal Distribution 46

3.4.1 Moments of Normal Distribution 48

Chapter 4: Curve Fitting & Correlation 54

4.1 Introduction 54

4.2 Principle of Least Squares 54

4.2.1 Fitting a straight line 55

4.2.2 Fitting a parabola 57

4.2.3 Change of Scale 59

4.3 Correlation 61

4.4 Karl Pearson Coefficient of Correlation...61

4.4.1 Shortcut Method for Karl Pearson Coefficient of Correlation 62

4.4.2 Coefficient of Correlation of Bivariate Frequency Distribution 63

4.4.3 Coefficient of Correlation by Rank differences 64

4.5 Linear Regression 68

4.5.1 Properties of Regression Coefficients 69

4.5.2 Angle between the Lines of Regression 70

Chapter 5: Sampling Distributions & Hypothesis Testing 78

5.1 Introduction 78

5.2 Sampling Distributions 80

5.3 Sampling of Attributes for large samples ($n > 30$) 81

5.3.1 Comparing Proportions of Large Samples from Two Different Populations in Terms of Attributes 83

5.4 Sampling Distributions of Sample means 84

Chapter 6: Tests of Significance for Small Samples 93

6.1 Introduction 93

6.2 Student's t-Distribution...93

6.3 Snedecor's F – test for Testing Equality of Two Population Variances 99

6.4 Fisher's Z Test for Testing Significance of Correlation Coefficient for Small Samples 101

Chapter 7: Tests of significance based on Chi-Square 104

7.1 Chi-Square (χ^2) Test 104

7.2 Chi-Square (χ^2) Test for Testing Significance of Sample Variance 104

7.3 Chi-Square (χ^2) Test for Testing Goodness of Fit 105

7.4 Chi-Square (χ^2) Test for Testing Independence of Attributes 107

Chapter 8: Analysis of Variance 112

8.1 Introduction 112

8.2 One-Way ANOVA 112

8.3 Two-Way ANOVA 119

Chapter 9: Numerical Solutions of Algebraic and Transcendental Equations 125

9.1 Introduction 125

9.2 Numerical methods to find roots of algebraic and transcendental equations 125

    9.2.1 Direct Iteration Method 125

    9.2.2 Bisection Method (or Bolzano Method) 127

    9.2.3 Regula- Falsi Method (Geometrical Interpretation) 130

    9.2.4 Newton-Raphson Method (Geometrical Interpretation) 133

        9.2.4.1 Generalized Newton's Method for Multiple Roots 137

        9.2.4.2 Convergence of Newton Raphson Method 138

    9.2.5 Secant Method 138

9.3 Iterative Methods for Solving Simultaneous Linear Equations 139

    9.3.1 Gauss-Jacobi's Iteration Method 140

    9.3.2 Gauss-Seidal's Iteration Method 142

Chapter 10: Finite Differences 145

10.1 Introduction 145

10.2 Shift or Increment Operator (E) 145

10.3 Differencing Operators 145

    10.3.1 Forward Difference Operator Δ 145

    10.3.2 Backward Difference Operator ∇ 147

    10.3.3 Central Difference Operator δ 147

    10.3.4 Averaging Operator μ 148

10.4   Missing values of Data 156

10.5 Finding Differences Using Factorial Notation 157

  10.5.1 Factorial Notation of a Polynomial 157

10.6 Series Summation Using Finite Differences 159

Chapter 11: Interpolation 163

11.1 Introduction 163

11.2 Interpolation within Equal Intervals 163

  11.2.1 Newton's Forward Interpolation Formula 163

  11.2.2 Newton's Backward Interpolation Formula 166

  11.2.3 Gauss's Forward and Backward Difference Formulae 170

  11.2.4 Stirling's Central Difference Formula 172

11.3 Interpolation with Unequal Intervals 177

  11.3.1 Newton's Divided Difference Method 177

  11.3.1.1 Newton's Divided Difference Formula 177

  11.3.2 Lagrange's Interpolation Formula 181

Chapter 12: Numerical Differentiation & Integration 185

12.1 Introduction 185

12.2 Numerical Differentiation 185

  12.2.1 Derivatives Using Newton's Forward Interpolation Formula 185

  12.2.2 Derivatives Using Newton's Backward Interpolation Formula 186

  12.2.3 Derivatives Using Stirling's Interpolation Formula 187

  12.2.4 Derivatives of Polynomial with Unequispaced x_i's 187

  12.2.5 Maxima and Minima of a Tabulated Function: 194

12.3 Numerical Integration 196

  12.3.1 Numerical Integration Using Trapezoidal Rule 197

  12.3.2 Numerical Integration Using Simpson's One-Third Rule 199

  12.3.3 Numerical Integration Using Simpson's Three- Eighth Rule 201

  12.3.4 Applications of Numerical Integration 202

Chapter 13: Numerical Solutions of Ordinary Differential Equations 208

13.1 Introduction 208

13.2 Single Step Methods to Solve Initial Value Problems 208

  13.2.1 Taylor Series Method 209

  13.2.2 Picard's Method of Successive Approximations 211

  13.2.3 Euler's Method 214

  13.2.4 Modified Euler's Method 215

13.2.5 Runge- Kutta's Method 218

13.3 Multistep Methods 221

13.3.1 Milne's Predictor Corrector Method 222

13.3.2 Adams- Bashforth Method 228

13.4 Simultaneous Differential Equations of First Order 232

13.5 Second Order Differential Equations 236

13.6 Boundary Value Problems (BVPs): 238

13.6.1 Finite Differences Method 238

Appendix 243

Scilab code to add two matrices 243

Scilab code to multiply two matrices 243

Scilab code to find matrix transpose 244

Scilab code to find the inverse of a matrix 245

Scilab code to find eigenvalues of a matrix 246

Scilab code to find mean, S.D. and first r moments about mean of a grouped data 247

Scilab code to find mean, S.D. and first r moments about mean of an ungrouped grouped data ... 248

Scilab code to find a line of regression 249

Scilab code to fit a straight line to given set of data points 249

Scilab code to fit a parabola to given set of data points 250

Scilab code to find roots of an equation using Bisection method 251

Scilab code to find roots of an equation using N-R method 251

Scilab code to find roots of an equation using Regula Falsi method 252

Scilab Code for numerical integration using Simpson's 1/3 rule 252

Scilab Code for numerical integration using Simpson's 3/8 rule 253

Scilab Code for solving initial value problems using Euler's method 254

Scilab Code for solving initial value problem using Runge Kutta method 254

Table1: Area Under Standard Normal Curve 256

Table2: Standard Normal Cumulative Probability Table 257

Table3: t-table 258

Table4: Chi-Square ($\chi 2$) Distribution Table 259

Table 5: Critical Values of the F Distribution with Alpha Level of 0.05 260

Chapter 1: Introduction to Inferential Statistics Numerical Methods

Introduction to Inferential Statistics
&
Numerical Methods

1.1 Inferential Statistics

Statistics is an important branch of mathematics dealing with data collection, analysis and interpretation. Statistics has vast scientific and research applications and can be divided into two parts: descriptive statistics and inferential statistics. Both of these are employed in data analysis and are of equal importance in practical applications.

By definition an inference is an educated guess or conclusion that is drawn from evidences and reasoning. Inferential statistics as the name suggests, involves drawing the conclusions from the statistical analysis that has been performed using descriptive statistics. Inferences are made about central tendency and variability or other aspects of a distribution and this is generally done through random sampling.

1.1.1 Measures of central tendencies

Statisticians often use common parameters to describe a set of observations which measure how the set is centered around a particular point on a line scale. This category of parameters is called measures of central tendencies and commonly used measures are mean, median and mode. The most reliable statistical parameter from this category is the mean or average.

1.1.1.1 Mean

The mean is the arithmetic average of a set of observations. The mean gives an idea where the center lies for a set of observations. The arithmetic mean is obtained by taking the sum of all the observations in the set and dividing by the total number of observations in the set.

Mean of a sample is said to be an unbiased estimator of the population mean and is highly useful in more advanced statistical interpretations drawn in inferential statistics. It is an inevitable parameter of inferential statistics.

For ungrouped data: Mean $(\bar{x}) = \frac{\sum_{i=1}^{n} x_i}{n} = \frac{x_1 + x_2 + \cdots + x_n}{n}$

For grouped data: Mean $(\bar{x}) = \frac{\sum f_i x_i}{\sum f_i}$, $\sum f_i = n$ **(In case of frequency distribution)**

Mean $(\bar{x}) = a + \frac{\sum f_i d_i}{\sum f_i} \times h$, $\sum f_i = n$ **(In case of class intervals)**

here a is the assumed mean and h is height of the class interval and $d_i = \frac{x_i - a}{h}$.

1.1.1.2 Median

Although mean is the most widely used measure of central tendency, it is not always appropriate to use it; especially when the set of observations has few one sided extreme (highest or lowest) values. Mean is strongly influenced by the few peripheral observations in skewed distributions. In such cases mean may not truthfully represent the central tendency of the set of observations because it has been raised (or lowered) by a few outlying scores. In such situations median may be a better measure of central tendency. The median value in a set of observations is that value which divides the set into equal halves when all the numbers have been arranged in ascending or descending order.

For ungrouped data: After arranging the data in ascending or descending order,

$$\text{Median} = \begin{cases} \text{middle term,} & \text{when } n \text{ is odd} \\ \text{mean of two middle terms,} & \text{when } n \text{ is even} \end{cases}$$

For grouped data: Median class is determined by class in which $\frac{n}{2}$ lies

$$\text{Median} = l + \frac{\frac{n}{2} - c}{f} \times h$$

where l is the lower limit of the median class, c is the cumulative frequency before the median class, f is the frequency of the median class, h is height of the class interval.

1.1.1.3 Mode

The mode is another measure of central tendency and is the most frequently occurring term in the set of observations. Mode of a set of observations may not be unique.

For ungrouped data: Mode of a raw data is the term which has highest frequency.

For ungrouped data: Modal class is the class having highest frequency

$$\text{Mode} = l + \frac{f_m - f_{m-1}}{(f_m - f_{m-1}) + (f_m - f_{m+1})} \times h$$

where l is the lower limit of the modal class, f_m is the frequency of the modal class, f_{m-1} is the frequency of the class before the modal class, f_{m+1} is the frequency of the class after the modal class, h is height of the class interval.

1.1.2 Measures of Variation (Dispersion)

Apart from measuring central tendencies to summarise a set of data, we need to measure the degree of variation (dispersion) of the set to determine its reliability. Measures of central tendency are useful in understanding how the observations cluster about a central value and measures of dispersion are useful in understanding how far the extreme (high or low) observations are scattered about the central value of the set.

Common measures of variation (dispersion) are: range, quartile deviation, standard deviation and variance.

1.1.2.1 Range

The range is the simplest of the measures of variation and describes the difference between the lowest value and the highest value in a set of observations.

Range = (largest value of the set) − (smallest value of the set)

For grouped data, the range is the difference between the upper bound of the highest class interval and the lower bound of the lowest class interval.

Range is very simple to use but it provides relatively little information on dispersion as it is highly influenced by an extremely large or an extremely small observation in the data set.

1.1.2.2 Quartile Deviation

If Q_1, Q_2, Q_3 are the quartiles which divide the data (arranged in ascending or descending order) into four equal parts, then Inter quartile range = $Q_3 - Q_1$.

Moreover

Q_1 is the median of the lower half of the data
Q_2 is the median of the whole set of data
Q_3 is the median of the upper half of the data

Quartile Deviation ($Q.D.$) is usually taken as measure of dispersion for the data whose central tendency is measured as median and is given by $Q.D. = \frac{1}{2}(Q_3 - Q_1)$.

1.1.2.3 Standard Deviation

Along with the mean, the standard deviation is an inevitable part of inferential statistics. Quartile deviation is used in association with the median whereas standard deviation is measured for data whose central tendency is taken as mean. The standard deviation gives an approximate picture of the average amount each number in a set varies from the central value (mean).

Standard deviation (σ) is defined as root mean square deviation, i.e. $\sigma = \sqrt{\frac{\sum(x_i - \bar{x})}{n}}$

In case of grouped data

In case of frequency distribution

$$\sigma = \sqrt{\frac{\sum f_i (x_i - \bar{x})}{\sum f_i}}, \quad \sum f_i = n$$

In case of class intervals

$$\sigma = \sqrt{\frac{\sum f_i d_i^2}{\sum f_i} - \left(\frac{\sum f_i d_i}{\sum f_i}\right)^2} \times h, \; d_i = \frac{x_i - a}{h}, \; h \text{ is height of the class interval}$$

1.1.2.4 Variance

Variance is the square of standard deviation. It is denoted by σ^2

1.2 Numerical Methods

There are mainly three types of techniques for problem solving: experimental, analytical, and numerical approximations. Experimental methodologies are workable for practical problems and are not only expensive and time consuming but also inflexible for parameter variations. Analytical methods provide exact solutions, but at times they are rigorous and complex; even inapplicable for many types of problems.

Numerical analysis puts in quantitative approximations to solve mathematical problems, providing sufficient accuracy and can be reckoned where analytical methods fail to provide results. Most numerical methods are iterative procedures; improving the solution in each step;

until the desired accuracy is achieved. Numerical procedures are well applicable in following situations:

- When presented with problems which cannot be solved using analytical methods, numerical methods attempt to replace them with nearby problems that can be solved more efficiently. For example numerical methods are well posed for solving complex algebraic equations and also they are apt for solutions of transcendental equations. Another good application is the use of interpolation formulae in deriving numerical differentiation & integration methods and also developing predictor corrector techniques for solving ordinary differential equations. Numerical techniques extend their applications in solving partial differential equations using finite differences and many more associable techniques.
- Numerical techniques are also associated with development of efficient techniques which can reduce the computation time taken by traditional methods. For instance the use of Gauss elimination method to solve a linear system of equations $AX = B$ containing more than three equations will require lesser time compared to traditional methods. Another good application is the use of finite difference method in solving boundary value problems which is worth replacing traditional techniques.
- Numerical analysis techniques are also concerned with stability of the solution; a concept referring to the sensitivity of a problem to small changes in the data values. Numerical methods for solving mathematical problems should be efficient for handling sensitivity issues; which sometimes attempt to destabilize the solution. For example Newton-Raphson method having quadratic convergence, tends to diverge for sensitivity issues and needs to be taken care off.

Numerical techniques being based on approximations; are also associated with allied errors and are to be modeled efficiently to keep errors within admissible range. While approximating any problem, it is prudent to understand the nature of the error in the computed solution. Moreover, understanding the error propagation allows evolutionary corrective techniques to improve the solution. Workable area is; what types of errors are propagated and how much tolerance is admissible in the solution domain.

1.2.1 Approximate Numerals

We have been using approximations for rational and irrational numbers traditionally; for instance $\frac{1}{3}, \sqrt{2}, \pi$ are all approximate numbers. Sometimes we may get exact number but with capped number of digits; concept of approximation comes into picture in all such concerns.

Approximations can be done in two ways:

i. **Truncation:** In case of truncation, digits are ignored (or chopped) after required number of figures.

ii. **Rounding:** Rounding off a number to given number of decimal places would mean to discard the digits after a specified place, and also to moderate the last digit at unit place.

For example truncating number 345.468 to 5 significant digits would mean to erase the figures after five digits, i.e. truncated number is 345.46, while rounding of a number to 5 significant digits means to ignore digits after 5 digits after rounding off current unit positioned digit; resulting in rounded off number as 345.47. Some more examples are given below:

7.4327596376 rounded to 3 decimal places gives 7.433 and truncation gives 7.432

32.7595376 rounded to 3 decimal places gives 32.760 and truncation gives 32.759

4327.6376 rounded to 1 decimal place gives 4327.6 and truncation also gives 4327.6

Truncation or rounding off procedures provide approximate numbers and induce some amount of errors as discussed below:

1.2.2 Types of Errors

Numerical errors arise from the use of approximations for desired results. Error propagation has inherent or computational reasons.

Inherent Errors: Many times errors are present in statement of the results; if they have been deduced on the basis of experimental data or taken up to a set precision; for example Euler's method to solve ordinary differential equations, which is equivalent to taking up two terms of the Taylor's series. These are also induced due to use of set tables such as log tables or statistical tables etc.

Round off or Truncation Errors: These errors arise due to rounding or truncating a number to specific number of digits and also when an infinite series is approximated by finite one.

For example Taylor's series expansion of a function $f(x)$ is given by:

$$f(x+h) = f(x) + hf'(x) + \frac{h^2}{2!}f''(x) + \cdots + \frac{h^n}{n!}f^{(n)}(x) + O(h^{n+1})$$

Here $O(h^{n+1})$ means that; if numerical addition is done till term containing h^n, we will end up with an error of order h^{n+1}; higher the value of h, greater the emanated error. In numerical computations, each step propagates a truncation error which is aggregated in subsequent iterations.

1.2.3 Calculation of Errors

We cannot exactly compute the errors associated with numerical methods; unless the true solution is known. But as a general rule; the degree of accuracy should be half a unit each side of the unit of measurement. For example a rope is measured as 112.53 meters long, with an accuracy of 0.01 of a meter, this would mean that it could have an error up to 0.005 meters either way.

∴ Length = 112.53 ± 0.005

This means that length could really be anywhere between 112.525 and 112.535 meters.

By definition; error in any result may be defined as:

True Value = Approximation + Error

or Error (E) = True Value − Approximation

Absolute Error: Absolute error is the magnitude of the error value.

Absolute Error = |Error|

Relative Error: It is comparative error to true value.

Relative Error = $\frac{|\text{Error}|}{|\text{True Value}|}$

Percentage Error: It is Relative error multiplied by 100.

∴ Percentage Error = $\frac{|\text{Error}|}{|\text{True Value}|} \times 100$

Chapter 2: Theory of Probability

Theory of Probability

2.1 Introduction

The numerical measure of certainty of an event is called probability. The probability of any event lies between 0 and 1. Probability of a sure event is 1 while that of an impossible event is 0.

Sample Space: The set of all possible outcomes associated with an experiment is called sample space. For example while tossing a coin, the sample space is $S = \{H, T\}$ and while tossing two coins $S = \{HH, HT, TH, TT\}$, whereas while rolling a die $S = \{1,2,3,4,5,6\}$.

Event: An event is the subset of a sample space. For example getting an odd number while rolling a die is an event $E = \{1,3,5\}$.

Mutually Exclusive Events: Two or more events $E_1, E_2, \ldots, E_n$ in a sample space are mutually exclusive if they have no point in common i.e. if $E_1 \cap E_2 \cap \ldots \cap E_n = \phi$. Getting an odd number and getting an even number while rolling a die are two mutually exclusive events.

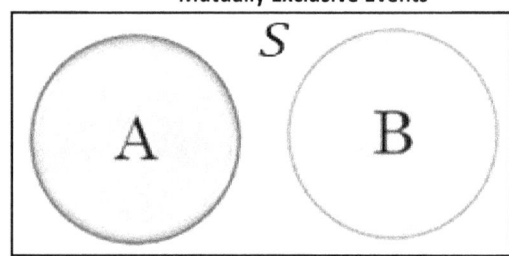

Exhaustive Events: Two or more events are called exhaustive events if at least one of them occurs when an experiment is performed. For example while tossing a coin; there are two exhaustive cases head and tail. Also while rolling a die; getting an odd number and getting an even number are two exhaustive events. Whereas getting a number less than two and getting a number greater than two, in case of rolling a die are non-exhaustive events.

Mutually Exclusive and Exhaustive Events:

If events $E_1, E_2, \ldots, E_n$ in a sample space are mutually exclusive and exhaustive then

$$P(E_1) + P(E_2) + \ldots + P(E_n) = 1$$

Equally Likely Events: Two or more events are equally likely if they have same probability of occurrence. Getting an odd number and getting an even number while rolling an unbiased die are two mutually exclusive and equally likely events.

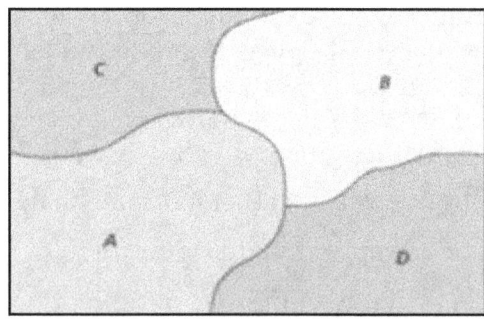

Independent Events: Two events are said to be independent if the occurrence or non-occurrence of one does not affect the probability of occurrence of the other. Mathematically the events E_1 and E_2 are independent if and only if $P(E_1 \cap E_2) = P(E_1) \times P(E_2)$.

If two events are independent; they cannot be mutually exclusive and vice-versa.

2.2 Mathematical Definition of Probability

If a trial results in n exhaustive, mutually exclusive and equally likely events and m of them are favorable to happening of an event E; then probability of happening of E is given by:

$$P(E) = \frac{\text{Favorable number of cases}}{\text{Exaustive number of cases}} = \frac{n(E)}{n(S)} = \frac{m}{n}$$

Example 1 Give an example of two events which are mutually exclusive but not exhaustive.

Solution: In an experiment of tossing two coins,

Let E_1: Getting two heads

E_2: Getting two tails

Then $S = \{HH, HT, TH, TT\}$

$E_1 = \{HH\}$ and $E_2 = \{TT\}$

$\therefore E_1 \cap E_2 = \phi$ but $P(E_1) + P(E_2) \neq 1$

Hence the events E_1 and E_2 are mutually exclusive but not exhaustive.

Example 2 Give an example of two events in an experiment of tossing two coins, which are mutually exclusive and exhaustive.

Solution: In an experiment of tossing two coins,

Let E_1: Getting at least one head

E_2: Getting two tails

Then $S = \{HH, HT, TH, TT\}$

$E_1 = \{HH, HT, TH\}$ and $E_2 = \{TT\}$

$\therefore E_1 \cap E_2 = \phi$ and $P(E_1) + P(E_2) = 1$

Hence the events E_1 and E_2 are mutually exclusive and exhaustive.

Example 3 An urn contains 5 white, 6 red and 4 black balls. Two balls are drawn at random. Find the probability that both are red. Also find the probability of one white and one black ball.

Solution: White Balls: 5, Red Balls: 6, Black Balls: 4

Let E: Both balls are red

$P(E) \equiv P(\text{red ball}) \text{ and } P(\text{red ball})$

$\Rightarrow P(E) = \frac{6}{15} \times \frac{5}{14} = \frac{1}{7}$

Let F: One white and one black ball are drawn

$P(F) \equiv \big(P(\text{white ball}) \text{ and } P(\text{black ball})\big) \text{ or } \big(P(\text{black ball}) \text{ and } P(\text{white ball})\big)$

$\Rightarrow P(F) = \frac{5}{15} \times \frac{4}{14} + \frac{4}{15} \times \frac{5}{14} = \frac{4}{21}$

Note: Questions in which replacement is not allowed can be attempted in a better manner using combinations.

Using combinations, $P(E) = \frac{6C_2}{15C_2} = \frac{6 \times 5}{15 \times 14} = \frac{1}{7}$

$P(F) = \frac{5C_1 \times 4C_1}{15C_2} = \frac{5 \times 4 \times 2!}{15 \times 14} = \frac{4}{21}$

Example 4 An urn contains 4 white, 5 red and 6 black balls. Three balls are drawn at random. Find the probability that balls are white, red and black.

Solution: Given that: white balls are 4, red balls are 5 and black balls are 6

Let E: Balls are white, red and black

$P(E) = \frac{4C_1 \times 5C_1 \times 6C_1}{15C_3} = \frac{4 \times 5 \times 6 \times 3!}{15 \times 14 \times 13} = \frac{24}{91}$

Example 5 Four cards are drawn without replacement from a well shuffled pack of 52 cards. Find the probability that
   (i)   All cards are spades
   (ii) There are two spades and two hearts
   (iii) All cards are black

Also compute the probabilities if four cards are drawn with replacement.

Solution: Let E_1: All cards are spades
E_2: There are two spades and two hearts
E_3: All cards are black

$$P(E_1) = \frac{13_{C_4}}{52_{C_4}} = \frac{13 \times 12 \times 11 \times 10 \times 4!}{52 \times 51 \times 50 \times 49 \times 4!} = \frac{11}{4165}$$

$$P(E_2) = \frac{13_{C_2} \times 13_{C_2}}{52_{C_4}} = \frac{13 \times 12 \times 13 \times 12 \times 4!}{52 \times 51 \times 50 \times 49 \times 2! \times 2!} = \frac{468}{20825}$$

$$P(E_3) = \frac{26_{C_4}}{52_{C_4}} = \frac{26 \times 25 \times 24 \times 23 \times 4!}{52 \times 51 \times 50 \times 49 \times 4!} = \frac{46}{833}$$

Again if the four cards are drawn with replacement

$$P(E_1) = \frac{13}{52} \times \frac{13}{52} \times \frac{13}{52} \times \frac{13}{52} = \frac{1}{256}$$

$$P(E_2) = \frac{13}{52} \times \frac{13}{52} \times \frac{13}{52} \times \frac{13}{52} \times \frac{4!}{2! \times 2!} = \frac{3}{128}$$

(Here $\frac{4!}{2! \times 2!}$ denotes the number of arrangements (permutations) of spades and hearts, which are SSHH or HHSS or SHSH or SHHS or HSHS or HSSH where S denotes Spade and H denotes Heart)

$$P(E_3) = \frac{26}{52} \times \frac{26}{52} \times \frac{26}{52} \times \frac{26}{52} = \frac{1}{16}$$

Example 6 A pair of dice is rolled. What is the probability of sum 7?

Solution: Let E: Getting a sum 7, when a pair of dice is rolled

$$S = \begin{Bmatrix} (1,1),(1,2),\dots,(1,6),(2,1),(2,2),\dots,(2,6),(3,1),(3,2),\dots,(3,6), \\ (4,1),(4,2),\dots,(4,6),(5,1),(5,2),\dots,(5,6),(6,1),(6,2),\dots,(6,6) \end{Bmatrix}$$

$E = \{(1,6),(2,5),(3,4),(4,3),(5,2),(6,1)\}$

$P(E) = \frac{n(E)}{n(S)} = \frac{6}{36} = \frac{1}{6}$

Example 7 Only 3 events A, B, C can happen. Given that chance of A is one-third that of B and odds against C are 2:1, find odds in favor of A.

Solution: Given $P(A) + P(B) + P(C) = 1$ ...①

Also $P(A) = \frac{1}{3} P(B) \Rightarrow P(B) = 3 P(A)$ ...②

And $P(C) = \frac{1}{3}$ ...③

Using ②, ③ in ①

$\Rightarrow P(A) + 3P(A) + \frac{1}{3} = 1$

$\Rightarrow 4P(A) = \frac{2}{3}$

$\Rightarrow P(A) = \frac{1}{6}$ and $P(\bar{A}) = \frac{5}{6}$

Hence odds in favour of A are 1:5

Example 8 A problem in mathematics is given to three students A, B and C whose chances of solving are $\frac{2}{3}, \frac{1}{2}$ and $\frac{1}{3}$ respectively. What is the probability that the problem will be solved?

Solution: Probability of A solving the problem $P(A) = \frac{2}{3} \Rightarrow P(\bar{A}) = \frac{1}{3}$

Probability of B solving the problem $P(B) = \frac{1}{2} \Rightarrow P(\bar{B}) = \frac{1}{2}$

Also probability of C solving the problem $P(C) = \frac{1}{3} \Rightarrow P(\bar{C}) = \frac{2}{3}$

Now probability that A, B and C do not solve the problem is $\frac{1}{3} \times \frac{1}{2} \times \frac{2}{3} = \frac{1}{9}$

∴ The probability that the problem is solved $= 1 - $ (Probability that problem is not solved)

$$= 1 - \frac{1}{9} = \frac{8}{9}$$

Example 9 A bag contains 50 tickets numbered from 1 to 50, out of which 5 are drawn at random and arranged in ascending order ($t_1 < t_2 < t_3 < t_4 < t_5$). Find the probability of t_4 carrying the number 45.

Solution: Exhaustive number of cases $= 50_{C_5}$.

To follow the given pattern, three tickets t_1, t_2, t_3 must be drawn out of tickets numbered from 1 to 44 with favorable number of cases $^{44}C_3$, then t_4 is drawn bearing number 25 which has only one favourable case $^1C_1$, and then t_5 is drawn out of remaining 5 tickets with favourable number of cases $^5C_1$.

∴ Required probability is $\frac{^{44}C_3 \times ^1C_1 \times ^5C_1}{50_{C_5}} = \frac{44 \times 43 \times 42 \times 5 \times 5!}{50 \times 49 \times 48 \times 47 \times 46 \times 3!} = 0.03$

Example 10 A has two shares in lottery in which there is 2 prizes and 3 blanks; B has three shares in a lottery in which there are 3 prizes and 6 blanks. Compare the probability of $A's$ success to that of $B's$ success.

Solution: Probability of A not getting a prize in two shares $= \frac{3_{C_2}}{5_{C_2}} = \frac{3 \times 2}{5 \times 4} = \frac{3}{10}$

∴ Probability of A getting a prize $= \frac{7}{10}$

Probability of B not getting a prize in three shares $= \frac{6_{C_3}}{9_{C_3}} = \frac{6 \times 5 \times 4}{9 \times 8 \times 7} = \frac{5}{21}$

∴ Probability of B getting a prize $= \frac{16}{21}$

Hence the probability of $A's$ success to that of $B's$ success $= \frac{7}{10} : \frac{16}{21}$

$$= 147 : 160$$

Example 11 What is the probability that a leap year selected at random will have 53 Mondays.

Solution: A leap year has 366 days, having 52 full weeks and 2 extra days.

∴ $S = \{$(Monday, Tuesday), (Tuesday, Wednesday), (Wednesday, Thursday),

(Thursday, Friday), (Friday, Saturday), (Saturday, Sunday), (Sunday, Monday)$\}$

Let E: The leap year selected will have 53 Mondays.

$E = \{(\text{Monday}, \text{Tuesday}), (\text{Sunday}, \text{Monday})\}$

$\therefore P(E) = \frac{n(E)}{n(S)} = \frac{2}{7}$

Example 12 A and B alternatively throw a die until one gets a success and wins the game, where success is defined as getting a six. What are their respective probabilities of winning if A takes the first trial?

Solution: If success (S) is getting a six, then $P(S) = \frac{1}{6}$, and if F denotes failure, then $P(F) = \frac{5}{6}$.

Now A can win in $1^{st}, 3^{rd}, 5^{th}, \cdots$ trials, i.e. getting S or FFS or FFFFS, $\cdots$

$\therefore$ A's probability of winning $P(A) = \frac{1}{6} + \frac{5}{6} \cdot \frac{5}{6} \cdot \frac{1}{6} + \frac{5}{6} \cdot \frac{5}{6} \cdot \frac{5}{6} \cdot \frac{5}{6} \cdot \frac{1}{6} + \cdots$

This is a G.P. with $a = \frac{1}{6}$ and $r = \frac{5}{6} \cdot \frac{5}{6} = \frac{25}{36}$

$\therefore P(A) = \frac{a}{1-r} = \frac{\frac{1}{6}}{1 - \frac{25}{36}} = \frac{6}{11}$, $P(B) = 1 - \frac{6}{11} = \frac{5}{11}$

2.3 Addition Law of Probability

Statement: If A and B be any two events, then

$P(A \cup B) = P(A) + P(B) - P(A \cap B)$

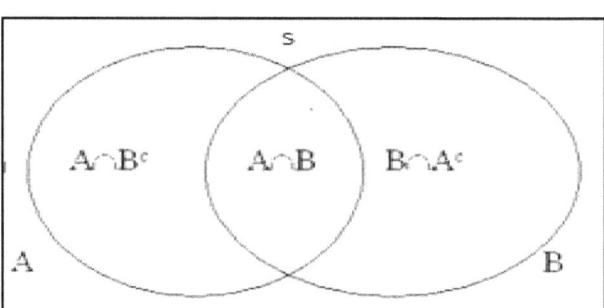

Proof: $A \cup B = A \cup (B \cap A^C)$

$\Rightarrow P(A \cup B) = P(A \cup (B \cap A^C))$

$= P(A) + P(B \cap A^C)$

($\because$ Both A and $B \cap A^C$ are disjoint)

$= P(A) + [P(B \cap A^C) + P(A \cap B)] - P(A \cap B)$

$= P(A) + P[(B \cap A^C) \cup (A \cap B)] - P(A \cap B)$

$= P(A) + P(B) - P(A \cap B)$ $\because (B \cap A^C) \cup (A \cap B) = B$

Hence proved

> If A and B are mutually exclusive events, $A \cap B = \phi$

$\Rightarrow P(A \cup B) = P(A) + P(B)$

If A, B, C are 3 events, then

$P(A \cup B \cup C) = P(A) + P(B) + P(C) - P(A \cap B) - P(B \cap C) - P(C \cap A) + P(A \cap B \cap C)$

Example 13 Find the probability of getting a spade or an ace when a card is drawn from a well shuffled pack of 52 cards.

Solution: Let A: Getting a spade, $P(A) = \frac{13}{52}$

B: Getting an ace, $P(B) = \frac{4}{52}$

$A \cap B$: Getting an ace of spade, $P(A \cap B) = \frac{1}{52}$

Now $P(A \cup B) = P(A) + P(B) - P(A \cap B)$
$$= \frac{13}{52} + \frac{4}{52} - \frac{1}{52} = \frac{16}{52} = \frac{4}{13}$$

Example14 Find the probability of getting neither heart nor king when a card is drawn from a well shuffled pack of 52 cards.

Solution: Let A: Getting a card of heart, $P(A) = \frac{13}{52}$

B: Getting a card of king, $P(B) = \frac{4}{52}$

$A \cap B$: Getting a king of heart, $P(A \cap B) = \frac{1}{52}$

Now $P(A \cup B) = P(A) + P(B) - P(A \cap B)$
$$= \frac{13}{52} + \frac{4}{52} - \frac{1}{52} = \frac{16}{52} = \frac{4}{13}$$

Again Probability of neither heart nor king is given by $P(A^C \cap B^C)$

$P(A^C \cap B^C) = P(A \cup B)^C = 1 - P(A \cup B)$
$$= 1 - \frac{4}{13} = \frac{9}{13}$$

Example15 Three newspapers A, B, C are published in a city and a survey on readers reveals the following information: 25% read A, 30% read B, 20% read C, 10% read both A and B, 5% read both A and C, 8% read both B and C, 3% read all three newspapers. For a person chosen at random, find the probability that he reads none of the newspapers.

Solution: $P(A) = \frac{25}{100}, P(B) = \frac{30}{100}, P(C) = \frac{20}{100}, P(A \cap B) = \frac{10}{100},$

$P(A \cap C) = \frac{5}{100}, P(B \cap C) = \frac{8}{100}, P(A \cap B \cap C) = \frac{3}{100}$

Now $P(A \cup B \cup C) = P(A) + P(B) + P(C) - P(A \cap B) - P(B \cap C) -$
$$P(C \cap A) + P(A \cap B \cap C)$$

$\Rightarrow P(A \cup B \cup C) = \frac{25}{100} + \frac{30}{100} + \frac{20}{100} - \frac{10}{100} - \frac{8}{100} - \frac{5}{100} + \frac{3}{100}$
$$= \frac{55}{100} = \frac{11}{20}$$

$\Rightarrow P(A \cup B \cup C)^C = 1 - P(A \cup B \cup C)$
$$= 1 - \frac{11}{20} = \frac{9}{20}$$

Example 16 Discuss and comment on following:

$P(A) = \frac{1}{4}, P(B) = \frac{1}{3}, P(C) = \frac{2}{3}$ are probabilities of three mutually exclusive events A, B, C.

Solution: Since A, B, C are mutually exclusive events,
$$P(A \cup B \cup C) = P(A) + P(B) + P(C)$$
$$\Rightarrow P(A \cup B \cup C) = \frac{1}{4} + \frac{1}{3} + \frac{2}{3} = \frac{5}{4} > 1$$

Which is not possible, hence it is a false statement.

2.4 Conditional Probability

The probability of occurrence of an event A, when event B has already occurred is called conditional probability of A and is denoted by $P(A|B)$.

If $P(A|B) = P(A)$, then event A is said to be independent of event B.

2.4.1 Multiplicative Law of Probability

Statement: The probability of simultaneous occurrence of two events is equal to the probability of event multiplied by conditional probability of the other, i.e. for two events A and B:
$$P(A \cap B) = P(A).P(B|A)$$

Proof: Suppose a trial results in n exhaustive, mutually exclusive and equally likely outcomes, m of them being favourable to the happening of event A,

then $P(A) = \frac{m}{n}$ ...①

Out of m outcomes, favorable to the happening of event A, let m_1 be favourable to happening of event B,

∴ Conditional probability of B given that A has already occurred is

$P(B|A) = \frac{m_1}{m}$ ...②

Again out of n exhaustive, mutually exclusive and equally likely outcomes, m_1 are favorable to happening of A and B

∴ Probability of simultaneous occurrence of A and B

$$= P(A \cap B) = \frac{m_1}{n}$$
$$= \frac{m_1}{n}.\frac{m}{m} = \frac{m}{n}.\frac{m_1}{m} = P(A).P(B|A) \quad \text{Using ① and ②}$$

$\therefore P(A \cap B) = P(A).P\left(\frac{B}{A}\right)$

- $P(A \cap B)$ is also written as $P(AB)$
- Interchanging A and B, $P(BA) = P(B).P(A|B)$

                  or $P(AB) = P(B).P(A|B)$ ∵ $A \cap B = B \cap A$

- If A and B are independent events, then $P(B|A) = P(B)$

  ∴ $P(AB) = P(A).P(B)$, if events A and B are independent

  Formulae for conditional probability:
  $$P(B|A) = \frac{P(A \cap B)}{P(A)}$$
  $$P(A|B) = \frac{P(A \cap B)}{P(B)}$$

Example 17 An unbiased coin is tossed thrice. In which of the following cases events A and B are independent

    (i) A: The first throw results in a tail
            B: The last throw results in a head

    (ii) A: The number of tails is two
            B: The last throw results in a tail

Solution: $S = \{HHH, HHT, HTH, HTT, TTT, TTH, THT, THH\}$

(i) $A = \{TTT, TTH, THT, THH\}$, $P(A) = \frac{4}{8} = \frac{1}{2}$

$B = \{HHH, HTH, TTH, THH\}$, $P(B) = \frac{4}{8} = \frac{1}{2}$

$A \cap B = \{TTH, THH\}$, $P(A \cap B) = \frac{2}{8} = \frac{1}{4}$

Now $P(A \cap B) = \frac{1}{4}$, and $P(A).P(B) = \frac{1}{4}$

$\therefore P(A \cap B) = P(A).P(B)$

Hence events A and B are independent.

(ii) $A = \{HTT, THT, TTH\}$, $P(A) = \frac{3}{8}$

$B = \{HHT, HTT, TTT, THT\}$, $P(B) = \frac{4}{8} = \frac{1}{2}$

$A \cap B = \{HTT, THT\}$, $P(A \cap B) = \frac{2}{8} = \frac{1}{4}$

Now $P(A \cap B) = \frac{1}{4}$, and $P(A).P(B) = \frac{3}{16}$

$\therefore P(A \cap B) \neq P(A).P(B)$

Hence the events A and B are not independent.

Example 18 If $P(A) = 0.3$, $P(B) = 0.5$ and $P(A|B) = 0.4$
Find (i) $P(A \cap B)$, (ii) $P(B|A)$, (iii) $P(A^C \cup B^C)$

Solution: (i) $P(A \cap B) = P(B).P\left(\frac{A}{B}\right) = 0.5 \times 0.4 = 0.2$

(ii) $P(B|A) = \frac{P(A \cap B)}{P(A)} = \frac{0.2}{0.3} = \frac{2}{3}$

(iii) $P(A^C \cup B^C) = P(A \cap B)^C$

$= 1 - P(A \cap B) = 1 - 0.2 = 0.8$

Example 19 Cards are dealt one by one from a well shuffled pack of 52 cards until an ace appears. Show that the probability that exactly n cards are dealt before the first ace appears is $\frac{4(51-n)(50-n)(49-n)}{52.51.50.49}$

Solution: Let A: Drawing n non ace cards

B: Drawing an ace in $(n+1)^{th}$ draw

$P(A) = \frac{^{48}C_n}{^{52}C_n} = \frac{48!}{n!(48-n)!} \cdot \frac{n!(52-n)!}{52!}$

$= \frac{(52-n)(51-n)(50-n)(49-n)}{52.51.50.49}$

Also $P(B|A) = \frac{^4C_1}{(52-n)C_1} = \frac{4}{52-n}$

$$\therefore P(A \cap B) = P(A).P(B|A) = \frac{4(51-n)(50-n)(49-n)}{52.51.50.49}$$

Example 20 Two dice are thrown and sum of numbers appearing is observed to be 6. Find the conditional probability that number 2 has appeared at least once.

Solution: Let A: Number 2 has appeared at least once

B: Sum of numbers on two dice is 6

Now $P(A|B) = \frac{P(A \cap B)}{P(B)}$

$A \equiv \{(1,2),(2,2),(3,2),(4,2),(5,2),(6,2),(2,1),(2,3),(2,4),(2,5),(2,6)\}$

$B \equiv \{(1,5),(2,4),(3,3),(4,2),(5,1)\}$

$A \cap B \equiv \{(2,4),(4,2)\}$

Now $P(A \cap B) = \frac{2}{36}$ and $P(B) = \frac{5}{36}$ $\therefore P(A|B) = \frac{2}{5}$

Example 21 A bag contains 7 red 5 black balls; another bag contains 5 red and 8 black balls. A ball is drawn from the first bag and without noticing its colour is put in the second bag and then a ball is drawn from second bag. Find the probability that ball drawn is red in colour.

Solution: Case 1 A: Red ball is drawn from first bag.

B: Red ball is drawn from second bag.

$$P(A \cap B) = P(A).P\left(\frac{B}{A}\right)$$

$$= \frac{7}{12} \cdot \frac{6}{14} = \frac{42}{168}$$

Case 2 C: Black ball is drawn from first bag.

B: Red ball is drawn from second bag.

Then $P(C \cap B) = P(C).P\left(\frac{B}{C}\right)$

$$= \frac{5}{12} \cdot \frac{5}{14} = \frac{25}{168}$$

Required probability $= P(A \cap B) + P(C \cap B)$

$$= \frac{42}{168} + \frac{25}{168} = \frac{67}{168}$$

2.5 Bay's Theorem

If $E_1, E_2, \ldots, E_n$ are n mutually exclusive and exhaustive events in a sample space such that $P(E_i) > 0$ for each $i, (i = 1,2,\ldots,n)$ and A is an arbitrary event for which $P(A) \neq 0$, then the conditional probability of occurrence of E_i, given that A has already occurred is given by

$P(E_i/A) = \frac{P(E_i)P(A/E_i)}{\sum P(E_i)P(A/E_i)}$, $i = 1,2,\ldots,n$

Proof: As $E_1, E_2, \ldots, E_n$ are n mutually exclusive and exhaustive events

$\therefore E_1 \cup E_2 \cup \ldots \cup E_n = S$

Now $A = A \cap S = A \cap (E_1 \cup E_2 \cup ... \cup E_n)$

$\Rightarrow A = (A \cap E_1) \cup (A \cap E_2) \cup (A \cap E_3) \cup ... \cup (A \cap E_n)$

By addition law of probability

$P(A) = P(A \cap E_1) + P(A \cap E_2) + P(A \cap E_3) + \cdots + P(A \cap E_n)$

$\because (A \cap E_1), (A \cap E_2), (A \cap E_3)$ are all mutually exclusive

$\Rightarrow P(A) = P(E_1)P(A/E_1) + P(E_2)P(A/E_2) + P(E_3)P(A/E_3) + \cdots + P(E_n)P(A/E_n)$

$= \sum P(E_i)P(A/E_i), \quad i = 1, 2, ..., n \quad ...①$

$\therefore P(E_i/A) = \frac{P(E_i \cap A)}{P(A)} = \frac{P(E_i)P(A/E_i)}{\sum P(E_i)P(A/E_i)} \quad$ using ①

Example 22 Three urns I, II, III contain 6 red and 4 black balls, 2 red and 6 black balls, and 1 red and 8 black balls respectively. An urn is chosen randomly and a ball is drawn from the urn. If the ball drawn is red, find the probability that it was drawn from urn I.

Solution: Let A: A Red ball is drawn

E_1: Urn I is chosen, $\quad P(E_1) = \frac{1}{3}, P(A/E_1) = \frac{6}{10} = \frac{3}{5}$

E_2: Urn II is chosen, $\quad P(E_2) = \frac{1}{3}, P(A/E_2) = \frac{2}{8} = \frac{1}{4}$

E_3: Urn III is chosen, $\quad P(E_3) = \frac{1}{3}, P(A/E_3) = \frac{1}{9}$

Required Probability $= P(E_1/A) = \dfrac{P(E_1)P(A/E_1)}{P(E_1)P(A/E_1) + P(E_2)P(A/E_2) + P(E_3)P(A/E_3)}$

$= \dfrac{\frac{1}{3} \cdot \frac{3}{5}}{\left(\frac{1}{3} \cdot \frac{3}{5}\right) + \left(\frac{1}{3} \cdot \frac{1}{4}\right) + \left(\frac{1}{3} \cdot \frac{1}{9}\right)} = \dfrac{108}{173}$

Example 23 Two urns I and II contain 3 red and 4 black balls, 2 red and 5 black balls respectively. A ball is transferred from urn I to urn II and then a ball is drawn from urn II. If the ball drawn is found to be red, find the probability that the ball transferred from urn I is red.

Solution: Let A: A Red ball is drawn from urn II

E_1: Ball transferred from urn I is red, $\quad P(E_1) = \frac{3}{7}, P(A/E_1) = \frac{3}{8}$

E_2: Ball transferred from urn I is black, $\quad P(E_2) = \frac{4}{7}, P(A/E_2) = \frac{2}{8}$

Required Probability $= P(E_1/A) = \dfrac{P(E_1)P(A/E_1)}{P(E_1)P(A/E_1) + P(E_2)P(A/E_2)}$

$= \dfrac{\frac{3}{7} \cdot \frac{3}{8}}{\left(\frac{3}{7} \cdot \frac{3}{8}\right) + \left(\frac{4}{7} \cdot \frac{2}{8}\right)} = \dfrac{9}{9+8} = \dfrac{9}{17}$

Example 24 An insurance company insured 2000 scooter drivers, 4000 car drivers and 6000 truck drivers. The probability of an accident involving a scooter driver is 0.01, a car driver is 0.03 and a truck driver is 0.15. If one of an insured person meets with an accident, what is the probability that he is a car driver?

Solution: Let A: An insured person meets with an accident

E_1: Person is an insured scooter driver

$$P(E_1) = \frac{2000}{2000+4000+6000} = \frac{2}{12}, \ P(A/E_1) = 0.01$$

E_2: Person is an insured car driver

$$P(E_2) = \frac{4000}{2000+4000+6000} = \frac{4}{12}, \ P(A/E_2) = 0.03$$

E_3: Person is an insured truck driver

$$P(E_3) = \frac{6000}{2000+4000+6000} = \frac{6}{12}, \ P(A/E_3) = 0.15$$

Required Probability $= P(E_2/A) = \dfrac{P(E_2)P(A/E_1)}{P(E_1)P(A/E_1)+P(E_2)P(A/E_2)+P(E_3)P(A/E_3)}$

$$= \frac{\frac{4}{12}(0.03)}{\frac{2}{12}(0.01)+\frac{4}{12}(0.03)+\frac{6}{12}(0.15)} = \frac{3}{26}$$

Example 25 Ram speaks truth 2 out of 3 times and Shyam 4 out of 5 times; they agree in an assertion that from a bag containing 6 balls of different colour, a red ball has been drawn. Find the probability that the statement is true.

Solution: Let A: Ram and Shyam agree in the assertion that a red ball has been drawn

E_1: Red ball is drawn. $P(E_1) = \frac{1}{6}$, $P(A/E_1) = \frac{2}{3} \cdot \frac{4}{5}$

E_2: Non Red ball is drawn. $P(E_1) = \frac{5}{6}$, $P(A/E_2) = \frac{1}{3} \cdot \frac{1}{5}$

Required Probability $= P(E_1/A) = \dfrac{P(E_1)P(A/E_1)}{P(E_1)P(A/E_1)+P(E_2)P(A/E_2)}$

$$= \frac{\frac{1}{6} \cdot \frac{2}{3} \cdot \frac{4}{5}}{\left(\frac{1}{6} \cdot \frac{2}{3} \cdot \frac{4}{5}\right)+\left(\frac{5}{6} \cdot \frac{1}{3} \cdot \frac{1}{5}\right)} = \frac{8}{8+5} = \frac{8}{13}$$

Example 26 A card from a deck of 52 cards is missing. 2 cards are drawn from the remaining deck of 51 cards and are found to be of spade. Find the probability that missing card is of spade.

Solution: Let A: 2 cards of spade are drawn from a deck of 51 cards.

E_1: Missing card is of spade, $\quad P(E_1) = \frac{13}{52}, \ P(A/E_1) = \frac{12_{C_2}}{51_{C_2}} = \frac{12 \times 11}{51 \times 50}$

E_2: Missing card is a non-spade, $P(E_2) = \frac{39}{52}, \ P(A/E_2) = \frac{13_{C_2}}{51_{C_2}} = \frac{13 \times 12}{51 \times 50}$

$\therefore P(E_1/A) = \dfrac{P(E_1)P(A/E_1)}{P(E_1)P(A/E_1)+P(E_2)P(A/E_2)} = \dfrac{\frac{13 \times 12 \times 11}{52 \times 51 \times 50}}{\frac{13 \times 12 \times 11}{52 \times 51 \times 50}+\frac{39 \times 13 \times 12}{52 \times 51 \times 50}} = \dfrac{11}{11+39} = \dfrac{11}{50}$

2.6 Random Variable and Probability Distributions

A random variable is often described as a variable whose values are determined by chance. The values taken by the variable may be countable or uncountable, based on which it is classified as discrete or continuous. Random variable is typically denoted using capital letters such as 'X'.

2.6.1 Discrete Probability Distributions

A discrete probability distribution '$P(X)$' describes the probability of occurrence of each value of a discrete random variable 'X'. A discrete random variable is a random variable that has countable values, such as a set of positive integers.

The discrete probability distribution '$P(X)$' of an experiment provides the corresponding probabilities to all possible values of the random variable 'X' associated with it such that $\sum P(X) = 1$

Example 27 Find the probability distribution of the number of aces when two cards are drawn at random with replacement from a well shuffled pack of 52 cards.

Solution: Let X be a random variable showing number of aces. Clearly X can take values 0, 1 or 2. If S denotes success i.e. getting an ace and F denotes failure i.e. getting a non-ace card, then $P(S) = \frac{4}{52} = \frac{1}{13}$ and $P(F) = \frac{12}{13}$

X	Event	$P(X)$
0	FF	$\frac{12}{13} \cdot \frac{12}{13} = \frac{144}{169}$
1	SF or FS	$\left(\frac{1}{13} \cdot \frac{12}{13}\right) + \left(\frac{12}{13} \cdot \frac{1}{13}\right) = \frac{24}{169}$
2	SS	$\frac{1}{13} \cdot \frac{1}{13} = \frac{1}{169}$

Example 28 Find the probability distribution of the number of successes in two tosses of a die, when success is defined as a number greater than 4.

Solution: Let X be a random variable showing number of successes in two tosses of a die. Clearly X can take values 0, 1 or 2. If S denotes success i.e. getting a number greater than 4 and F denotes failure, then $P(S) = \frac{2}{6} = \frac{1}{3}$ and $P(F) = \frac{2}{3}$

X	Event	$P(X)$
0	FF	$\frac{2}{3} \cdot \frac{2}{3} = \frac{4}{9}$
1	SF or FS	$\left(\frac{1}{3} \cdot \frac{2}{3}\right) + \left(\frac{2}{3} \cdot \frac{1}{3}\right) = \frac{4}{9}$
2	SS	$\frac{1}{3} \cdot \frac{1}{3} = \frac{1}{9}$

Example 29 3 bad articles are mixed with 7 good ones. Find the probability distribution of number of bad articles if 3 are drawn at random without replacement from the lot.

Solution: Let X be a random variable showing number of bad articles. Clearly X can take values 0, 1, 2 or 3.

X	Event	$P(X)$
0	0 Bad 3 Good	$\dfrac{7_{C_3}}{10_{C_3}} = \dfrac{210}{720}$
1	1 Bad 2 Good	$\dfrac{3_{C_1}.7_{C_2}}{10_{C_3}} = \dfrac{378}{720}$
2	2 Bad 1 Good	$\dfrac{3_{C_2}.7_{C_1}}{10_{C_3}} = \dfrac{126}{720}$
3	3 Bad 0 Good	$\dfrac{3_{C_3}}{10_{C_3}} = \dfrac{6}{720}$

Note: Combination is being used as articles are drawn without replacement.

2.6.2 Mathematical Expectation

If X be a random variable which can assume any one of the values $x_1, x_2, \cdots, x_n$ with respective probabilities $p_1, p_2, \cdots, p_n$; then the mathematical expectation of X usually called as expected value of X, denoted by $E(X)$ is defined as:

$$E(X) = p_1 x_1 + p_2 x_2 + \cdots + p_n x_n = \sum p_i x_i \; ; \text{where } \sum p_i = 1$$

Physical interpretation of $E(X)$

If $\bar{x}$ denotes mean of set of observations $x_1, x_2, \cdots, x_n$ with respective frequencies $f_1, f_2, \cdots, f_n$; then $\bar{x} = \dfrac{\sum f_i x_i}{N}$, $N = \sum f_i$

$$\Rightarrow \bar{x} = \dfrac{f_1}{N} x_1 + \dfrac{f_2}{N} x_2 + \cdots + \dfrac{f_n}{N} x_n \qquad \cdots \text{①}$$

Now out of total N cases, f_i are favorable to x_i

$$\therefore P(X = x_i) = \dfrac{f_i}{N} = p_i \; , i = 1, 2, \cdots, n$$

$$\Rightarrow \dfrac{f_1}{N} = p_1, \dfrac{f_2}{N} = p_2, \cdots, \dfrac{f_n}{N} = p_n \qquad \cdots \text{②}$$

$$\therefore \bar{x} = p_1 x_1 + p_2 x_2 + \cdots + p_n x_n = \sum p_i x_i \qquad \text{using ② in ①}$$

Hence $\bar{x} = E(X)$

Hence mathematical expectation of a random variable is nothing but its arithmetic mean.

∴ We conclude Mean $(\bar{x}) = E(X) = \sum p_i x_i$

Similarly Variance $(\sigma^2) = \dfrac{\sum f_i x_i^2}{N} - \left(\dfrac{\sum f_i x_i}{N}\right)^2$

$$= \sum p_i x_i^2 - (\sum p_i x_i)^2 = E(X^2) - (E(X))^2$$

Example 30 What is the expected number of heads appearing when a fair coin is tossed 3 times.

Solution: Let X be a random variable showing number of heads. Clearly X can take values $0, 1, 2, 3$.

x_i	Event	p_i	$p_i x_i$
0	TTT	$\frac{1}{8}$	0
1	HTT, THT, TTH	$\frac{3}{8}$	$\frac{3}{8}$
2	HHT, HTH, THH	$\frac{3}{8}$	$\frac{6}{8}$
3	HHH	$\frac{1}{8}$	$\frac{3}{8}$

$\therefore E(X) = \sum p_i x_i = \frac{12}{8} = 1.5$

Example 31 A man draws 2 balls from a bag containing 3 white and 5 black balls. If he receives Rs 70 for every white ball he draws and Rs. 35 for every black ball, what is his expectation?

Solution: Following table shows the amount received by the man for each event:

Event	Probability (p_i)	Amount (x_i)	$p_i x_i$
2 black balls	$\frac{5_{C_2}}{8_{C_2}} = \frac{10}{28}$	$35 + 35 = 70$	$\frac{10}{28} \times 70$
1 white 1 black ball	$\frac{3_{C_1} \times 5_{C_1}}{8_{C_2}} = \frac{15}{28}$	$70 + 35 = 105$	$\frac{15}{28} \times 105$
2 white balls	$\frac{3_{C_2}}{8_{C_2}} = \frac{3}{28}$	$70 + 70 = 140$	$\frac{3}{28} \times 140$

$\therefore E(X) = \sum p_i x_i = \frac{700}{28} + \frac{1575}{28} + \frac{420}{28} = \frac{385}{4} = 96.25$

Example 32 For a random variable X, the probability mass function is

$f(x) = kx$, for $x = 1, 2, \cdots, n$

$= 0$, otherwise

Find expectation of X.

Solution: Here $f(x)$ denotes probability mass function

$\therefore \sum_{x=1}^{n} f(x) = \sum_{x=1}^{n} kx = k \sum_{x=1}^{n} x = 1$

$\Rightarrow k \frac{n(n+1)}{2} = 1$

$\Rightarrow k = \frac{2}{n(n+1)}$

$$\therefore E(X) = \sum_{x=1}^{n} xf(x) = \sum_{x=1}^{n} xkx = \sum_{x=1}^{n} x^2 \frac{2}{n(n+1)} = \frac{2}{n(n+1)} \sum_{x=1}^{n} x^2$$

$$= \frac{2}{n(n+1)}(1^2 + 2^2 + \cdots + n^2) = \frac{2}{n(n+1)} \frac{n(n+1)(2n+1)}{6} = \frac{(2n+1)}{3}$$

Example 33 A random variable X has the following probability function:

X	-2	-1	0	1	2	3
P(X)	k	0.1	0.3	2k	0.2	K

Calculate mean and variance.

Solution: For any probability function $\sum p_i = 1$

$\Rightarrow k + 0.1 + 0.3 + 2k + 0.2 + k = 1$

$\Rightarrow k = 0.1$

x_i	p_i	$p_i x_i$	$p_i x_i^2$
-2	0.1	-0.2	0.4
-1	0.1	-0.1	0.1
0	0.3	0	0
1	0.2	0.2	0.2
2	0.2	0.4	0.8
3	0.1	0.3	0.9
		$\sum p_i x_i = 0.6$	$\sum p_i x_i^2 = 2.4$

$\therefore$ Mean $= \sum p_i x_i = 0.6$

Variance $(\sigma^2) = \sum p_i x_i^2 - (\sum p_i x_i)^2$

$= 2.4 - 0.36 = 2.04$

2.6.2 Continuous Probability Distributions

The probability distribution $P(X)$ associated with a continuous random variable X is called a continuous distribution. A continuous random variable is having a set of infinite and uncountable values, for example set of real numbers in the interval (0,1) is uncountable.

If X be a continuous random variable taking values in the interval $[a, b]$, the function $f(x)$ is said to be the Probability Density Function (PDF) of X, if it satisfies the following properties:

i. $f(x) \geq 0 \quad \forall x \in X$ in $[a, b]$.
ii. Total area under the probability curve is one, i.e. $P(a \leq x \leq b) = 1$.

For two distinct points c and d in the interval $[a, b]$; $P(c \leq x \leq d)$ is givenby area under the probability curve between the ordinates $x = c$ and $x = d$, i.e. $P(c \leq x \leq d) = \int_c^d f(x)dx$.

Also $F(x) = P(X \leq x) = \int_{-\infty}^{x} f(x)dx$ is called cumulative distribution function or simply distribution function.

Example 34 Find whether the following is a probability density function:
$$f(x) = \begin{cases} x, & 0 \leq x \leq 1 \\ 2x, & 1 < x \leq 2 \end{cases}$$

Solution: For $f(x)$ to be a probability density function, $\int_{-\infty}^{\infty} f(x)\, dx = 1$

$$\int_{-\infty}^{\infty} f(x)\, dx = \int_{-\infty}^{0} f(x)\, dx + \int_{0}^{1} f(x)\, dx + \int_{1}^{2} f(x)\, dx + \int_{2}^{\infty} f(x)\, dx$$

$$= \int_{-\infty}^{0} 0\, dx + \int_{0}^{1} x\, dx + \int_{1}^{2} 2x\, dx + \int_{2}^{\infty} 0\, dx$$

$$= 0 + \frac{1}{2}[x^2]_0^1 + [x^2]_1^2$$

$$= \frac{1}{2}[1-0] + [4-1] = \frac{7}{2} \neq 1$$

Hence $f(x)$ is not a probability density function.

Example 35 Let X be a random variable with PDF given by
$$f(x) = kx^4, \ |x| \leq 1$$
$$= 0, \text{ otherwise}$$

   i.   Find the value of the constant k

   ii.   Find $E(X)$ and $Var(X)$

   iii. Find $P(X) \geq \frac{1}{2}$

Solution: i. For $f(x)$ to be PDF, $\int_{-\infty}^{\infty} f(x)\, dx = 1$

$$\Rightarrow \int_{-1}^{1} kx^4\, dx = 1$$

$$\Rightarrow \frac{k}{5}[x^5]_{-1}^{1} = 1 \quad \Rightarrow k = \frac{5}{2}$$

ii. $E(X) = \int_{-\infty}^{\infty} xf(x)\, dx$

$$= \int_{-1}^{1} x\, kx^4\, dx$$

$$= k \int_{-1}^{1} x^5\, dx = 0 \quad (\because x^5 \text{ is an odd function})$$

Also $Var(X) = E(X^2) - (E(X))^2$

$$= \int_{-\infty}^{\infty} x^2 f(x)\, dx - 0$$

$$= k \int_{-1}^{1} x^6\, dx = \frac{5}{7}$$

iii. $P(X) \geq \frac{1}{2} = \int_{\frac{1}{2}}^{1} kx^4\, dx = \frac{k}{5}[x^5]_{\frac{1}{2}}^{1} = \frac{31}{64}$

Example 36 Show that the function $f(x)$ defined as
$$f(x) = \begin{cases} e^{-x}, & x \geq 0 \\ 0, & x < 0 \end{cases}$$, is a probability density function and find the probability that the variate X having $f(x)$ as density function will lie in the interval (1,2). Also find the probability distribution function $F(2)$.

Solution: We have $\int_{-\infty}^{\infty} f(x)dx = \int_{-\infty}^{0} 0\, dx + \int_{0}^{\infty} e^{-x}\, dx = -[e^{-x}]_{0}^{\infty} = 1$

∴ $f(x)$ is a probability density function.

Also $P(1 \leq x \leq 2) = \int_{1}^{2} e^{-x}\, dx = -[e^{-x}]_{1}^{2} = 2.33$

Again $F(2) = \int_{-\infty}^{2} f(x)dx = \int_{-\infty}^{0} 0\, dx + \int_{0}^{2} e^{-x}\, dx = -[e^{-x}]_{0}^{2} = 1 - e^{-2}$

$= 1 - 0.135 = 0.865$

2.7 Moments

The expected values $E(x - a), E((x - a)^2), E((x - a)^3), \ldots, E((x - a)^r)$ are called moments about any point a.

Thus r^{th} moment about any point 'a' of any distribution is denoted by μ_r'
and is given by: $\mu_r' = \sum p_i (x_i - a)^r = \frac{1}{N}\sum f_i (x_i - a)^r$, where $N = \sum f_i$

In particular r^{th} moment about mean $\bar{x}$ is given by $\mu_r = \frac{1}{N}\sum f_i (x_i - \bar{x})^r$

Some important results:

- $\mu_0 = \mu_0' = 1$
- $\mu_1 = \frac{1}{N}\sum f_i (x_i - \bar{x}) = \frac{1}{N}\sum f_i x_i - \frac{\bar{x}}{N}\sum f_i = \bar{x} - \bar{x} = 0$
- $\mu_2 = \frac{1}{N}\sum f_i (x_i - \bar{x})^2 = \frac{1}{N}\sum f_i x_i^2 + \frac{\bar{x}^2}{N}\sum f_i - 2\frac{\bar{x}}{N}\sum f_i x_i$

  $= \frac{1}{N}\sum f_i x_i^2 + \bar{x}^2 - 2\bar{x}^2 = \frac{1}{N}\sum f_i x_i^2 - \bar{x}^2 = \sigma^2$
- $\mu_1' = \frac{1}{N}\sum f_i (x_i - a) = \frac{1}{N}\sum f_i x_i - \frac{a}{N}\sum f_i = \bar{x} - a$

Relation between μ_r' and μ_r

$\mu_r = \frac{1}{N}\sum f_i (x_i - \bar{x})^r$

$= \frac{1}{N}\sum f_i ((x_i - a) - (\bar{x} - a))^r$

$= \frac{1}{N}\sum f_i (d - \mu_1')^r$

by putting $x_i - a = d$

$\Rightarrow \mu_r = \frac{1}{N}\sum f_i \left(d^r - r_{C_1} d^{r-1}\mu_1' + r_{C_2} d^{r-2}\mu_1'^2 + \cdots + (-1)^r \mu_1'^r\right)$

$= \frac{1}{N}\sum f_i d^r - r_{C_1}\mu_1'\frac{1}{N}\sum f_i d^{r-1} + r_{C_2}\mu_1'^2\frac{1}{N}\sum f_i d^{r-2} + \cdots + (-1)^r \mu_1'^r \frac{1}{N}\sum f_i$

$\Rightarrow \mu_r = \mu_r' - r_{C_1}\mu_{r-1}'\mu_1' + r_{C_2}\mu_{r-2}'\mu_1'^2 + \cdots + (-1)^r \mu_1'^r$

In particular

$\mu_2 = \mu_2' - 2\mu_1'^2 + \mu_0'\mu_1'^2 = \mu_2' - \mu_1'^2$ as $\mu_0' = 1$

$\mu_3 = \mu_3' - 3\mu_2'\mu_1' + 3\mu_1'^3 - \mu_0'\mu_1'^3 = \mu_3' - 3\mu_2'\mu_1' + 2\mu_1'^3$

Similarly $\mu_4 = \mu_4' - 4\mu_3'\mu_1' + 6\mu_2'\mu_1'^2 - 3\mu_1'^4$

2.8 Moment Generating Function (MGF)

The moment generating function of the variable x about a point $x = a$ is defined as the expected value of $e^{t(x-a)}$ and is denoted by $M_a(t)$.

$\therefore M_a(t) = E[e^{t(x-a)}]$.

$\Rightarrow M_a(t) = \sum p_i\, e^{t(x_i - a)}$ ...①

$= \sum p_i + t \sum p_i (x_i - a) + \frac{t^2}{2!} \sum p_i (x_i - a)^2 + \ldots + \frac{t^r}{r!} \sum p_i (x_i - a)^r + \ldots$

$(\because e^x = 1 + x + \frac{x^2}{2!} + \ldots + \frac{x^r}{r!} + \ldots)$

$\Rightarrow M_a(t) = 1 + t\mu_1' + \frac{t^2}{2!}\mu_2' + \ldots + \frac{t^r}{r!}\mu_r' + \ldots$ ...②

$\because p_i = \frac{f_i}{N}$

where μ_r' is the moment of order r about a point a

Hence $\mu_r' =$ coefficient of $\frac{t^r}{r!}$

Thus $M_a(t)$ generates moments and therefore it is called moment generating function.

Again rewriting ① as $M_a(t) = e^{-at} \sum p_i\, e^{tx_i}$

$\Rightarrow M_a(t) = e^{-at} M_0(t)$

Thus (MGF about the point a) $= e^{-at}$ (MGF about origin)

Again if $f(y)$ be density function of a continuous variate Y, then the moment generating function of the continuous probability distribution about $y = a$ is given by:

$M_a(t) = \int_{-\infty}^{\infty} e^{t(y-a)} f(y)\, dy$

We can also generate moments by differentiating r times w.r.t. t and then putting $t = 0$

i.e. $\mu_r' = \left[\frac{d^r}{dt^r} M_a(t)\right]_{t=0}$ ...③

Thus the moments about any point $x = a$ can be found using ② or more conveniently using ③

If a moment-generating function exists for a random variable X, then:
- $M(0) = 1$
- $M'(0) = E(X),\ M''(0) = E(X^2),\ M'''(0) = E(X^3), \ldots$
- The mean and variance of X can be found by evaluating the first and second derivatives of the moment-generating function at $t = 0$.

i.e. $\bar{x} = E(X) = M'(0),\ \sigma^2 = E(X^2) - (E(X))^2 = M''(0) - [M'(0)]^2$

Example 37 Find the moment generating function for the probability distribution given by number of heads appearing when a fair coin is tossed 3 times and hence find mean and variance.

Solution: Let X be a random variable showing number heads. Clearly X can take values $0, 1, 2, 3$. Probability distribution is given by:

x_i	0	1	2	3
p_i	$\frac{1}{8}$	$\frac{3}{8}$	$\frac{3}{8}$	$\frac{1}{8}$

$$M(t) = E(e^{tX}) = \sum p_i e^{tx_i}$$

$$\Rightarrow M(t) = \frac{1}{8} + \frac{3}{8}e^t + \frac{3}{8}e^{2t} + \frac{1}{8}e^{3t}$$

$$M'(t) = \frac{3}{8}e^t + \frac{6}{8}e^{2t} + \frac{3}{8}e^{3t}$$

$\therefore$ Mean $= M'(0) = \frac{3}{8} + \frac{6}{8} + \frac{3}{8} = \frac{12}{8} = 1.5$

Also $M''(t) = \frac{3}{8}e^t + \frac{12}{8}e^{2t} + \frac{9}{8}e^{3t}$

$\Rightarrow M''(0) = \frac{3}{8} + \frac{12}{8} + \frac{9}{8} = \frac{24}{8} = 3$

$\therefore$ Variance $= M''(0) - [M'(0)]^2 = 3 - (1.5)^2 = 0.75$

Example 38 Let X be a random variable with PDF given by

$$f(x) = \begin{cases} ke^{-kx}, & x \in (0, \infty) \\ 0, & \text{otherwise} \end{cases}$$

Find moment generating function of $f(x)$, hence find mean and variance.

Solution: $M(t) = E(e^{tX}) = \int_{-\infty}^{\infty} e^{tx} f(x)\, dx$

$$= \int_{-0}^{\infty} e^{tx} ke^{-kx}\, dx = k \int_{-0}^{\infty} e^{-(k-t)x}\, dx = k \left[\frac{e^{-(k-t)}}{t-k} \right]_0^{\infty} = \frac{k}{k-t}$$

Also $M'(t) = \frac{k}{(k-t)^2}$

$\therefore$ Mean $= M'(0) = \frac{1}{k}$

Also $M''(t) = \frac{2k}{(k-t)^3} \Rightarrow M''(0) = \frac{2}{k^2}$

$\therefore$ Variance $= M''(0) - [M'(0)]^2 = \frac{2}{k^2} - \frac{1}{k^2} = \frac{1}{k^2}$

2.9 Skewness

Skewness is a measure of the asymmetry of the probability distribution of a random variable about its mean. In a symmetrical distribution, mean, mode and median coincide.

The skewness value can be positive or negative.

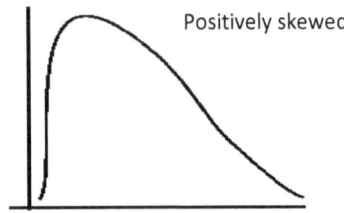

Positively skewed

Mass of distribution is concentrated to left

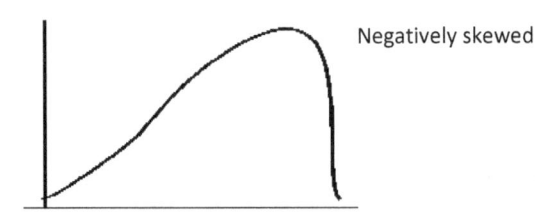

Negatively skewed

Mass of distribution is concentrated to the right

$$\text{Coefficient of skewness} = \frac{\text{Mean} - \text{Mode}}{\text{Standard Deviation}}$$

$\therefore S_k = \frac{M-M_o}{\sigma}$, S_k is called Karl Pearson's coefficient of skewness and lies between -3 and $+3$.

If mode is ill defined, then using $M_o = 3M_d - 2M$

$S_k = \frac{3(M-M_d)}{\sigma}$

Karl Pearson defined the following four coefficients based upon the first four moments about the mean:

$$\beta_1 = \frac{\mu_3^2}{\mu_2^3}, \quad \gamma_1 = \sqrt{\beta_1}$$

$$\beta_2 = \frac{\mu_4}{\mu_2^2}, \quad \gamma_2 = \beta_2 - 3$$

β_1 gives a measure of departure from symmetry and β_2 is associated with skewness.

2.10 Kurtosis

Kurtosis measures the degree of peakness of a distribution and is given by β_2

If $\beta_2 > 3$, the curve is peaked or leptokurtic

If $\beta_2 = 3$, the curve is normal or mesokurtic

If $\beta_2 < 3$, the curve is flat topped or platykurtic

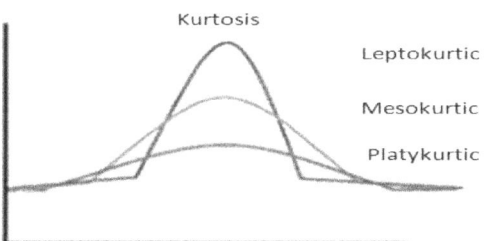

Example 39 Calculate the first four moments of the following distribution about the mean and hence find β_1 and β_2.

x	0	1	2	3	4	5	6	7	8
f	1	8	28	56	70	56	28	8	1

Solution: Taking 4 as assumed mean, let us first calculate moments about $x = 4$

$\mu_r' = \frac{1}{N}\Sigma f(x-a)^r = \frac{1}{N}\Sigma f d^r$, where $d = x - 4$

x	f	$d = x-4$	fd	fd^2	fd^3	fd^4
0	1	-4	-4	16	-64	256
1	8	-3	-24	72	-216	648
2	28	-2	-56	112	-224	448
3	56	-1	-56	56	-56	56
4	70	0	0	0	0	0
5	56	1	56	56	56	56
6	28	2	56	112	224	448
7	8	3	24	72	216	648
8	1	4	4	16	64	256
	N= 256		0	512	0	2816

$\mu_1' = \frac{1}{N}\Sigma f d = 0, \quad \mu_2' = \frac{1}{N}\Sigma f d^2 = \frac{512}{256} = 2,$

$\mu_3' = \frac{1}{N}\Sigma f d^3 = 0, \quad \mu_4' = \frac{1}{N}\Sigma f d^4 = \frac{2816}{256} = 11$

Moments about mean:

$\mu_1 = 0$

$\mu_2 = \mu_2' - \mu_1'^2 = 2$

$\mu_3 = \mu_3' - 3\mu_2'\mu_1' + 2\mu_1'^3 = 0$

$\mu_4 = \mu_4' - 4\mu_3'\mu_1' + 6\mu_2'\mu_1'^2 - 3\mu_1'^4 = 11$

$\beta_1 = \dfrac{\mu_3^2}{\mu_2^3} = 0$, $\beta_2 = \dfrac{\mu_4}{\mu_2^2} = \dfrac{11}{4} = 2.75$

Exercise 2

1. In a single throw of two dice, what is the probability of getting six as a product of numbers on two dice?
2. A bag contains 4 black, 2 red and 3 blue pens. If 2 pens are drawn at random from the pack and then another pen is drawn without replacement, what is the probability of drawing 2 black and 1 blue pens?
3. A purse contains 2 silver and 4 copper coins. A second purse contains 4 silver and 3 copper coins. If a coin is pulled out at random from one of the two purses, what is the probability that coin is a silver coin?
4. A problem in mathematics is given to three students A, B and C whose chances of solving are $\dfrac{1}{2}, \dfrac{1}{3}$ and $\dfrac{1}{4}$ respectively. What is the probability that the problem will not be solved?
5. If four whole numbers taken at random are multiplied together, find the probability that the last digit in the product is 1,3,7 or 9.
6. A speaks truth in 60% and B in 75% cases of the cases. In what percentage of cases they are likely to contradict each other in stating the same fact.
7. A and B take turns in throwing two dice and the first to throw 10 wins the game. If A has the first throw, find B's chances of winning.
8. A bag contains 10 white and 3 black balls. Another bag contains 3 white and 5 black balls. Two balls are transferred from the first bag and placed in the second and then one ball is drawn from the second bag. What is the probability that it is a white ball?
9. In each of a set of games it is 2 to 1 in favour the winner of the previous game. What is the probability that who wins the first game shall win at least 3 games out of the next four games?
10. A and B take turns in throwing two dice with A having the first trial. The first to throw 10 being awarded a prize. Find their expectations if the prize money is Rs 460.
11. In a lottery m tickets are drawn out of n tickets numbered from 1 to n. Show that the expectation of sum of numbers drawn is $m\left(\dfrac{n+1}{2}\right)$.
12. Find the probability distribution of number of kings drawn when two cards are drawn one by one, without replacement, from a deck of 52 cards.

13. A random variable X has the following probability distribution:

X	0	1	2	3	4	5
$P(X)$	0.1	K	0.2	$2K$	0.3	K

Find i. K ii. $P(X \leq 1)$ iii. $P(X > 3)$

14. Let X be a random variable with PDF given by
$$f(x) = \begin{cases} \frac{1}{4}, & |x| \leq 2 \\ 0, & \text{otherwise} \end{cases}$$

Obtain the values of i. $P(X < 1)$ ii. $P(|X| > 1)$ iii. $P(2X + 3) > 5)$

15. For the function $f(x)$ defined by $f(x) = ce^{-x}$, $0 \leq x \leq \infty$, find the value of c which changes $f(x)$ to a probability density function.

16. If X be a random variable with PDF given by
$$f(x) = \begin{cases} e^{-x}, & x \in (0, \infty) \\ 0, & \text{otherwise} \end{cases}$$, find moment generating function of $f(x)$.

17. Find $E(X^3)$ using the moment generating function $(1 - 2t)^{-10}$

Answers

1. $\frac{1}{9}$
2. $\frac{1}{14}$
3. $\frac{19}{42}$
4. $\frac{1}{4}$
5. $\frac{16}{625}$
6. 45%
7. $\frac{11}{23}$
8. $\frac{59}{130}$
9. $\frac{4}{9}$
10. Rs.240, Rs. 220
12. $X = 0$; $P(X) = \frac{188}{221}$, $X = 1$; $P(X) = \frac{32}{221}$, $X = 2$; $P(X) = \frac{1}{221}$
13. i. 0.1   ii. 0.2   iii. 0.4
14. i. $\frac{3}{4}$ ii. $\frac{1}{2}$ iii. $\frac{1}{4}$
15. 1
16. $\frac{1}{1-t}$
17. 10,560.

Chapter 3: Theoretical Probability Distributions

Theoretical Probability Distributions

3.1 Introduction

Probability distributions are either discrete or continuous, depending on whether they define probabilities for discrete or continuous variables. Here we shall confine our studies to Binomial, Poisson and Normal distributions of which Binomial and Poisson distributions are discrete distributions whereas Normal distribution is a continuous probability distribution.

3.2 Binomial Distribution

A series of independent trials which result in one of the two mutually exclusive outcomes 'success' or 'failure' such that the probability of the success (or failure) in each trials is constant, then such repeated independent trials are called as 'Bernoulli trials'. A discrete random variable which results in only one of the two possible outcomes (success or failure) is called Binomial variable.

Let there be n independent finite trials in an experiment such that

i. Each trial has only two possible outcomes success and failure
ii. Probability of success (p) and probability of failure (q) are constant for all the trials and $p + q = 1$.

Then if a random variable X denotes the number of successes in n trials, then

$P(X = r) = {}^nC_r \, p^r q^{n-r}$ or $P(r) = {}^nC_r \, q^{n-r} p^r$

∴ Binomial distribution may be given as $(q + p)^r$

3.2.1 Mean of Binomial Distribution

Mean $= \sum_{r=0}^{n} rP(r)$

$= \sum_{r=0}^{n} r \, {}^nC_r \, q^{n-r} p^r$

$= {}^nC_1 \, q^{n-1} p^1 + 2 \, {}^nC_2 \, q^{n-2} p^2 + 3 \, {}^nC_3 \, q^{n-3} p^3 + \cdots + n \, {}^nC_n \, q^0 p^n$

$= n \, q^{n-1} p + \frac{2n(n-1)}{2!} q^{n-2} p^2 + \frac{3n(n-1)(n-2)}{3!} q^{n-3} p^3 + \cdots + np^n$

$= np \left[q^{n-1} + (n-1) q^{n-2} p + \frac{(n-1)(n-2)}{2!} q^{n-3} p^2 + \cdots + p^{n-1} \right]$

$= np \left[{}^{n-1}C_0 \, q^{n-1} + {}^{n-1}C_1 \, q^{n-2} p + {}^{n-1}C_2 \, q^{n-3} p^2 + \cdots + {}^{n-1}C_{n-1} \, p^{n-1} \right]$

$= np \, (q + p)^{n-1}$

$= np \qquad \because q + p = 1$

3.2.2 Variance of Binomial Distribution

Variance $= \sum_{r=0}^{n} r^2 P(r) - (\text{mean})^2$

Now $\sum_{r=0}^{n} r^2 P(r) = \sum_{r=0}^{n} r^2 \, {}^nC_r \, q^{n-r} p^r$

$$= {}^nC_1 q^{n-1}p + 2^2 {}^nC_2 q^{n-2}p^2 + 3^2 {}^nC_3 q^{n-3}p^3 + \cdots + n^2 {}^nC_n p^n$$

$$= n q^{n-1}p + \frac{4n(n-1)}{2!} q^{n-2}p^2 + \frac{9n(n-1)(n-2)}{3!} q^{n-3}p^3 + \cdots + n^2 p^n$$

$$= np\left[q^{n-1} + 2(n-1) q^{n-2}p + \frac{3(n-1)(n-2)}{2!} q^{n-3}p^2 + \cdots + np^{n-1}\right]$$

$$= np\left[q^{n-1} + (n-1)q^{n-2}p + \frac{(n-1)(n-2)}{2!} q^{n-3}p^2 + \cdots + p^{n-1}\right]$$

$$+ np\left[(n-1)q^{n-2}p + (n-1)(n-2) q^{n-3}p^2 + \cdots + (n-1)p^{n-1}\right]$$

$$= np\left[{}^{n-1}C_0 q^{n-1} + {}^{n-1}C_1 q^{n-2}p + {}^{n-1}C_2 q^{n-3}p^2 + \cdots + {}^{n-1}C_{n-1} p^{n-1}\right]$$

$$+ np(n-1)p\left[{}^{n-2}C_0 q^{n-2} + {}^{n-2}C_1 q^{n-3}p + \cdots + {}^{n-2}C_{n-2}p^{n-2}\right]$$

$$= np[(q+p)^{n-1} + (n-1)p(q+p)^{n-2}]$$

$$= np[1 + (n-1)p] \qquad \because q + p = 1$$

$\therefore$ Variance $= np[1 + np - p] - n^2p^2 = np[q + np] - n^2p^2 \quad \because 1 - p = q$

$$= npq$$

$\therefore$ For Binomial distribution

Mean $= np$

$\mu_1 = 0$

$\mu_2 = \sigma^2 = npq \qquad \because \mu_2 = variance (\sigma^2)$

Similarly $\mu_3 = npq(q - p)$

$\mu_4 = npq[1 + 3pq(n - 2)]$

$\beta_1 = \frac{\mu_3^2}{\mu_2^3} = \frac{[npq(q-p)]^2}{(npq)^3} = \frac{(q-p)^2}{npq}$, $\qquad \gamma_1 = \frac{(q-p)}{\sqrt{npq}}$

$\beta_2 = \frac{\mu_4}{\mu_2^2} = \frac{1 + 3pq(n-2)}{npq} = 3 + \frac{1 - 6pq}{npq}$, $\quad \gamma_2 = \beta_2 - 3 = \frac{1 - 6pq}{npq}$

Remarks:

- Binomial Distribution is symmetrical if $\beta_1 = 0$ i.e. if $\frac{(q-p)^2}{npq} = 0$ or $p = q = \frac{1}{2}$
- Binomial Distribution is positively skewed if $\gamma_1 > 0$ i.e. if $q - p > 0$ or $1 - 2p > 0$ or $p < \frac{1}{2}$
- Binomial Distribution is negatively skewed if $p > \frac{1}{2}$
- Since $0 < q < 1$, $\therefore$ for binomial distribution $npq < np$ i.e. vaiance $<$ mean.

3.2.3 Moment Generating Function of Binomial Distribution

Moment Generating Function (MGF) about origin is expected value of e^{tr}

$M_0(t) = E(e^{tr})$

$= \sum p(r)e^{tr}$

$= \sum {}^nC_r p^r q^{n-r} e^{tr}$

$$= \sum {}^nC_r (pe^t)^r q^{n-r}$$
$$= (q + pe^t)^n$$

Now $\left[\frac{d}{dt} M_0(t)\right]_{t=0} = [n(q + pe^t)^{n-1} pe^t]_{t=0} = n(q + p)^{n-1} p = np$

$\Rightarrow \mu_1' = np$

Again $M_a(t) = e^{-at} M_0(t)$

$\therefore M_m(t) = e^{-npt} (q + pe^t)^n$ where $m = np$ denotes mean of the distribution

$$= (qe^{-pt} + pe^{qt})^n$$

$$= \left[(q+p) + (-qpt + pqt) + \left(\frac{qp^2t^2}{2!} + \frac{pq^2t^2}{2!}\right) + \left(\frac{-qp^3t^3}{3!} + \frac{pq^3t^3}{3!}\right) + \left(\frac{qp^4t^4}{4!} + \frac{pq^4t^4}{4!}\right) + \cdots\right]^n$$

$$= \left[1 + \left(pq\frac{t^2}{2!} + pq(q^2-p^2)\frac{t^3}{3!} + pq(q^3+p^3)\frac{t^4}{4!} + \cdots\right)\right]^n$$

$$= 1 + n\left(pq\frac{t^2}{2!} + pq(q^2-p^2)\frac{t^3}{3!} + pq(q^3+p^3)\frac{t^4}{4!} + \cdots\right)$$

$$+ \frac{n(n-1)}{2!}\left(pq\frac{t^2}{2!} + pq(q^2-p^2)\frac{t^3}{3!} + pq(q^3+p^3)\frac{t^4}{4!} + \cdots\right)^2 + \cdots$$

$\therefore M_m(t) = 1 + npq\frac{t^2}{2!} + npq(q-p)\frac{t^3}{3!} +$

$\quad [npq(q^2 + p^2 - pq) + 3n(n-1)p^2q^2]\frac{t^4}{4!} + \cdots \quad \because (q+p) = 1$

Now $M_m(t) = 1 + \mu_1 t + \mu_2 \frac{t^2}{2!} + \mu_3 \frac{t^3}{3!} + \mu_4 \frac{t^4}{4!}$

$\therefore 1 + \mu_1 t + \mu_2 \frac{t^2}{2!} + \mu_3 \frac{t^3}{3!} + \mu_4 \frac{t^4}{4!} = 1 + npq\frac{t^2}{2!} + npq(q-p)\frac{t^3}{3!} +$

$\quad [npq(q^2 + p^2 - pq) + 3n(n-1)p^2q^2]\frac{t^4}{4!} + \cdots$

Comparing coefficients of different powers of t on both sides, we get

$\mu_1 = 0$

$\mu_2 = npq$

$\mu_3 = npq(q - p)$

$\mu_4 = npq(q^2 + p^2 - pq) + 3n(n-1)p^2q^2$

$\quad = npq(1 - 3pq) + 3n(n-1)p^2q^2 \quad \because q^2 + p^2 = (q+p)^2 - 2pq$

$\quad = npq(1 - 3pq + 3npq - 3pq)$

$\quad = npq[1 + 3pq(n-2)]$

Example1 Show that Variance of a binomial distribution is less than or equal to $\frac{n}{4}$.

Solution: Variance (σ^2) of a binomial distribution is npq

$\therefore \sigma^2 = npq = np(1-p) = np - np^2 = f(p)$ say

For $f(p)$ to be maximum

$f'(p) = 0$ and $f''(p) < 0$

Now $f'(p) = n - 2np = 0 \Rightarrow p = \frac{1}{2}$

Also $f''(p) = -2n < 0$

$\therefore f(p)$ is maximum at $p = \frac{1}{2}$

Thus maximum variance is at $p = \frac{1}{2}$, $q = \frac{1}{2}$

i.e. maximum variance $= npq = n.\frac{1}{2}.\frac{1}{2} = \frac{n}{4}$ $\therefore$ Variance $\leq \frac{n}{4}$

Example 2 6 dice are thrown 729 times. How many times would you expect at least three dice to show 1 or 2?

Solution: Here the Binomial Distribution (B.D.) is given by $N(q+p)^n$

Where $p = \frac{2}{6} = \frac{1}{3}$, $q = \frac{2}{3}$, $n = 6$, $N = 729$

$\therefore$ B.D. is given by $729\left(\frac{2}{3} + \frac{1}{3}\right)^6$ and if X is random variable showing number of successes, then

$P(X \geq 3) = 729\left[{}^6C_3 \left(\frac{2}{3}\right)^3 \left(\frac{1}{3}\right)^3 + {}^6C_4 \left(\frac{2}{3}\right)^2 \left(\frac{1}{3}\right)^4 + {}^6C_5 \left(\frac{2}{3}\right)^1 \left(\frac{1}{3}\right)^5 + {}^6C_6 \left(\frac{1}{3}\right)^6\right]$

$= \frac{729}{3^6}[160 + 60 + 12 + 1] = 233$

Example 3 A die is tossed 3 times. Find mean and variance of number of successes if getting 5 or 6 is considered as success.

Solution: Here $p = \frac{2}{6} = \frac{1}{3}$, $q = \frac{2}{3}$, $n = 3$

$\therefore$ Mean $= np = \frac{3}{3} = 1$, Variance $= npq = 3 \times \frac{1}{3} \times \frac{2}{3} = \frac{2}{3}$

Example 4 If the sum of mean and variance of Binomial Distribution is 4.8 for 5 trials. Find the distribution.

Solution: Given $np + npq = 4.8$, $n = 5$

$\Rightarrow np(1+q) = 4.8$

$\Rightarrow 5(1-q)(1+q) = 4.8$ $\because n = 5$

$\Rightarrow 1 - q^2 = 9.6 \Rightarrow q^2 = \frac{1}{25} \Rightarrow q = \frac{1}{5}$

$\therefore q = \frac{1}{5}$, $p = \frac{4}{5}$ and distribution is given by $\left(\frac{1}{5} + \frac{4}{5}\right)^5$

Example 5 For a binomial distribution; mean is 4 and standard deviation is $\sqrt{2}$. Find the distribution.

Solution: Given $np = 4$, $npq = 2$

$\Rightarrow \frac{npq}{np} = \frac{2}{4} \Rightarrow q = \frac{2}{4} = \frac{1}{2} \therefore p = \frac{1}{2}$

Also $np = 4$ or $\frac{n}{2} = 4 \Rightarrow n = 8$

∴ The distribution is given by $\left(\frac{1}{2} + \frac{1}{2}\right)^8$

Example6 Find the expected number of the defective bulbs in a lot of 100 bulbs; if one out of 5 bulbs is defective. Also find the standard deviation, coefficient of skewness γ_1 and determine whether the distribution curve is leptokurtic, mesokurtic or platykurtic.

Solution: We have $p = \frac{1}{5} = 0.2$, $q = 1 - 0.2 = 0.8$, $n = 1000$

Expected number of defective bulbs $= np = 100 \times 0.2 = 20$

Also standard deviation $= \sqrt{npq} = \sqrt{100 \times 0.2 \times 0.8} = 4$

$\gamma_1 = \frac{(q-p)}{\sqrt{npq}} = \frac{(0.8-0.2)}{4} = 0.15$

$\beta_2 = 3 + \frac{1-6pq}{npq} = 3 + \frac{1-6(0.2)(0.8)}{16} = 3.0025$

∴ The curve is a bit leptokurtic.

Example7 If the probability of success of an event is $\frac{1}{20}$; how many trials are required in order that the probability of getting at least one success, is just greater than $\frac{1}{2}$?

Solution: Here $p = \frac{1}{20}$, $q = \frac{19}{20}$

Let n be the required number of trials such that the probability of getting at least one success, is just greater than $\frac{1}{2}$

i.e. $P(X \geq 1) > \frac{1}{2}$

$\Rightarrow 1 - P(X = 0) > \frac{1}{2}$

$\Rightarrow P(X = 0) < 1 - \frac{1}{2}$

$\Rightarrow {}^nC_0 \left(\frac{19}{20}\right)^n \left(\frac{1}{20}\right)^0 < \frac{1}{2}$

$\Rightarrow \left(\frac{19}{20}\right)^n < \frac{1}{2}$

$\Rightarrow n \log_{10} \frac{19}{20} < \log_{10} \frac{1}{2}$

$\Rightarrow n > \frac{\log_{10}\frac{1}{2}}{\log_{10}\frac{19}{20}}$ ∵ $\log_{10} \frac{19}{20} < 0$

$\Rightarrow n > \frac{-0.3010}{-0.02228} = 13.5099$ ∴ $n = 14$

Example 8 The probability of a man hitting a target is 1/3. How many times must he take the shot so that the probability of hitting the target at least once is less than 90%?

Solution: Here $p = \frac{1}{3}, q = \frac{2}{3}$, Let n be the number of shots so that the probability of hitting the target, at least once, is less than 90%.

i.e. $P(X \geq 1) < \frac{9}{10}$

$\Rightarrow 1 - P(X = 0) < \frac{9}{10}$

$\Rightarrow P(X = 0) > 1 - \frac{9}{10}$

$\Rightarrow {}^nC_0 \left(\frac{2}{3}\right)^n \left(\frac{1}{3}\right)^0 > \frac{1}{10}$

$\Rightarrow \left(\frac{2}{3}\right)^n > \frac{1}{10}$

$\Rightarrow n \log_{10} \frac{2}{3} > \log_{10} \frac{1}{10}$

$\Rightarrow n < \frac{\log_{10} \frac{1}{10}}{\log_{10} \frac{2}{3}} \qquad \because \log_{10} \frac{2}{3} < 0$

$\Rightarrow n < \frac{-1}{-0.1761} = 5.6786 \qquad \therefore n = 5$

Example 9 Assuming that half the population are consumers of chocolates, so that the chances of an individual being consumer is $\frac{1}{2}$ and assuming that each of the 25 surveyors takes 10 individuals to see whether they are consumers. How many surveyors would you expect to report that three or less people were consumers?

Solution: The probability of an individual to be consumer $(p) = \frac{1}{2}, \therefore q = \frac{1}{2}$

Also $n = 10$, $N = 25$

$\therefore$ B.D. is given by $25 \left(\frac{1}{2} + \frac{1}{2}\right)^{10}$ and if X is random variable showing number of successes, then

$P(X \leq 3) = 25 \left[{}^{10}C_0 \left(\frac{1}{2}\right)^{10} + {}^{10}C_1 \left(\frac{1}{2}\right)^9 \left(\frac{1}{2}\right)^1 + {}^{10}C_2 \left(\frac{1}{2}\right)^8 \left(\frac{1}{2}\right)^2 + {}^{10}C_3 \left(\frac{1}{2}\right)^7 \left(\frac{1}{2}\right)^3 \right]$

$= 25 \left(\frac{1}{2}\right)^{10} \left[1 + 10 + \frac{10 \times 9}{2!} + \frac{10 \times 9 \times 8}{3!}\right] = 25(0.171875) = 4.3$

So we can expect four surveyors to report that three or less people were consumers.

Example 10: Fit a binomial distribution to the following data and compare theoretical frequencies with actual ones

x	0	1	2	3	4	5	6	7	8	9
f	6	20	28	12	8	6	0	0	0	0

Solution: Mean of the given distribution $= \frac{\Sigma fx}{\Sigma f}$, $\Sigma f = 80$

$$= \frac{0+20+56+36+32+30+0}{80} = \frac{87}{40} = 2.175$$

Let mean of binomial distribution to be fitted $= np = 2.175$

Also $n = 10 \therefore p = 0.2175$ $q = 1 - 0.2175 = 0.7825$

$\therefore$ B.D. is given by $80(0.7825 + 0.2175)^{10}$

Theoretical frequencies using binomial distribution are given in the table below:

x	$P(r) = {}^nC_r\, q^{n-r} p^r$	Theoretical frequencies $(f) = 80 \times P(r)$
0	${}^{10}C_0 (0.7825)^{10}(0.2175)^0 = 0.086$	$6.9 = 7\,(\text{say})$
1	${}^{10}C_1 (0.7825)^9 (0.2175)^1 = 0.239$	$19.1 = 19\,(\text{say})$
2	${}^{10}C_2 (0.7825)^8 (0.2175)^2 = 0.299$	$23.9 = 24\,(\text{say})$
3	${}^{10}C_3 (0.7825)^7 (0.2175)^3 = 0.22$	$17.8 = 18\,(\text{say})$
4	${}^{10}C_4 (0.7825)^6 (0.2175)^4 = 0.11$	$8.6 = 9\,(\text{say})$
5	${}^{10}C_5 (0.7825)^5 (0.2175)^5 = 0.04$	$2.9 = 3\,(\text{say})$
6	${}^{10}C_6 (0.7825)^4 (0.2175)^6 = 0.008$	$0.66 = 0\,(\text{say})$
7	${}^{10}C_7 (0.7825)^3 (0.2175)^7 = 0.001$	$0.11 = 0\,(\text{say})$
8	${}^{10}C_8 (0.7825)^2 (0.2175)^8 = 0$	0
9	${}^{10}C_9 (0.7825)^1 (0.2175)^9 = 0$	0

3.3 Poisson Distribution

Result: Poisson distribution with $P(r) = \frac{e^{-\lambda} \lambda^r}{r!}$ is a limiting case of Binomial distribution, under the conditions **i.** $n \to \infty$ **ii.** $p \to 0$ **iii.** $np = \lambda$ is finite

Proof: In a Binomial distribution

$$P(r) = {}^nC_r\, q^{n-r} p^r$$

$$= {}^nC_r (1-p)^{n-r} p^r$$

$$= {}^nC_r \left(1 - \frac{\lambda}{n}\right)^{n-r} \left(\frac{\lambda}{n}\right)^r \qquad \because np = \lambda$$

$$= \frac{n(n-1)(n-2)\cdots(n-(r-1))}{r!} \left(1 - \frac{\lambda}{n}\right)^{n-r} \left(\frac{\lambda}{n}\right)^r$$

$$= \frac{1\left(1-\frac{1}{n}\right)\left(1-\frac{2}{n}\right)\cdots\left(1-\frac{(r-1)}{n}\right)}{\left(1-\frac{\lambda}{n}\right)^r r!} \left(1 - \frac{\lambda}{n}\right)^n \lambda^r$$

Taking limit as $n \to \infty$

$$P(r) = \frac{\lambda^r}{r!} \lim_{n \to \infty} \left(1 - \frac{\lambda}{n}\right)^n$$

$$= \frac{\lambda^r}{r!} \lim_{n \to \infty} \left[\left(1 - \frac{\lambda}{n}\right)^{\frac{-n}{\lambda}}\right]^{-\lambda}$$

$$\therefore P(r) = \frac{e^{-\lambda} \lambda^r}{r!} \qquad \because \lim_{n \to \infty} \left(1 + \frac{1}{x}\right)^x = e$$

3.3.1 Mean of Poisson Distribution

$$\text{Mean} = \sum_{r=0}^{\infty} rP(r) = \sum_{r=0}^{\infty} \frac{re^{-\lambda}\lambda^r}{r!}$$

$$= \frac{e^{-\lambda}\lambda^1}{1!} + \frac{2e^{-\lambda}\lambda^2}{2!} + \frac{3e^{-\lambda}\lambda^3}{3!} + \frac{4e^{-\lambda}\lambda^4}{4!} + \cdots$$

$$= e^{-\lambda}\lambda\left(1 + \lambda + \frac{\lambda^2}{2!} + \frac{\lambda^3}{3!} + \cdots\right)$$

$$= e^{-\lambda}\lambda\, e^{\lambda} = \lambda$$

3.3.2 Variance of Poisson Distribution

$$\text{Variance} = \sum_{r=0}^{n} r^2 P(r) - (\text{mean})^2$$

Now $\sum_{r=0}^{n} r^2 P(r) = \sum_{r=0}^{\infty} \frac{r^2 e^{-\lambda}\lambda^r}{r!}$

$$= \frac{e^{-\lambda}\lambda^1}{1!} + \frac{2^2 e^{-\lambda}\lambda^2}{2!} + \frac{3^2 e^{-\lambda}\lambda^3}{3!} + \frac{4^2 e^{-\lambda}\lambda^4}{4!} + \cdots$$

$$= e^{-\lambda}\lambda\left(1 + 2\lambda + \frac{3\lambda^2}{2!} + \frac{4\lambda^3}{3!} + \cdots\right)$$

$$= e^{-\lambda}\lambda\left[\left(1 + \lambda + \frac{\lambda^2}{2!} + \frac{\lambda^3}{3!} + \cdots\right) + \left(\lambda + \lambda^2 + \frac{3\lambda^3}{3!} + \cdots\right)\right]$$

$$= e^{-\lambda}\lambda\left[\left(1 + \lambda + \frac{\lambda^2}{2!} + \frac{\lambda^3}{3!} + \cdots\right) + \lambda\left(1 + \lambda + \frac{\lambda^2}{2!} + \frac{\lambda^3}{3!} + \cdots\right)\right]$$

$$= e^{-\lambda}\lambda[e^{\lambda} + \lambda e^{\lambda}]$$

$$\Rightarrow \sum_{r=0}^{n} r^2 P(r) = e^{-\lambda}\lambda e^{\lambda}[1 + \lambda] = \lambda + \lambda^2$$

$\therefore$ Variance $= \lambda + \lambda^2 - \lambda^2 = \lambda$

3.3.3 Moments of Poisson Distribution

Since Poisson distribution is a limiting case of binomial distribution, therefore mean and moments may be obtained from Binomial distribution by taking $np = \lambda$, $p \to 0$ and $q \to 1$ as limit $n \to \infty$

Mean $= \lim_{n \to \infty} np = \lambda$,

$$\mu_1 = 0$$

$$\mu_2 = \sigma^2 = \lim_{n \to \infty} npq = \lambda \quad \because \mu_2 = variance\,(\sigma^2)$$

Similarly $\mu_3 = \lim_{n \to \infty} npq(q-p) = \lambda.1(1-0) = \lambda$

$$\mu_4 = \lim_{n \to \infty} npq[1 + 3pq(n-2)]$$

$$= \lim_{n \to \infty} npq[1 + 3npq - 6pq]$$

$$= \lambda.1[1 + 3\lambda.1 - 6.0.1]$$

$$= \lambda + 3\lambda^2$$

$$\beta_1 = \frac{u_3^2}{u_2^3} = \frac{\lambda^2}{\lambda^3} = \frac{1}{\lambda}, \quad \beta_2 = \frac{\mu_4}{u_2^2} = \frac{\lambda + 3\lambda^2}{\lambda^2} = \frac{1}{\lambda} + 3$$

Example11: If the standard deviation of a Poisson variate X is $\sqrt{3}$, then find the probability that X is strictly positive.

Solution: Here variance $(\lambda) = 3$

$$P(X = r) = \frac{e^{-\lambda}\lambda^r}{r!} = \frac{e^{-3}3^r}{r!}, \quad r = 0, 1, 2, 3, \cdots$$

The probability that X is strictly positive is:

$$P(X > 0) = 1 - P(X = 0)$$
$$= 1 - \frac{e^{-3}3^0}{0!} = 1 - e^{-3}$$

Example12: *i.* If the probability of a bad reaction from a certain injection is 0.001, determine the chance that out of 1000 individuals more than two will have bad reaction.

ii. A manufacturer who produces medicine bottles finds that 0.1% of the bottles are defective. The bottles are packed in boxes containing 500 bottles. Find the probability that in 100 such boxes how many are expected to contain (*a*) no defective bottle (*b*) at least two defective bottles.

Solution: *i.* Here $p = 0.001$, $n = 1000$ $\lambda = np = 1000 \times 0.001 = 1$

$$P(X = r) = \frac{e^{-\lambda}\lambda^r}{r!} = \frac{e^{-1}(1)^r}{r!}$$

Probability that more than two individuals will have bad reaction is given by:

$$P(X > 2) = 1 - P(X \leq 2) = \{1 - [P(X = 0) + P(X = 1) + P(X = 2)]\}$$

$$= \left\{1 - e^{-1}\left[\frac{(1)^0}{0!} + \frac{(1)^1}{1!} + \frac{(1)^2}{2!}\right]\right\} = 1 - e^{-1}\left[\frac{5}{2}\right] = 1 - 0.9197 = 0.0803$$

ii. Here $p = 0.1\% = \frac{0.1}{100} = 0.001$, $n = 500$ $\lambda = np = 500 \times 0.001 = 0.5$

$$P(X = r) = \frac{e^{-\lambda}\lambda^r}{r!} = \frac{e^{-0.5}(0.5)^r}{r!}$$

(*a*) Probability of zero defective bottles in a box of 500 bottles is given by:

$$P(X = 0) = \frac{e^{-0.5}(0.5)^0}{0!} = e^{-0.5} = 0.6065$$

∴ Number of boxes having no defective bottle out of 100 boxes
$$= 100 \times 0.6065 = 60.65 \text{ approx}$$

(*b*) Probability of at least 2 defective bottles in a box of 500 is given by:

$$P(X \geq 2) = 1 - P(X < 2) = \{1 - [P(X = 0) + P(X = 1)]\}$$

$$= \left\{1 - e^{-0.5}\left[\frac{(0.5)^0}{0!} + \frac{(0.5)^1}{1!}\right]\right\} = 1 - e^{-0.5}[1.5] = 0.0902$$

∴ Number of boxes having at least 2 defective bottles out of 100 boxes
$$= 100 \times 0.0902 = 9.02 \text{ approx}$$

Example13 In a certain factory producing tyres, there is a small chance of 1 in 500 tyres to be defective. The tyres are supplied in lots of 10. Using Poisson distribution, calculate the approximate number of lots containing   *i.* no defective *ii.* at least one defective tyre in a consignment of 10,000 lots.

Solution: Here $p = \frac{1}{500}$, $n = 10$ $\lambda = np = \frac{10}{500} = 0.02$

$$P(X = r) = \frac{e^{-\lambda}\lambda^r}{r!} = \frac{e^{-0.02}(0.02)^r}{r!}$$

i. Probability of no defective tyre in a lot is given by:

$$P(X = 0) = \frac{e^{-0.02}(0.02)^0}{0!} = e^{-0.02} = 0.9802$$

∴ Number of lots containing no defective tyre = $10000 \times 0.9802 = 9802$ approx

ii. Probability of at least one defective tyre in a lot is given by:

$$1 - P(X = 0) = 1 - 0.9802 = 0.0198$$

∴ Number of lots containing at least one defective tyre

$$= 10000 \times 0.0198 = 198 \text{ Approx}$$

Example14: A skilled typist kept a record of his mistakes made per day during 300 working days. Fit a Poisson distribution to compare theoretical frequencies with actual ones

Mistakes per day	0	1	2	3	4	5	6
Number of days	143	90	42	12	9	3	1

Solution: Mean of the given distribution = $\frac{\sum fx}{\sum f}$, $\sum f = 300$

$$= \frac{0+90+84+36+36+15+6}{300} = \frac{89}{100} = 0.89 = \lambda$$

Mistakes per day	$P(r) = \frac{e^{-\lambda}\lambda^r}{r!}$	Theoretical frequency $300 \times P(r)$
0	$\frac{e^{-(0.89)}(0.89)^0}{0!} = 0.411$	123.3=123 (say)
1	$\frac{e^{-(0.89)}(0.89)^1}{1!} = 0.365$	109.5=110 (say)
2	$\frac{e^{-(0.89)}(0.89)^2}{2!} = 0.163$	48.9=49 (say)
3	$\frac{e^{-(0.89)}(0.89)^3}{3!} = 0.048$	14.4=14 (say)
4	$\frac{e^{-(0.89)}(0.89)^4}{4!} = 0.011$	3.3=3 (say)
5	$\frac{e^{-(0.89)}(0.89)^5}{5!} = 0.002$	0.6=1 (say)
6	$\frac{e^{-(0.89)}(0.89)^6}{6!} = 0.0003$	0.09=0 (say)

Example15: The distribution of number of road accidents per day in a city is Poisson with mean 5. Find the number of days in a year when there will be

i. at most 2 accidents *ii.* between 3 and 5 accidents

Solution: Here $\lambda = np = 5$

$$P(X = r) = \frac{e^{-\lambda}\lambda^r}{r!} = \frac{e^{-5}(5)^r}{r!}$$

i. Probability of at most 2 accidents per day is given by:

$$P(X \leq 2) = P(X = 0) + P(X = 1) + P(X = 2)$$

$$= e^{-5}\left[\frac{(5)^0}{0!} + \frac{(5)^1}{1!} + \frac{(5)^2}{2!}\right] = 0.1247$$

∴ Number of days in a year having at most 2 accidents per day

$$= 365 \times 0.1247 = 45.5 \text{ approx}$$

ii. Probability of 3 to 5 accidents per day is given by:

$$P(3 \leq X \leq 5) = P(X = 3) + P(X = 4) + P(X = 5)$$

$$= e^{-5}\left[\frac{(5)^3}{3!} + \frac{(5)^4}{4!} + \frac{(5)^5}{5!}\right] = 0.4913$$

∴ Number of days in a year having 3 to 5 accidents per day

$$= 365 \times 0.4913 = 179.3 \text{ approx}$$

3.4 Normal Distribution

The normal distribution developed by Gauss is a continuous distribution and is very useful in practical applications. It can be considered as the limiting form of the Binomial Distribution when the number of trials (n), is very large and neither p nor q is very small. The probability curve of a normal variate x with mean μ and standard deviation σ is given by:

$$p(x) = \frac{1}{\sigma\sqrt{2\pi}} e^{-\frac{1}{2}\left(\frac{x-\mu}{\sigma}\right)^2}, -\infty < x < \infty$$

Any normal variate x with mean μ and standard deviation σ is changed to a standard normal variate $z = \frac{x-\mu}{\sigma}$, and hence the probability density function of z is given by: $\phi(z) = \frac{1}{\sqrt{2\pi}} e^{-\frac{1}{2}z^2}$, $-\infty < z < \infty$

The normal distribution with mean μ and variance σ^2 is denoted by $N(\mu, \sigma^2)$.

Adjoining figure shows a normal distribution curve for standard normal variate z.

Properties of Normal Curve:

- The graph of $p(x)$ or $\phi(z)$ is a bell shaped curve.
- Since the distribution is symmetrical, mean, mode and median coincide at $x = \mu$ or $z = 0$. Also $\beta_1 = 0 \Rightarrow \gamma_1 = 0$ and $\beta_2 = 3$ $\Rightarrow \gamma_2 = 0$

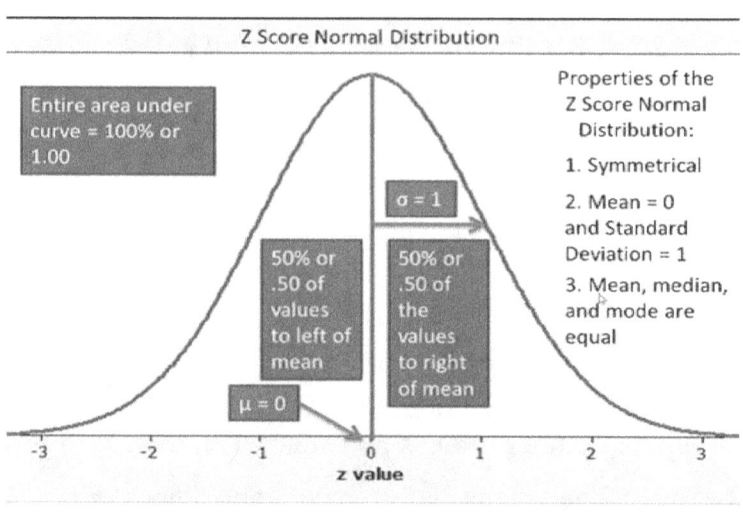

- The ordinate at $x = \mu$ or $z = 0$, divides the whole area into two equal parts. Also since the total area under the probability curve is 1, area to the right of the ordinate as well as to the left of the ordinate at $x = \mu$ or $z = 0$ is 0.5.
- Area under the curve between the ordinates $a < z < b$ gives the probability of variate z taking values between a and b. Area is concentrated more towards the middle and goes on decreasing on the either sides of the curve, i.e. tails, but never becomes zero. The curve never intersects x-axis at any finite point. i.e. x-axis is its asymptote.
- Since the curve is symmetrical about mean. The first quartile Q_1 and the third quartile Q_3 lie at the same distance on the two sides of the mean μ. The distance of any quartile from μ is $0.6745\,\sigma$ units. Thus

  $Q_1 = \mu - 0.6745\sigma$ or $\mu - \frac{2}{3}\sigma$

  $Q_3 = \mu + 0.6745\sigma$ or $\mu + \frac{2}{3}\sigma$

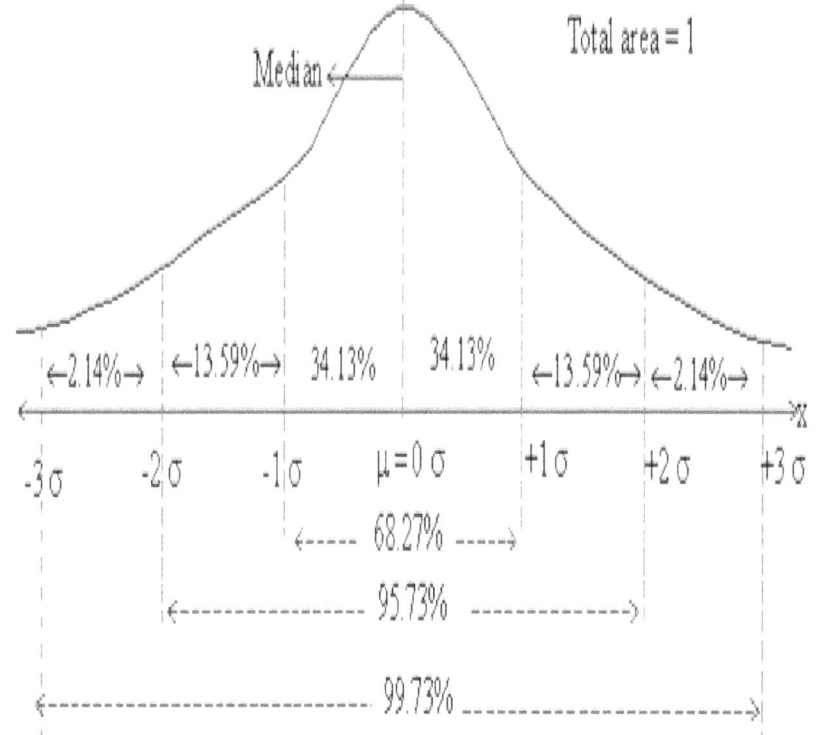

- Area under the normal curve is distributed as follows:

  Area between $x = \mu - \sigma$ and $x = \mu + \sigma$ is 18.27%

  Area between $x = \mu - 2\sigma$ and $x = \mu + 2\sigma$ is 95.45%

  Area between $x = \mu - 3\sigma$ and $x = \mu + 3\sigma$ is 99.73%

- Points of inflection are $\mu \pm \sigma$
- Quartile deviation (QD) is

  $\frac{1}{2}(Q_3 - Q_1)\,\frac{2}{3}\sigma$

- Mean deviation (MD) is $\frac{4}{5}\sigma$

  $\therefore QD:MD:SD \equiv 10:12:15$

- Since the distribution is symmetrical, all the moments of odd order about mean are zero, i.e. $\mu_{2n+1} = 0$, $n = 1, 2, 3, \cdots$
- Moments of even order are given by:

  $\mu_{2n} = 1.3.5\ldots(2n-1)\sigma^{2n}$, $n = 1, 2, 3, \cdots$

Putting $n = 1$ and 2; $\mu_2 = \sigma^2$, $\mu_4 = 3\sigma^4$, $\beta_1 = \frac{u_3^2}{u_2^3} = 0$, $\beta_2 = \frac{\mu_4}{u_2^2} = \frac{3\sigma^4}{\sigma^4} = 3$

Using Normal Distribution tables: We first convert the variate x into standard normal variate z; using the relation $z = \frac{x-\mu}{\sigma}$ and find the limits of z-score corresponding to the given limits of the variate x. Normal Distribution tables are broadly of two types; either of the two may be used to determine the area (probability) for the standard normal variate z usually given in the range $(-3.49 \leq z \leq 3.49)$.

The first type of table gives the area covered by standard normal variate z between the ordinates 0 to z as shown in adjoining figure. This table covers more than 0.499 units of area on

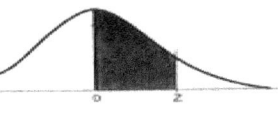

positive side of the curve on the whole. If value of z is negative, we may use the symmetrical property of the normal curve, i.e area in the region $z < -a$ is same as area $z > a$ where a is any positive number within the range $(0, 3.49)$.

The second table gives the area covered by standard normal variate z from $-\infty$ to z as shown in the given figure. This table covers more than 0.998 units of area on the whole.

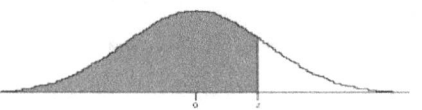

3.4.1 Moments of Normal Distribution

Result: Odd order moments about mean are zero

Proof: r^{th} moment about the mean $\bar{x}$ of a continuous distribution is given by:

$$\mu_r = \int_{-\infty}^{\infty} \frac{f(x-\bar{x})^r}{n} dx$$

∴ Odd order moments of normal distribution with mean μ are given by:

$$\mu_{2n+1} = \int_{-\infty}^{\infty} (x-\mu)^{2n+1} p(x) dx \quad \because \frac{f}{n} = p(x)$$

$$= \frac{1}{\sigma\sqrt{2\pi}} \int_{-\infty}^{\infty} (x-\mu)^{2n+1} e^{-\frac{1}{2}\left(\frac{x-\mu}{\sigma}\right)^2} dx \quad \because p(x) = \frac{1}{\sigma\sqrt{2\pi}} e^{-\frac{1}{2}\left(\frac{x-\mu}{\sigma}\right)^2}$$

$$= \frac{\sigma^{2n+1}}{\sqrt{2\pi}} \int_{-\infty}^{\infty} z^{2n+1} e^{-\frac{1}{2}z^2} dz \text{ Putting } \frac{x-\mu}{\sigma} = z, \; dx = \sigma \, dz$$

$$= 0, \; z^{2n+1} e^{-\frac{1}{2}z^2} \text{ being an odd function of } z.$$

Result: Even order moments about mean are given by:

$$\mu_{2n} = 1.3.5 \ldots (2n-1)\sigma^{2n}, \; n = 1, 2, 3, \cdots$$

Proof: $\mu_{2n} = \int_{-\infty}^{\infty} (x-\mu)^{2n} p(x) dx$

$$= \frac{1}{\sigma\sqrt{2\pi}} \int_{-\infty}^{\infty} (x-\mu)^{2n} e^{-\frac{1}{2}\left(\frac{x-\mu}{\sigma}\right)^2} dx \quad \because p(x) = \frac{1}{\sigma\sqrt{2\pi}} e^{-\frac{1}{2}\left(\frac{x-\mu}{\sigma}\right)^2}$$

$$= \frac{\sigma^{2n}}{\sqrt{2\pi}} \int_{-\infty}^{\infty} z^{2n} e^{-\frac{1}{2}z^2} dz \text{ Putting } \frac{x-\mu}{\sigma} = z \Rightarrow dx = \sigma \, dz$$

$$= \frac{2\sigma^{2n}}{\sqrt{2\pi}} \int_{0}^{\infty} z^{2n} e^{-\frac{1}{2}z^2} dz, \; z^{2n} e^{-\frac{1}{2}z^2} \text{ being even function of } z.$$

$$= \frac{2\sigma^{2n}}{\sqrt{2\pi}} \int_{0}^{\infty} (2t)^n e^{-t} (2t)^{-\frac{1}{2}} dt \text{ Putting } \frac{z^2}{2} = t \Rightarrow z \, dz = dt$$

$$= \frac{2^n \sigma^{2n}}{\sqrt{\pi}} \int_{0}^{\infty} e^{-t} t^{n-\frac{1}{2}} dt$$

$$\mu_{2n} = \frac{2^n \sigma^{2n}}{\sqrt{\pi}} \Gamma\left(n + \frac{1}{2}\right) \qquad \because \Gamma n = \int_{0}^{\infty} e^{-t} t^{n-1} dt$$

Again changing n to $(n-1)$

$$\mu_{2n-2} = \frac{2^{n-1} \sigma^{2n-2}}{\sqrt{\pi}} \Gamma\left(n - \frac{1}{2}\right)$$

$$\therefore \frac{\mu_{2n}}{\mu_{2n-2}} = 2\sigma^2 \frac{\Gamma\left(n+\frac{1}{2}\right)}{\Gamma\left(n-\frac{1}{2}\right)} = 2\sigma^2 \frac{\left(n-\frac{1}{2}\right)\Gamma\left(n-\frac{1}{2}\right)}{\Gamma\left(n-\frac{1}{2}\right)} = 2\sigma^2 \left(n - \frac{1}{2}\right)$$

$$\therefore \mu_{2n} = \sigma^2 (2n-1)\mu_{2n-2}$$

$$\Rightarrow \mu_{2n} = \sigma^2(2n-1)\sigma^2(2n-3)\mu_{2n-4}$$
$$= \sigma^2(2n-1)\sigma^2(2n-3)\sigma^2(2n-5)\mu_{2n-6}$$
$$\vdots$$
$$\therefore \mu_{2n} = 1.3.5\ldots(2n-5)(2n-3)(2n-1)\sigma^{2n}, n = 1, 2, 3, \cdots$$

Example 16 For a normal distribution the mean 20 and the standard deviation 15, find *i.* Q_1 and Q_3 *ii.* Mean deviation *iii.* the inter quartile range.

Solution: *i.* For a normal distribution
$$Q_1 = \mu - \frac{2}{3}\sigma = 20 - \frac{2}{3}(15) = 10$$
$$Q_3 = \mu + \frac{2}{3}\sigma = 20 + \frac{2}{3}(15) = 30$$

ii. Mean deviation is $\frac{4}{5}\sigma = \frac{4}{5}(15) = 12$

iii. The inter quartile range is $Q_3 - Q_1 = 20$

Example 17 Find p, mean and the standard deviation of the normal distribution given by $y = pe^{-\left(\frac{x^2}{8} - x + 2\right)}$

Solution: Rewriting $\left(\frac{x^2}{8} - x + 2\right)$ as $\frac{1}{2}\left(\frac{x^2 - 8x + 16}{4}\right) = \frac{1}{2}\left(\frac{x-4}{2}\right)^2$

$$\therefore y = pe^{-\frac{1}{2}\left(\frac{x-4}{2}\right)^2}$$

Comparing with $y = \frac{1}{\sigma\sqrt{2\pi}}e^{-\frac{1}{2}\left(\frac{x-\mu}{\sigma}\right)^2}$, we get $\mu = 4, \sigma = 2, p = \frac{1}{\sigma\sqrt{2\pi}} = \frac{1}{2\sqrt{2\pi}}$

Example 18 If X is a random variable with mean 20 and standard deviation 5, find the probabilities that (i) $15 \leq X \leq 25$ (ii) $|X - 20| > 5$

Solution: Given that X is a random variable with mean $\mu = 20$ and S.D. $\sigma = 5$.

i.e. $X \sim N(100, 25)$ and $z = \frac{X - \mu}{\sigma} = \frac{X - 20}{5} \Rightarrow X = 20 + 5z$

(i) $P(15 \leq X \leq 25)$

$P(-1 < z < 1)$

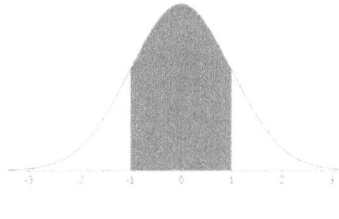

$= P(15 < 20 + 5z < 25)$
$= P\left(\frac{15-20}{5} < z < \frac{25-20}{5}\right)$
$= P(-1 < z < 1) = 2P(0 < z < 1)$
$= 2(0.3413)$ Using normal table
$= 0.6826$

(ii) $P(|X - 20| > 5) = 1 - P(|X - 20| \leq 5)$

Now $P(|X - 20| \leq 5) = P(-5 \leq X - 20 \leq 5) = P(15 \leq X \leq 25)$
$= 0.6826$ using part (i)

$\therefore P(|X - 20| > 5) = 1 - 0.6826 = 0.3174$

Example 19 Fifteen hundred candidates appeared in a certain examination, having maximum marks as 100. It was found that the marks are normally distributed with mean 55 and standard deviation as 10.5. Determine approximately the number of candidates who were passed with distinction, i.e. 75% and above marks.

Solution: Let the random variable X denote the marks obtained out of 100.

Then X is a random variable with mean $\mu = 55$ and S.D. $\sigma = 10.5$.

i.e. $X \sim N(55, 10.5^2)$ and $z = \frac{X-\mu}{\sigma} = \frac{X-55}{10.5}$

$\Rightarrow X = 55 + 10.5z$

$P(X \geq 75)$

$= P(55 + 10.5z \geq 75)$

$= P\left(z \geq \frac{75-55}{10.5}\right) = P(z \geq 1.9)$

$= 0.5 - P(0 < z < 1.9)$

$= 0.5 - 0.4713 = 0.0287$

$P(z \geq 1.9)$

∴Number of candidates who passed with distinction is:

$1500 \times 0.0287 = 43.05$ i.e. 43 approximately

Example 20 The daily wages of 1000 workers are normally distributed with mean 100$ and standard deviation 5$. Estimate the number of workers whose daily wages will be:

(i) between 100$ and 105$ (ii) between 96$ and 105$

(iii) more than 110$ (iv) more than 110$

(v) Also estimate the daily wages of 100 highest paid workers.

Solution: Let the random variable X denote the daily wages of workers in dollars.

Then X is a random variable with mean $\mu = 100$ and S.D. $\sigma = 5$.

i.e. $X \sim N(100, 25)$ and $z = \frac{X-\mu}{\sigma} = \frac{X-100}{5}$

$\Rightarrow X = 100 + 5z$

(i) $P(100 < X < 105)$

$= P(100 < 100 + 5z < 105)$

$= P\left(\frac{100-100}{5} < z < \frac{105-100}{5}\right)$

$= P(0 < z < 1) = 0.3413$ using Z table

(ii) $P(96 < X < 105)$

$= P(96 < 100 + 5z < 105)$

$= P\left(\frac{96-100}{5} < z < \frac{105-100}{5}\right)$

$= P(-0.8 < z < 1)$

$P(0 < z < 1)$

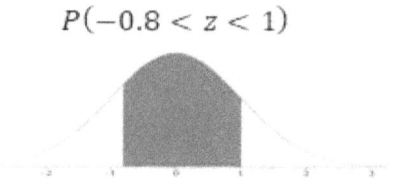

$P(-0.8 < z < 1)$

$$= P(0 < z < 0.8) + P(0 < z < 1)$$
$$= 0.2881 + 0.3413 = 0.6294 \text{ using Z table}$$

(iii) $P(X > 110)$
$$= P(100 + 5z > 110)$$
$$= P\left(z > \frac{110-100}{5}\right)$$
$$= P(z > 2)$$
$$= 0.5 - P(0 < z < 2)$$
$$= 0.5 - 0.4772 = 0.0228 \text{ using Z table}$$

$P(z > 2)$

(iv) $P(X < 92)$
$$= P(100 + 5z < 92)$$
$$= P\left(z < \frac{92-100}{5}\right)$$
$$= P(z < -1.6)$$
$$= P(z > 1.6) \text{ (Symmetry)}$$
$$= 0.5 - P(0 < z < 1.6)$$
$$= 0.5 - 0.4452 = 0.0548 \text{ using Z table}$$

$P(z < -1.6)$

(v) Proportion of 100 highest paid workers is $\frac{100}{1000} = \frac{1}{10} = 0.1$

To determine $X = r$ such that $P(X > r) = 0.1$

When $X = r$, $z = \frac{r-100}{5} = z_1$ (say)

$\therefore P(z > z_1) = 0.1$

$\Rightarrow P(0 < z < z_1) = 0.5 - 0.1 = 0.4$

From normal distribution table, $z_1 = 1.28$ approx.

$\therefore z = \frac{r-100}{5} = 1.28$

$\Rightarrow r = 100 + 5 \times 1.28 = 106.4$

Hence the lowest daily wages of 100 highest paid workers are 106.4\$

Example 21 If in a normal distribution 7% of the items are under 35 and 89% are under 63. What are the mean and standard deviations of the distribution?

Solution: Let X be normally distributed with mean μ and standard deviation σ such that $z = \frac{X-\mu}{\sigma}$.
Also given that $P(X < 35) = 0.07$ and $P(X < 63) = 0.89$
Let z_1 be a standard normal variate corresponding to $X = 35$,

$\therefore z_1 = \frac{35-\mu}{\sigma}$ and $P(z_1 < 35) = 0.07$

$\Rightarrow P(-z_1 > 35) = 0.07$

$\Rightarrow P(0 < -z_1 < z) = 0.5 - 0.07 = 0.43$

$\Rightarrow -z_1 = 1.48$ From normal distribution table

or $z_1 = \frac{35-\mu}{\sigma} = -1.48$

$\Rightarrow \mu - 1.48\sigma = 35$ ... ①

Again let z_2 be a standard normal variate corresponding to $X = 63$

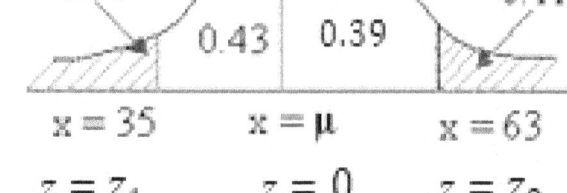

$\therefore z_2 = \frac{63-\mu}{\sigma}$ and $P(z_2 < 63) = 0.89$

$\Rightarrow P(0 < z_2 < z) = 0.89 - 0.5 = 0.39$

$\Rightarrow z_2 = \frac{63-\mu}{\sigma} = 1.23$ from normal distribution table

$\Rightarrow \mu + 1.23\sigma = 63$ ... ②

Solving ① and ②, we get: $\mu = 109.6$, $\sigma = 50.4$

Example 21: Fit a normal curve to the following distribution

x	2	4	6	8	10
f	1	4	6	4	1

Solution: Let x be normally distributed with mean μ and standard deviation σ such that $z = \frac{X-\mu}{\sigma}$.

$$\mu = \frac{\Sigma fx}{\Sigma f} = \frac{2+16+36+32+10}{16} = 6$$

Also $\sigma = \sqrt{\frac{\Sigma fx^2}{\Sigma f} - \left(\frac{\Sigma fx}{\Sigma f}\right)^2} = \sqrt{\frac{(4+64+216+256+100)}{16} - 36} = \sqrt{40 - 36} = 2$

Again x has to be a continuous variable to follow normal distribution, therefore taking x as mid value of an interval (x_1, x_2)

x	(x_1, x_2)	(z_1, z_2)	$P(z) \equiv$ Area under the curve in the interval (z_1, z_2)	Theoretical frequency $16 \times P(z)$
2	(1, 3)	(-2.5, -1.5)	0.0606	0.9696
4	(3, 5)	(-1.5, -0.5)	0.2417	3.8672
6	(5, 7)	(-0.5, 0.5)	0.3829	6.1264
8	(7, 9)	(0.5, 1.5)	0.2417	3.8672
10	(9, 11)	(1.5, 2.5)	0.0606	0.9696

Exercise 3

1. In 256 sets of 12 tosses of a coin, in how many cases can one expect 8 heads and 4 tails?
2. In a precision bombing attack, there are 50% chances that any one bomb will hit the target. If two direct hits are required to destroy the target completely, how many bombs must be dropped to provide a 99% or more chances to completely destroy the target.

3. Comment on the statement: For a binomial distribution mean is 5 and standard deviation is 3.
4. If the probability of a defective item is 0.02, find the probability that at most 5 defective items will be found out in a box of 200 items.
5. Six coins are tossed 6400 times. Using Poisson distribution, find the approximate probability of getting six heads 2 times.
6. In a certain factory making a machine part, probability of it being defective is 0.002. If the part is supplied in packs of 10, use Poisson distribution to calculate the approximate number of packets containing no defective and one defective machine part in a consignment of 10,000 packets.
7. A car hire firm has 2 cars, which are hired on daily basis. The number of demands for a car on each day follows Poisson distribution with mean 1.5. Calculate the proportion of days on which neither car is used and the proportion of days on which some demand is refused, given that $e^{-1.5} = 0.2231$.
8. If in a Poisson distribution $P(X = 0) = P(X = 1) = a$, show that $a = \frac{1}{e}$.
9. If X is a random variable with mean 30 and standard deviation 5, find the probabilities that (i) $X > 45$ (ii) $|X - 30| \geq 5$
10. If in a normal distribution 31% of the items are under 45 and 8% are over 64. What are the mean and standard deviation of the distribution?
11. In an intelligence test given to 1000 children, the average score is 42 with the standard deviation 24. Find $i.$ the number of children whose score exceeds 60

    $ii.$ the number of children whose score lie between 20 and 40.
12. Assuming mean height of the soldiers to be 68.22 inches with a variance of 10.8 square inches, how many soldiers in the regiment of 1000 would you expect to be over 6 feet tall, given that area under the standard normal curve between $z = 0$ to $z = 1.15$ is 0.3746.
13. Fit a binomial distribution for the given data:

x	0	1	2	3	4	5	6	7	8
f	0	5	9	22	25	26	14	4	1

Answers

1. 31
2. 11
3. Variance cannot be greater than mean in case of a binomial distribution.
4. 0.7845
5. $5000 \, e^{-100}$
6. 9802, 196
7. 0.2231, 0.1913
9. 0.0014, 0.3174
10. 50, 10
11. 227, 287
12. 125
13. 0.4, 3.3, 11.6, 23.2, 29, 23.2, 11.6, 3.3, 0.4

Chapter 4: Curve Fitting & Correlation

Curve Fitting & Correlation

4.1 Introduction

The process of constructing an approximate curve $y = f(x)$, which fit best to a given discrete set of points (x_i, y_i), $i = 1, 2, 3, \ldots, n$ is called curve fitting. Curve fitting and interpolation are closely associated procedures. In interpolation, the fitted function should pass through all given data points; whereas curve fitting methodologically fits a unique curve to the data points, which may or may not lie on the fitted curve. The difference between interpolation and curve fitting; while attempting to fit a linear function; is illustrated in the adjoining figure.

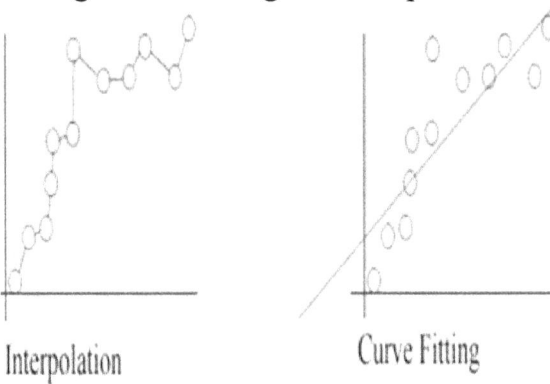

Interpolation   Curve Fitting

4.2 Principle of Least Squares

The principle of least squares is one of the most popular methods for finding the curve of best fit to a given data set (x_i, y_i), $i = 1, 2, 3, \ldots, n$.

Let $y = f(x)$ be the equation of the curve to be fitted to the given set of points $P_1(x_1, y_1)$, $P_2(x_2, y_2), \cdots, P_n(x_n, y_n)$.

Then at a point $x = x_i$, the observed value of the ordinate is $P_i M_i = y_i$ say and let the expected (theoretical) value be $f(x_i)$, shown by $L_i M_i$ in the adjoining figure.

The difference between the observed and expected values is the error $e_i = P_i L_i$

Then $e_1 = y_1 - f(x_1)$

$e_2 = y_2 - f(x_2)$

$\vdots$

$e_n = y_n - f(x_n)$

Squaring each error e_i (to take care of negative errors) and adding, we get

$E = e_1^2 + e_2^2 + \cdots + e_n^2 = \sum_{i=1}^{n} e_i^2$

$= \sum_{i=1}^{n}(y_i - f(x_i))^2$. The curve of best fit is that for which E is minimum. This is called the Principle of least squares.

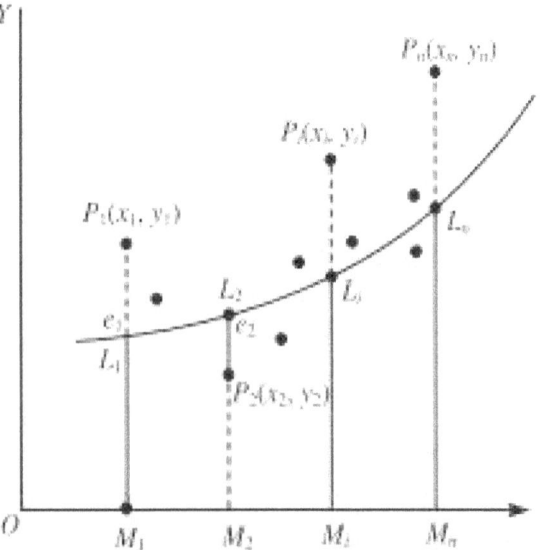

4.2.1 Fitting a straight line

Let $y = ax + b$ be the straight line to be fitted to the given set of data points (x_1, y_1), $(x_2, y_2), \ldots, (x_n, y_n)$.

Then $e_i = y_i - f(x_i) = y_i - (ax_i + b), \quad i = 1, 2, \ldots, n$

$$\therefore E = \sum_{i=1}^{n} e_i^2 = \sum_{i=1}^{n}(y_i - (ax_i + b))^2$$

Now by the principle of least square, for the curve of best fit, E is minimum

$$\therefore \frac{\partial E}{\partial a} = 0 \text{ and } \frac{\partial E}{\partial b} = 0$$

$$\therefore \frac{\partial E}{\partial a} = 0 \quad \text{and} \quad \frac{\partial E}{\partial b} = 0$$

$\Rightarrow \sum_{i=1}^{n} 2(y_i - (ax_i + b))(-x_i) = 0$ $\quad\quad \Rightarrow \sum_{i=1}^{n} 2(y_i - ax_i - b)(-1) = 0$

$\Rightarrow \sum_{i=1}^{n}(y_i x_i - ax_i^2 - bx_i) = 0$ $\quad\quad \Rightarrow \sum_{i=1}^{n}(y_i - ax_i - b) = 0$

$\Rightarrow \sum_{i=1}^{n} y_i x_i = a\sum_{i=1}^{n} x_i^2 + b\sum_{i=1}^{n} x_i$ $\quad \Rightarrow \sum_{i=1}^{n} y_i = a\sum_{i=1}^{n} x_i + nb$

$\quad\quad\quad\quad\quad\quad\quad \ldots \text{①}$ $\quad\quad\quad\quad\quad\quad\quad\quad\quad\quad\quad \ldots \text{②}$

Since x_i, y_i are known set of values, equation ① and ② are two equation with variables a and b, ignoring the suffices, ② and ① can be rewritten as

$$\sum y = a\sum x + nb \quad \quad \ldots \text{③}$$

and $\quad \sum xy = a\sum x^2 + b\sum x \quad \quad \ldots \text{④}$

③ and ④ are known as Normal equation for fitting a straight line $y = ax + b$

> If the equation of line is taken as $y = a + bx$
> we get normal equations as: $\sum y = na + b\sum x$ and $\sum xy = a\sum x + b\sum x^2$

Example 1 By the method of least squares, find a straight line that best fits the following data points.

x	0	1	2	3	4
y	1.0	2.9	4.8	6.7	8.6

Solution: Let line of best fit be given by $y = ax + b$ $\quad \ldots \text{①}$

Normal equations are given by:

$$\sum y = a\sum x + nb \quad \quad \ldots \text{②}$$

and $\quad \sum xy = a\sum x^2 + b\sum x \quad \quad \ldots \text{③}$

Calculating $\sum x, \sum y, \sum xy$ and $\sum x^2$

x	y	xy	x^2
0	1.0	0	0
1	2.9	2.9	1
2	4.8	9.6	4
3	6.7	20.1	9
4	8.6	34.4	16
$\sum x = 10$	$\sum y = 24$	$\sum xy = 67.0$	$\sum x^2 = 30$

Substituting values of $\sum x, \sum y, \sum xy$ and $\sum x^2$ in ② and ③

$$\Rightarrow 24 = 10a + 5b \quad \ldots ④$$

and $\quad 67 = 30a + 10b \quad \ldots ⑤$

Solving ④ and ⑤, we get $a = 1.9$ and $b = 1$

Substituting in ①, line of best fit is $y = 1.9x + 1$

Example 2 Fit a straight line to following data

x	0	1	2	3	4
y	1.0	1.8	3.3	4.5	6.3

Solution: Let line of best fit be given by $y = ax + b \quad \ldots ①$

Normal equations are given by:

$$\sum y = a \sum x + nb \quad \ldots ②$$

and $\quad \sum xy = a \sum x^2 + b \sum x \quad \ldots ③$

Calculating $\sum x, \sum y, \sum xy$ and $\sum x^2$

x	y	xy	x^2
0	1.0	0	0
1	1.8	1.8	1
2	3.3	6.6	4
3	4.5	13.5	9
4	6.3	25.2	16
$\sum x = 10$	$\sum y = 16.9$	$\sum xy = 47.1$	$\sum x^2 = 30$

Substituting values of $\sum x, \sum y, \sum xy$ and $\sum x^2$ in ② and ③

$$\Rightarrow 16.9 = 10a + 5b \quad \ldots ④$$

and $\quad 47.1 = 30a + 10b \quad \ldots ⑤$

Solving ④ and ⑤, we get $a = \frac{133}{100} = 1.33$ and $b = \frac{18}{25} = 0.72$

Substituting in ①, line of best fit is $y = 1.33x + 0.72$

Example 3 If F is the force required to lift a load W, by means of a pulley, fit a linear expression $F = a + bW$ against the following data:

W	50	70	100	120
F	12	15	21	25

Solution: Line for best fit is given as $\quad F = a + bW \quad$ ... ①

Normal equations are given by:

$$\sum F = na + b\sum W \quad \text{... ②}$$

and $\quad \sum WF = a\sum W + b\sum W^2 \quad$ ... ③

W	F	WF	W^2
50	12	600	2500
70	15	1050	4900
100	21	2100	10000
120	25	3000	14400
$\sum W = 340$	$\sum F = 73$	$\sum WF = 6750$	$\sum W^2 = 31800$

Substituting values of $\sum W, \sum F, \sum WF$ and $\sum W^2$ in ② and ③

$$\Rightarrow 73 = 4a + 340b \quad \text{... ④}$$

and $\quad 6750 = 340a + 31800b \quad$ ... ⑤

Solving ④ and ⑤, we get $a = 2.2759$ and $b = 0.1879$

Substituting in ①, line of best fit is $F = 2.2759 + 0.1879W$

4.2.2 Fitting a parabola

Let $y = ax^2 + bx + c$ be the parabola to be fitted to the given set of data points (x_1, y_1), $(x_2, y_2), \ldots, (x_n, y_n)$.

Then $\quad e_i = y_i - f(x_i) = y_i - (ax_i^2 + bx_i + c), \quad i = 1, 2, \ldots, n$

$$\therefore E = \sum_{i=1}^{n} e_i^2 = \sum_{i=1}^{n}(y_i - (ax_i^2 + bx_i + c))^2$$

Now by the principle of least square, for the curve of best fit, E is minimum

$$\therefore \frac{\partial E}{\partial a} = 0, \quad \frac{\partial E}{\partial b} = 0 \text{ and } \frac{\partial E}{\partial c} = 0$$

Solving we get normal equations as:

$$\sum y = a\sum x^2 + b\sum x + nc$$

$$\sum xy = a\sum x^3 + b\sum x^2 + c\sum x$$

$$\sum x^2 y = a\sum x^4 + b\sum x^3 + c\sum x^2$$

Example 4 Fit a parabola $y = ax^2 + bx + c$ to the given data

x	10	12	15	23	20
y	14	17	23	25	21

Solution: Let the parabola of best fit be given by $y = ax^2 + bx + c$ ... ①

Normal equations are given by:

$$\sum y = a\sum x^2 + b\sum x + nc \quad \ldots ②$$

$$\sum xy = a\sum x^3 + b\sum x^2 + c\sum x \quad \ldots ③$$

$$\sum x^2y = a\sum x^4 + b\sum x^3 + c\sum x^2 \quad \ldots ④$$

x	y	xy	x^2	x^2y	x^3	x^4
10	14	140	100	1400	1000	10000
12	17	204	144	2448	1728	20736
15	23	345	225	5175	3375	50625
23	25	575	529	13225	12167	279841
20	21	420	400	8400	8000	160000
$\sum x = 80$	$\sum y = 100$	$\sum xy = 1684$	$\sum x^2 = 1398$	$\sum x^2y = 30684$	$\sum x^3 = 26270$	$\sum x^4 = 521202$

Substituting values of $\sum x, \sum y, \sum xy$ and $\sum x^2$ in ② and ③ and ④

$$\Rightarrow 100 = 1398a + 80b + 5c \quad \ldots ⑤$$

$$1684 = 26270a + 1398b + 80c \quad \ldots ⑥$$

$$30648 = 521202a + 26270b + 1398c \quad \ldots ⑦$$

Solving ⑤ ⑥ and ⑦, we get $a = -0.07$, $b = 3.01$, $c = -8.73$

Substituting in ①, parabola of best fit is $y = -0.07 x^2 + 3.01x - 8.73$

Example 5 Fit a *2nd* parabola to the given data

x	1	3	4	6	8	9	11	14
y	1	2	4	4	5	7	8	9

Solution: Let the parabola of best fit be given by $y = ax^2 + bx + c$ ... ①

x	y	xy	x^2	x^2y	x^3	x^4
1	1	1	1	1	1	1
3	2	6	9	18	27	81
4	4	16	16	64	64	256
6	4	24	36	144	216	1296
8	5	40	64	320	512	4096
9	7	63	81	567	729	6561
11	8	88	121	968	1331	14641
14	9	126	196	1764	2744	38416
$\sum x = 56$	$\sum y = 40$	$\sum xy = 364$	$\sum x^2 = 524$	$\sum x^2y = 3846$	$\sum x^3 = 5624$	$\sum x^4 = 65348$

Normal equations are given by:

$$\sum y = a \sum x^2 + b \sum x + nc \qquad \ldots ②$$

$$\sum xy = a \sum x^3 + b \sum x^2 + c \sum x \qquad \ldots ③$$

$$\sum x^2 y = a \sum x^4 + b \sum x^3 + c \sum x^2 \qquad \ldots ④$$

Substituting values of $\sum x, \sum y, \sum xy$ and $\sum x^2$ in ② and ③ and ④

$$\Rightarrow 40 = 524a + 56b + 8c \qquad \ldots ⑤$$

$$364 = 5624a + 524b + 56c \qquad \ldots ⑥$$

$$3846 = 65348a + 5624b + 524c \qquad \ldots ⑦$$

Solving ⑤ ⑥ and ⑦ , we get

$a = \frac{103}{11229} = 0.009$, $b = \frac{8672}{11229} = 0.77$, $c = \frac{4375}{22458} = 0.195$

Substituting in ①, parabola of best fit is $y = 0.009 x^2 + 0.77x + 0.195$

4.2.3 Change of Scale

If the data values are equispaced (with height (h)) and quite large for computation, simplification may be done by origin shifting as given below:

- When number of observations (n) is odd, take the origin at middle value of the table; say (x_0) and substitute $u = \frac{x-x_0}{h}$
- y values if small; may be left unchanged; or we can shift them at average value of y data $v = \frac{y-y_0}{h}$
- When number of observations (n) is even, take the origin as mean of two middle values, with new height $\frac{h}{2}$ and substitute $u = \frac{x-x_0}{h/2}$

Example 6 The weight of a calf taken at end of every month is given below. Fit a straight line using the method of least squares. Also compute monthly growth rate.

x	1	2	3	4	5	6	7	8	9	10
y	52.5	58.7	65.0	70.2	75.4	81.1	87.2	95.5	102.2	108.4

Solution: Here $n = 10$ is even, $\therefore$ taking origin at $\frac{5+6}{2} = 5.5$ and new height as

$\frac{h}{2} = 0.5$ $\therefore u = \frac{x-5.5}{0.5}$ and let $v = y$

Let line of best fit $y = ax + b$ be transformed to $v = Au + B$ ...①

Normal equations are given by $\sum v = a \sum u + nb$ ...②

and $\sum uv = a \sum u^2 + b \sum u$ ...③

Calculating $\sum u, \sum u^2, \sum v$ and $\sum uv$

x	$u = \dfrac{x - 5.5}{0.5}$	$v = y$	uv	u^2
1	-9	52.5	-472.5	81
2	-7	58.7	-410.9	49
3	-5	65.0	-325.0	25
4	-3	70.2	-210.6	9
5	-1	75.4	-75.4	1
6	1	81.1	81.1	1
7	3	87.2	261.6	9
8	5	95.5	477.5	25
9	7	102.2	715.4	49
10	9	108.4	975.6	81
	$\sum u = 0$	$\sum v = 796.2$	$\sum uv = 1016.8$	$\sum u^2 = 330$

Substituting values of $\sum u, \sum v, \sum uv$ and $\sum u^2$ in ② and ③

$\Rightarrow 796.2 = 10B$ and $1016.8 = 330a$

$\therefore A = 3.081$ and $B = 79.62$

Substituting in ①, line of best fit is $v = 3.081u + 79.62$

$\Rightarrow y = 3.081\left(\dfrac{x-5.5}{0.5}\right) + 79.62$

$\therefore$ Line of best fit is $y = 6.162x + 45.729$

Average growth rate per month is given by: $\dfrac{dy}{dx} = 6.162$

Example 6 Fit a 2nd degree parabola for the following data:

x	1929	1930	1931	1932	1933	1934	1935	1936	1937
y	352	356	357	358	360	361	361	360	359

Solution: Since number of observations is odd and $h = 1$,

taking $x_0 = 1933, y_0 = 357, u = x - 1933, v = y - 357$

The equation $y = ax^2 + bx + c$ is transformed to $v = Au^2 + Bu + C$ ... ①

Normal equations are

$$\sum v = A\sum u^2 + B\sum u + 9c \qquad ...②$$
$$\sum uv = A\sum u^3 + B\sum u^2 + c\sum u \qquad ...③$$
$$\sum u^2 v = A\sum u^4 + B\sum u^3 + c\sum u^2 \qquad ...④$$

Calculating $\sum u, \sum u^2, \sum u^3, \sum u^4, \sum v, \sum uv$ and $\sum u^2 v$

x	u	y	v	uv	u^2	u^2v	u^3	u^4
1929	-4	352	-5	20	16	-80	-64	256
1930	-3	356	-1	3	9	-9	-27	81
1931	-2	357	0	0	4	0	-8	16
1932	-1	358	1	-1	1	1	-1	1
1933	0	360	3	0	0	0	0	0
1934	1	361	4	4	1	4	1	1
1935	2	361	4	8	4	16	8	16
1936	3	360	3	9	9	27	27	81
1937	4	359	2	8	16	32	64	256
	$\sum u = 0$		$\sum v = 11$	$\sum uv = 51$	$\sum u^2 = 60$	$\sum u^2 v = -9$	$\sum u^3 = 0$	$\sum u^4 = 708$

Substituting $\sum u, \sum u^2, \sum u^3, \sum u^4, \sum v, \sum uv$ and $\sum u^2 v$ in ② and ③ and ④

$$\Rightarrow 11 = 60A + 9C \qquad \ldots ⑤$$

$$51 = 60B \qquad \ldots ⑥$$

$$-9 = 708A + 60C \qquad \ldots ⑦$$

Solving ⑤ ⑥ and ⑦, we get $A = \frac{-247}{924}$ and $B = \frac{17}{20}$, $C = \frac{694}{231}$

Substituting in ①, parabola of best fit is $v = \frac{-247}{924} x^2 + \frac{17}{20} x + \frac{694}{231}$

$$\Rightarrow y - 357 = \frac{-247}{924}(x - 1933)^2 + \frac{17}{20}(x - 1933) + \frac{694}{231}$$

$$\Rightarrow y = -0.267 x^2 + 1034.29 x - 1000106.41$$

4.3 Correlation

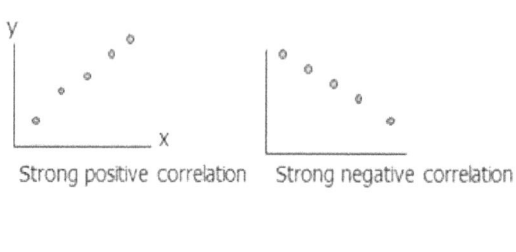

Correlation is a measure of association between two variables; which may be dependent or independent. Whenever two variables x and y are so related; that increase in one is accompanied by an increase or decrease in the other, then the variables are said to be correlated. Coefficient of correlation (r) lies between -1 and $+1$, i.e. $-1 \leq r \leq 1$.

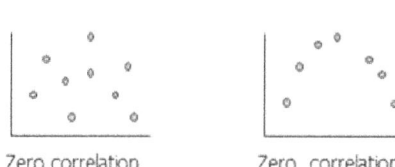

If r is zero; no correlation between two variables, positive correlation ($0 < r \leq +1$); when both variables increase or decrease simultaneously, and negative correlation ($-1 \leq r < 0$); when increase in one is associated with decrease in other variable and vice-versa.

4.4 Karl Pearson Coefficient of Correlation

Coefficient of correlation (r) between two variables x and y is defined as

$$r = \frac{Covariance\ (x,y)}{\sqrt{Variance\ (x)}\sqrt{Variance\ (y)}} = \frac{\sum d_x d_y}{\sqrt{(\sum d_x^2)(\sum d_y^2)}} = \frac{\rho}{\sigma_x \sigma_y}$$

where $d_x = x - \bar{x}$, $d_y = y - \bar{y}$, $\bar{x}, \bar{y}$ are means of x and y data values.

$\rho = Cov(x,y) = \frac{\Sigma d_x d_y}{n}$ is the covariance between the variables x and y.

Also $\sigma_x = \sqrt{\frac{\Sigma d_x^2}{n}}$ and $\sigma_y = \sqrt{\frac{\Sigma d_y^2}{n}}$

Example 8 If $Cov(x,y) = 10$, $var(x) = 25$, $var(y) = 9$ find coefficient of correlation.

Solution: $r = \frac{Cov(x,y)}{\sqrt{Var(x)}\sqrt{Var(y)}} = \frac{10}{\sqrt{25}\sqrt{9}} = \frac{10}{5\times 3} = 0.67$

Example 9 Calculate coefficient of correlation from the following data:

x	9	8	7	6	5	4	3	2	1
y	15	16	14	13	11	12	10	8	9

Solution: Karl Pearson coefficient of correlation (r) is given by: $r = \frac{\Sigma d_x d_y}{\sqrt{(\Sigma d_x^2)(\Sigma d_y^2)}}$

where $d_x = x - \bar{x}$, $d_y = y - \bar{y}$, $\bar{x}, \bar{y}$ are means of x and y data values.

x	d_x $(x-\bar{x})$	d_x^2	y	d_y $(y-\bar{y})$	d_y^2	$d_x d_y$
9	4	16	15	3	9	12
8	3	9	16	4	16	12
7	2	4	14	2	4	4
6	1	1	13	1	1	1
5	0	0	11	-1	1	0
4	-1	1	12	0	0	0
3	-2	2	10	-2	4	4
2	-3	9	8	-4	16	12
1	-4	16	9	-3	9	12
$\Sigma x = 45$ $\bar{x} = 5$		$\Sigma d_x^2 = 60$	$\Sigma y = 108$ $\bar{y} = 12$		$\Sigma d_y^2 = 60$	$\Sigma d_x d_y = 57$

$\therefore r = \frac{\Sigma d_x d_y}{\sqrt{(\Sigma d_x^2)(\Sigma d_y^2)}} = \frac{57}{\sqrt{60 \times 60}} = \frac{57}{60} = 0.95$

4.4.1 Shortcut Method for Karl Pearson Coefficient of Correlation

We can also find Karl Pearson Coefficient of Correlation by taking assumed means as shown:

If we take $d_x = x - a$; $d_y = y - b$, a and b are assumed means of x and y data values

Then $r = \frac{\Sigma d_x d_y - \frac{\Sigma d_x \Sigma d_y}{n}}{\sqrt{(\Sigma d_x^2) - \frac{1}{n}(\Sigma d_x)^2}\sqrt{(\Sigma d_y^2) - \frac{1}{n}(\Sigma d_y)^2}}$

➤ If x_i's are equispaced with height h, we can take $d_x = \frac{x-a}{h}$

Similarly if y_i's are equispaced with height k, then $d_y = \frac{y-b}{k}$

Example 10 Calculate coefficient of correlation from the following data:

x	1	3	5	7	8	10
y	8	12	15	17	18	20

Solution: Let $d_x = x - 7$, $d_y = y - 15$

$$r = \frac{\sum d_x d_y - \frac{\sum d_x \sum d_y}{n}}{\sqrt{(\sum d_x^2) - \frac{1}{n}(\sum d_x)^2}\sqrt{(\sum d_y^2) - \frac{1}{n}(\sum d_y)^2}}$$

Calculating $\sum d_x, \sum d_y, \sum d_x^2, \sum d_y^2$ and $\sum d_x d_y$

x	d_x (x−7)	d_x^2	y	d_y (y−15)	d_y^2	$d_x d_y$
1	-6	36	8	-7	49	42
3	-4	16	12	-3	9	12
5	-2	4	15	0	0	0
7	0	0	17	2	4	0
8	1	1	18	3	9	3
10	3	9	20	5	25	15
	$\sum d_x = -8$	$\sum d_x^2 = 66$		$\sum d_y = 0$	$\sum d_y^2 = 96$	$\sum d_x d_y = 72$

$$\therefore r = \frac{72 - \frac{(-8)(0)}{6}}{\sqrt{(66) - \frac{1}{6}(-8)^2}\sqrt{(96) - \frac{1}{6}(0)^2}} = 0.9879$$

4.4.2 Coefficient of Correlation of Bivariate Frequency Distribution

If given data is in the form of a bivariate frequency distribution,

Then $r = \dfrac{\sum f d_x d_y - \frac{\sum f d_x \sum f d_y}{n}}{\sqrt{(\sum f d_x^2) - \frac{1}{n}(\sum f d_x)^2}\sqrt{(\sum f d_y^2) - \frac{1}{n}(\sum f d_y)^2}}$, $n = \sum f$

Example 11 Following table gives a bivariate distribution showing frequency of marks obtained according to age by a group of 52 students in an intelligent test:

Marks \ Age	16-18	18-20	20-22	22-24
10-20	2	1	1	
20-30	3	2	3	2
30-40	3	4	5	6
40-50	2	2	3	4
50-60	-	1	2	2
60-70	-	1	2	1

Compute the correlation between marks and age of the students.

Solution: Let marks obtained by the students be denoted by x and age by y, then coefficient of correlation (r) for the bivariate frequency distribution is given by:

$$r = \frac{\sum f d_x d_y - \frac{\sum f d_x \sum f d_y}{n}}{\sqrt{(\sum f d_x^2) - \frac{1}{n}(\sum f d_x)^2}\sqrt{(\sum f d_y^2) - \frac{1}{n}(\sum f d_y)^2}}, n = \sum f$$

Let $d_x = \frac{x-a}{10}$, $d_y = \frac{y-b}{2}$, where a & b denote assumed mean classes

Here a is taken as $30 - 40$, b is taken as $18 - 20$

Also quantities in brackets denote $d_x d_y$ for each cell.

$\therefore f d_x d_y$ for each cell is obtained by multiplying frequency of each cell with $d_x d_y$ and added across rows or columns to get $f d_x d_y$

y \ x	16-18	18-20	20-22	22-24	F	d_x	$f d_x$	$f d_x^2$	$f d_x d_y$
10-20	2 (2)	1 (0)	1 (-2)	-	4	-2	-8	16	2
20-30	3 (1)	2 (0)	3 (-1)	2 (-2)	10	-1	-10	10	-4
30-40	3 (0)	4 (0)	5 (0)	6 (0)	18	0	0	0	0
40-50	2 (-1)	2 (0)	3 (1)	4 (2)	11	1	11	11	9
50-60	-	1 (0)	2 (2)	2 (4)	5	2	10	20	12
60-70	-	1 (0)	2 (3)	1 (6)	4	3	12	36	12
F	10	11	16	15	**52**	Totals	15	93	31
d_y	-1	0	1	2	Totals				
$f d_y$	-10	0	16	30	36				
$f d_y^2$	10	0	16	60	86				
$f d_x d_y$	5	0	8	18	31				

$$\therefore r = \frac{31 - \frac{(15)(36)}{52}}{\sqrt{(93) - \frac{1}{52}(15)^2}\sqrt{(86) - \frac{1}{52}(36)^2}} = \frac{20.6154}{(9.4166)(7.8152)} = 0.2801$$

Thus there is a weak positive correlation between marks and age of the students.

4.4.3 Coefficient of Correlation by Rank differences

Rank correlation is used for attributes (like beauty, intelligence etc.) which cannot be measured quantitatively but can be provided with comparative ranks.

Spearman's Rank Correlation in given by: $r = 1 - \frac{6 \sum D^2}{n(n^2 - 1)}$, where $D = R_1 - R_2$

Repeated or Tied Ranks

If two or more observations in a data are equal, each observation is provided with an average rank and a correction factor is applied to correlation formula given as:

Correction Factor (C.F.) = $\sum m(m^2 - 1)$, m is the number of times each observation is repeated.

Spearman's Rank Correlation for repeated (tied) ranks is given by:

$$r = 1 - \frac{6\left(\sum D^2 + \frac{1}{12}C.F.\right)}{n(n^2-1)}, \text{ where } D = R_1 - R_2$$

Example12 Calculate the coefficient of correlation from the following data; given ranks of 10 students in English and Mathematics.

Rank in English	3	1	5	4	2	6	8	10	9	7
Rank in Mathematics	2	4	3	1	5	10	7	9	8	6

Solution: Since comparative ranks are given; instead of marks, using Spearman's Rank Correlation is given by: $r = 1 - \frac{6\sum D^2}{n(n^2-1)}$, where $D = R_1 - R_2$

Rank in English R_1	Rank in Mathematics R_2	$D = R_1 - R_2$	D^2
3	2	1	1
1	4	-3	9
5	3	2	4
4	1	3	9
2	5	-3	9
6	10	-4	16
8	7	1	1
10	9	1	1
9	8	1	1
7	6	1	1
			$\sum D^2 = 52$

$$\therefore r = 1 - \frac{6(52)}{10(10^2-1)} = 0.6848$$

Example13 Eight competitors in a beauty contest got marks (out of 10) by three judges as given below:

Judge A   9   6   5   10   3   1   4   2

Judge B   3   5   8   4   7   10   2   1

Judge C   6   4   9   8   1   2   3   10

Use rank correlation to discuss which pair of judges has the nearest approach to common tastes in beauty.

Solution: Since instead of ranks; marks are given by the three judges, converting the given data to comparative ranks for the eight competitors

Judge A		Judge B		Judge C		D_{AB}	D_{AB}^2	D_{BC}	D_{BC}^2	D_{AC}	D_{AC}^2
Marks	Rank	Marks	Rank	Marks	Rank						
9	2	3	6	6	4	-4	16	2	4	-2	4
6	3	5	4	4	5	-1	1	-1	1	-2	4
5	4	8	2	9	2	2	4	0	0	2	4
10	1	4	5	8	3	-4	16	2	4	-2	4
3	6	7	3	1	8	3	9	-5	25	-2	4
1	8	10	1	2	7	7	49	-6	36	1	1
4	5	2	7	3	6	-2	4	1	1	-1	1
2	7	1	8	10	1	-1	1	7	49	6	36

Here D_{AB} = Rank by Judge A − Rank by Judge B, also $\sum D_{AB}^2 = 100$

Similarly D_{BC} = Rank by Judge B − Rank by Judge C, also $\sum D_{BC}^2 = 120$

D_{AC} = Rank by Judge A − Rank by Judge C, also $\sum D_{AC}^2 = 58$

Rank Correlation between judges A and B is given by:

$$r_{AB} = 1 - \frac{6\sum D_{AB}^2}{n(n^2-1)} = 1 - \frac{6(100)}{8(8^2-1)} = -0.1905$$

Rank Correlation between judges B and C is given by:

$$r_{BC} = 1 - \frac{6\sum D_{BC}^2}{n(n^2-1)} = 1 - \frac{6(120)}{8(8^2-1)} = -0.4286$$

Rank Correlation between judges A and C is given by:

$$r_{AC} = 1 - \frac{6\sum D_{AC}^2}{n(n^2-1)} = 1 - \frac{6(58)}{8(8^2-1)} = 0.3095$$

Therefore Judges A and C have the nearest approach to common tastes in beauty, while Judges B and C have most different beauty tastes.

Example14: Obtain rank correlation coefficient for following marks in economics (x) and Mathematics (y) out of 25 for eight students.

x	20	24	12	20	10	12	24	20
y	18	19	16	22	14	16	19	12

Solution: Converting data into ranks: Ranks of x as R_x, Ranks of y as R_y

x	R_x	y	R_y	$D = R_x - R_y$	D^2
20	4	18	4	0	0
24	1.5	19	2.5	-1	1
12	6.5	16	5.5	1	1
20	4	22	1	3	9
10	8	14	7	1	1
12	6.5	16	5.5	1	1
24	1.5	19	2.5	-1	1
20	4	12	8	-4	16
					$\sum D^2 = 30$

Correction Factor $= \sum m(m^2 - 1)$, m is the number of times each data value is repeated

$\therefore$ C.F. $= 2(2^2 - 1) + 3(3^2 - 1) + 2(2^2 - 1) + 2(2^2 - 1) + 2(2^2 - 1) = 48$

Spearman's Rank Correlation for repeated ranks is given by:

$$r = 1 - \frac{6\left(\sum D^2 + \frac{1}{12}\text{C.F.}\right)}{n(n^2-1)}, \text{ where } D = R_x - R_y$$

$$\therefore r = 1 - \frac{6\left(30 + \frac{48}{12}\right)}{8(8^2-1)} = \frac{25}{42} = 0.595$$

Example 15 Obtain rank correlation coefficient for following data

x	68	64	75	50	64	80	75	40	55	64
y	62	58	68	45	81	60	68	48	50	70

Solution: Converting data into ranks: Ranks of x as R_x, Ranks of y as R_y

X	R_x	Y	R_y	$D = R_x - R_y$	D^2
68	4	62	5	-1	1
64	6	58	7	-1	1
75	2.5	68	3.5	-1	1
50	9	45	10	-1	1
64	6	81	1	5	25
80	1	60	6	-5	25
75	2.5	68	3.5	-1	1
40	10	48	9	1	1
55	8	50	8	0	0
64	6	70	2	4	16
					$\sum D^2 = 72$

Correction Factor (C.F.) $= \sum m(m^2 - 1)$, m is the number of times each data value is repeated

$\therefore$ C.F. $= 2(2^2 - 1) + 3(3^2 - 1) + 2(2^2 - 1) = 36$

Spearman's Rank Correlation for repeated ranks is given by:

$r = 1 - \frac{6\left(\sum D^2 + \frac{1}{12}\text{C.F.}\right)}{n(n^2-1)}$, where $D = R_x - R_y$ $\therefore r = 1 - \frac{6\left(72 + \frac{36}{12}\right)}{10(10^2-1)} = \frac{6}{11} = 0.545$

4.5 Linear Regression

Regression describes the functional relationship between dependent and independent variables; which helps us to make estimates of one variable from the other. Correlation quantifies the association between the two variables; whereas linear regression finds the best line that predicts y from x and also x from y. The difference between correlation and regression is illustrated in the adjoining figure.

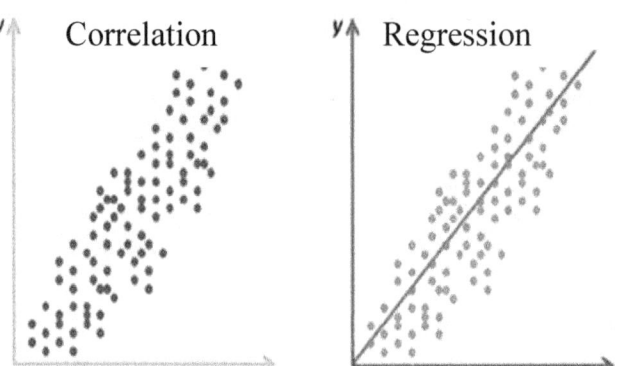

4.5.1 Lines of Regression:
If we plot the observations of the linear regression between two variables, actually two straight lines can approximately be drawn through the scatter diagram. One line estimates values of y for specified values of x (known as line of regression of y on x); and other predicts values of x from given values of y (called line of regression of x on y).

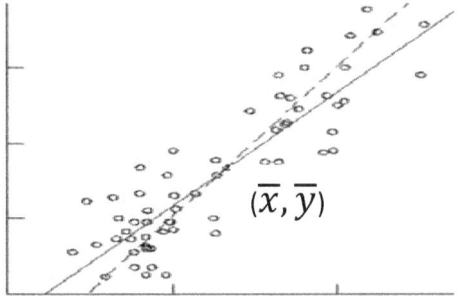

Let line of regression of y on x be represented by

$$y = a + bx \quad \ldots \text{①}$$

Normal equations as derived by the method of least Square are:

$$\sum y = an + b \sum x \quad \ldots \text{②}$$

and $\sum xy = a \sum x + b \sum x^2 \quad \ldots \text{③}$

Dividing ② by n, we get

$$\frac{\sum y}{n} = a + b \frac{\sum x}{n} \Rightarrow \bar{y} = a + b\bar{x}$$

Where $\bar{x}$ and $\bar{y}$ are the means of x series and y series. This shows that $(\bar{x}, \bar{y})$ lies on the line of regression given by ①.

Again as $(\bar{x}, \bar{y})$ satisfies ①, shifting the origin to $(\bar{x}, \bar{y})$ in equation ③, we get

$$\sum (x - \bar{x})(y - \bar{y}) = a \sum (x - \bar{x}) + b \sum (x - \bar{x})^2$$

$$\Rightarrow \sum (x - \bar{x})(y - \bar{y}) = b \sum (x - \bar{x})^2 \qquad \because \sum (x - \bar{x}) = 0$$

$$\Rightarrow b = \frac{\sum (x - \bar{x})(y - \bar{y})}{\sum (x - \bar{x})^2} = \frac{\sum d_x d_y}{\sum d_x^2} \quad \ldots \text{④}$$

Again $r = \dfrac{\sum d_x d_y}{\sqrt{(\sum d_x^2)(\sum d_y^2)}} = \dfrac{\sum d_x d_y}{n \sqrt{\frac{\sum d_x^2}{n}} \sqrt{\frac{\sum d_y^2}{n}}} = \dfrac{\sum d_x d_y}{n \sigma_x \sigma_y} \qquad \because \sigma_x = \sqrt{\dfrac{\sum d_x^2}{n}}, \sigma_y = \sqrt{\dfrac{\sum d_y^2}{n}}$

Here σ_x, σ_y are standard deviations of x and y data points respectively

$$\Rightarrow \sum d_x d_y = nr\sigma_x \sigma_y \qquad \ldots ⑤$$

Using ⑤ in ④, we get

$$b = \frac{nr\sigma_x\sigma_y}{\sum d_x^2} = \frac{r\sigma_x\sigma_y}{\sigma_x^2}$$

$\Rightarrow b = \frac{r\sigma_y}{\sigma_x}$ which is slope of line of regression line of y on x

$\therefore b_{yx} = \frac{r\sigma_y}{\sigma_x}$, b_{yx} denotes slope of line of regression line of y on x.

Thus line of regression of y on x given by ①, passes through $(\bar{x}, \bar{y})$ and is having slope

$$b_{yx} = \frac{r\sigma_y}{\sigma_x}$$

∴ Equation of line of regression of y on x is given by $y - \bar{y} = b_{yx}(x - \bar{x})$

Similarly line of regression of x on y is given by: $x - \bar{x} = b_{xy}(y - \bar{y})$

where $b_{xy} = \frac{r\sigma_x}{\sigma_y}$ is slope of line of regression line of x on y

Here b_{xy} and b_{yx} are known coefficients of regression and are connected by the relation:

$$b_{xy} b_{yx} = \left(\frac{r\sigma_x}{\sigma_y}\right)\left(\frac{r\sigma_y}{\sigma_x}\right) = r^2$$

4.5.1 Properties of Regression Coefficients

- As $\sqrt{b_{xy} b_{yx}} = r$, the coefficient of correlation is the geometric mean between the two regression coefficients.
- Since $\frac{b_{xy} + b_{yx}}{2} \geq \sqrt{b_{xy} b_{yx}} = r$, ∴ arithmetic mean of the two regression coefficients is greater than or equal to the correlation coefficient (r).
- If there is a perfect correlation between the two variables under consideration, then $b_{xy} = b_{yx} = r$; and the two lines of regression coincide. Converse is also true, i.e. if two lines of regression coincide, then there is a perfect correlation; $r = \pm 1$.
- Since $b_{xy} b_{yx} = r^2 > 0$, the signs of both regression coefficients b_{xy} and b_{yx} and coefficient of correlation (r) must be same; either all three negative or all positive.
- ∵ $b_{xy} b_{yx} = r^2 \leq 1$, if one of the regression coefficients is greater than unity, other must be less than unity.
- Point of intersection of two lines of regression is $(\bar{x}, \bar{y})$, Where $\bar{x}$ and $\bar{y}$ are the means of x series and y series.
- If both lines of regression cut each other at right angle, there is no correlation between the two variables; i.e. $r = 0$.

Example 16 Prove that arithmetic mean of coefficients of regression is greater than the coefficient of correlation.

Solution: We know that $b_{xy} = \dfrac{r\,\sigma_x}{\sigma_y}$ and $b_{yx} = \dfrac{r\,\sigma_y}{\sigma_x}$

To prove $\dfrac{b_{xy}+b_{yx}}{2} > r$

or $\dfrac{1}{2}\left[\dfrac{r\,\sigma_x}{\sigma_y} + \dfrac{r\,\sigma_y}{\sigma_x}\right] > r$

or $\dfrac{1}{2}\left[\dfrac{\sigma_x^2 + \sigma_y^2}{\sigma_x \sigma_y}\right] > 1$

or $\left[\dfrac{\sigma_x^2 + \sigma_y^2}{\sigma_x \sigma_y}\right] - 2 > 0$

or $\sigma_x^2 + \sigma_y^2 - 2\sigma_x \sigma_y > 0$

or $[\sigma_x - \sigma_y]^2 > 0$

which is true

Note : A.M. $= r$ if $b_{xy} = b_{yx} = r = \pm 1$

4.5.2 Angle between the Lines of Regression

If θ be the acute angle between the two regression lines for two variables x and y, then $\tan\theta = \dfrac{1-r^2}{r}\dfrac{\sigma_x \sigma_y}{\sigma_x^2 + \sigma_y^2}$

Proof: The two lines of regression are given by:

$y - \bar{y} = \dfrac{r\,\sigma_y}{\sigma_x}(x - \bar{x})$ ... ①

and $x - \bar{x} = \dfrac{r\,\sigma_x}{\sigma_y}(y - \bar{y})$ ... ②

If m_1 and m_2 are slopes of lines ① and ②, then

$\tan\theta = \dfrac{m_2 - m_1}{1 + m_1 m_2}$, where $m_1 = \dfrac{r\,\sigma_y}{\sigma_x}$, $m_2 = \dfrac{\sigma_y}{r\,\sigma_x}$

$\Rightarrow \tan\theta = \dfrac{\dfrac{\sigma_y}{r\,\sigma_x} - \dfrac{r\,\sigma_y}{\sigma_x}}{1 + \dfrac{r\,\sigma_y}{\sigma_x}\dfrac{\sigma_y}{r\,\sigma_x}} = \dfrac{\left(\dfrac{1}{r} - r\right)\dfrac{\sigma_y}{\sigma_x}}{1 + \dfrac{\sigma_y^2}{\sigma_x^2}} = \dfrac{1-r^2}{r}\dfrac{\sigma_x \sigma_y}{\sigma_x^2 + \sigma_y^2}$ ... ③

➢ When $r = 0$, $\tan\theta = \infty \Rightarrow \theta = \dfrac{\pi}{2}$ from ③

∴ when $r = 0$, the two lines of regression are perpendicular to each other.

➢ When $r = \pm 1$, $\tan\theta = 0 \Rightarrow \theta = 0$ from ③

∴ when $r = \pm 1$, the two lines of regression are coincident

Example 17 Find the correlation coefficient between x and y, when the two lines of regression are given by: $2x - 9y + 6 = 0$ and $x - 2y + 1 = 0$

Solution: Let the line of regression of x on y be $2x - 9y + 6 = 0$ ... ①

Then the line of regression of y on x is $x - 2y + 1 = 0$ ... ②

Now ① $\Rightarrow x = \frac{9}{2}y - 3 \quad \therefore b_{xy} = \frac{9}{2}$

Also ② $\Rightarrow y = \frac{1}{2}x + \frac{1}{2} \quad \therefore b_{yx} = \frac{1}{2}$

$\therefore r = \sqrt{b_{xy} b_{yx}} = \sqrt{\frac{9}{2} \times \frac{1}{2}} = \frac{3}{2}$, which is not possible as $-1 \leq r \leq 1$

So our choice of regression lines is incorrect.

$\therefore$ Line of regression of x on y is $x - 2y + 1 = 0$

$\Rightarrow x = 2y - 1 \quad \therefore b_{xy} = 2$

Also line of regression of y on x is $2x - 9y + 6 = 0$

$\Rightarrow y = \frac{2}{9}x + \frac{2}{3} \quad \therefore b_{yx} = \frac{2}{9}$

$\therefore r = \sqrt{b_{xy} b_{yx}} = \sqrt{2 \times \frac{2}{9}} = \frac{2}{3}$

Example 18 The regression equations calculated from a given set of observations for two random variables are: $x = -0.4y + 6.4$ and $y = -0.6x + 4.6$

Calculate $\bar{x}, \bar{y}$ and r.

Solution: The two equations of regression are:

$x = -0.4y + 6.4 \quad \quad \text{... ①}$

$y = -0.6x + 4.6 \quad \quad \text{... ②}$

$\Rightarrow b_{xy} = -0.4$ and $b_{yx} = -0.6$

$\therefore r^2 = b_{xy} b_{yx} = 0.24$

$\Rightarrow r = \pm 0.49$

We know that the signs of b_{xy}, b_{yx} and r must be same

$\therefore r = -0.49$

Again we know that the point of intersection of two regression lines is $(\bar{x}, \bar{y})$

Therefore solving ① and ②, we get $\bar{x} = 6$, $\bar{y} = 1$

Example 19 Find the regression line of y on x from the following data:

x	1	3	4	6	8	9	11	14
y	1	2	4	4	5	7	8	9

Also estimate the value of y, when $x = 10$

Solution: Let line of regression of y on x be:

$y = a + bx \quad \quad \text{... ①}$

Then normal equations are given by:

$\sum y = an + b \sum x \quad \quad \text{... ②}$

and $\sum xy = a \sum x + b \sum x^2 \quad \quad \text{... ③}$

Calculating $\sum x, \sum y, \sum xy$ and $\sum x^2$

x	y	x^2	xy
1	1	1	1
3	2	9	6
4	4	16	16
6	4	36	24
8	5	64	40
9	7	81	63
11	8	121	88
14	9	196	126
$\sum x = 56$	$\sum y = 40$	$\sum x^2 = 524$	$\sum xy = 364$

Substituting values of $\sum x, \sum y, \sum xy$ and $\sum x^2$ in ② and ③

$$\Rightarrow 40 = 8a + 56b \qquad \ldots ④$$

and $\quad 364 = 56a + 524b \qquad \ldots ⑤$

Solving ④ and ⑤, we get $a = \frac{6}{11}$ and $b = \frac{7}{11}$

Substituting in ①, line of regression of y on x is $y = \frac{6}{11} + \frac{7}{11}x$

$\Rightarrow 7x - 11y + 6 = 0$

Also at $x = 10$, $y = \frac{76}{11}$

Example 20 From a partially destroyed lab data, following results were retrieved:

Lines of regression are:

$x = 0.45y + 5.35$ and $y = 0.8x + 6.6$, $\sigma_x^2 = 9$

Find $\bar{x}, \bar{y}, \sigma_y$ and r for the existing data.

Solution: The two equations of regression are:

$x = 0.45y + 5.35 \qquad \ldots ①$

$y = 0.8x + 6.6 \qquad \ldots ②$

We know that the point of intersection of two regression lines is $(\bar{x}, \bar{y})$

Therefore solving ① and ②, we get $\bar{x} = 13$, $\bar{y} = 17$

Again ① $\Rightarrow b_{xy} = 0.45$ and $b_{yx} = 0.8$

$\therefore r^2 = b_{xy}b_{yx} = 0.36$

$\Rightarrow r = \pm 0.6$

We know that the signs of b_{xy}, b_{yx} and r must be same

$\therefore r = 0.6$

Also $b_{yx} = \frac{r\sigma_y}{\sigma_x} \Rightarrow 0.8 = \frac{(0.6)\sigma_y}{3} \Rightarrow \sigma_y = \frac{0.8 \times 3}{0.6} = 4$

Example 21 Following data depicts the statistical values of rainfall and production of wheat in a region for a specified time period.

	Mean	Standard Deviation
Production of Wheat (kg. per unit area)	10	8
Rainfall (cm)	8	2

Estimate the production of wheat when rainfall is 9cm if correlation coefficient between production and rainfall is given to be 0.5.

Solution: Let the variables x and y denote production and rainfall respectively.

Given that $\bar{x} = 10$, $\bar{y} = 8$ also $\sigma_x = 8$, $\sigma_y = 2$

Now equation of regression of x on y is given by:

$$x - \bar{x} = \frac{r\sigma_x}{\sigma_y}(y - \bar{y})$$

$$\Rightarrow x - 10 = \frac{(0.5)8}{2}(y - 8)$$

$$\Rightarrow x = 2y - 6$$

$\therefore$ When rainfall is 9cm, production of wheat is estimated to be

$2(9) - 6 = 12$ kg. per unit area

Example 22 Find the coefficient of correlation and the lines of regression for the data given below:

$n = 18, \sum x = 12, \sum y = 18, \sum x^2 = 60, \sum y^2 = 96$ and $\sum xy = 48$

Solution: $\bar{x} = \frac{\sum x}{n} = \frac{12}{18} = 0.67$, $\bar{y} = \frac{\sum y}{n} = \frac{18}{18} = 1$

$\sigma_x^2 = \frac{\sum x^2}{n} - \left(\frac{\sum x}{n}\right)^2 = \frac{60}{18} - \left(\frac{12}{18}\right)^2 = 2.89 \therefore \sigma_x = 1.7$

$\sigma_y^2 = \frac{\sum y^2}{n} - \left(\frac{\sum y}{n}\right)^2 = \frac{96}{18} - \left(\frac{18}{18}\right)^2 = 4.33 \therefore \sigma_y = 2.08$

$$r = \frac{\sum xy - \frac{\sum x \sum y}{n}}{\sqrt{(\sum x^2) - \frac{1}{n}(\sum x)^2}\sqrt{(\sum y^2) - \frac{1}{n}(\sum y)^2}}$$

$$= \frac{48 - \frac{(12)(18)}{18}}{\sqrt{(60) - \frac{1}{18}(12)^2}\sqrt{(96) - \frac{1}{18}(18)^2}} = \frac{36}{(7.2)(8.83)} = 0.57$$

$$b_{xy} = \frac{r\,\sigma_x}{\sigma_y} = \frac{(0.57)(1.7)}{2.08} = 0.47 \quad , \quad b_{yx} = \frac{r\,\sigma_y}{\sigma_x} = \frac{(0.57)(2.08)}{1.7} = 0.7$$

Equations of lines of regression are:

$$y - \bar{y} = b_{yx}(x - \bar{x}) \quad , \quad x - \bar{x} = b_{xy}(y - \bar{y})$$

$$\Rightarrow y - 1 = 0.7(x - 0.67) \text{ and } x - 0.67 = 0.47(y - 1)$$

$$\Rightarrow y = 0.7x + 0.53 \qquad \text{and } x = 0.47y + 0.2$$

Example 23 Marks obtained by 11 students in statistics papers are given below:

Paper I	60	65	68	70	75	85	80	45	55	56	58
Paper II	62	64	65	70	74	90	82	56	50	48	60

Calculate the coefficient of correlation and the equations of lines of regression.

Solution: Let marks obtained in paper I be denoted by x and marks obtained in paper II be denoted by y.

Let $A_x = 65$, $A_y = 70$ $\therefore$ $d_x = x - 65$, $d_y = y - 70$

Calculating $\sum d_x, \sum d_y, \sum d_x^2, \sum d_y^2$ and $\sum d_x d_y$

x	d_x $(x - 65)$	d_x^2	y	d_y $(y - 70)$	d_y^2	$d_x d_y$
60	-5	25	62	-8	64	40
65	0	0	64	-6	36	0
68	3	9	65	-5	25	-15
70	5	25	70	0	0	0
75	10	100	74	4	16	40
85	20	400	90	20	400	400
80	15	225	82	12	144	180
45	-20	400	56	-14	196	280
55	-10	100	50	-20	400	200
56	-9	81	48	-22	484	198
58	-7	49	60	-10	100	70
	$\sum d_x = 2$	$\sum d_x^2 = 1414$		$\sum d_y = -49$	$\sum d_y^2 = 1865$	$\sum d_x d_y = 1393$

Karl Pearson coefficient of correlation (r) is given by:

$$r = \frac{\sum d_x d_y - \frac{\sum d_x \sum d_y}{n}}{\sqrt{(\sum d_x^2) - \frac{1}{n}(\sum d_x)^2}\sqrt{(\sum d_y^2) - \frac{1}{n}(\sum d_y)^2}}$$

$$\therefore \quad r = \frac{1393 - \frac{(2)(-49)}{11}}{\sqrt{(1414) - \frac{1}{11}(2)^2}\sqrt{(1865) - \frac{1}{11}(-49)^2}} = \frac{1401.9091}{(37.5984)(40.5799)} = 0.9188$$

Now $\bar{x} = A_x + \frac{\Sigma d_x}{n} = 65 + \frac{2}{11} = 65.1818$

$\bar{y} = A_y + \frac{\Sigma d_y}{n} = 70 + \frac{-49}{11} = 65.5455$

Also $\sigma_x = \sqrt{\frac{\Sigma d_x^2}{n} - \left(\frac{\Sigma d_x}{n}\right)^2} = \sqrt{\frac{1414}{11} - \left(\frac{2}{11}\right)^2} = 11.3363$

and $\sigma_y = \sqrt{\frac{\Sigma d_y^2}{n} - \left(\frac{\Sigma d_y}{n}\right)^2} = \sqrt{\frac{1865}{11} - \left(\frac{-49}{11}\right)^2} = 12.2353$

$\therefore b_{xy} = \frac{r \sigma_x}{\sigma_y} = \frac{(0.9188)(11.3363)}{12.2353} = 0.8513$

$b_{yx} = \frac{r \sigma_y}{\sigma_x} = \frac{(0.9188)(12.2353)}{11.3363} = 0.9917$

Equations of lines of regression are:

$y - \bar{y} = b_{yx}(x - \bar{x})$, $x - \bar{x} = b_{xy}(y - \bar{y})$

$\Rightarrow y - 65.55 = 0.99(x - 65.18)$ and $x - 65.18 = 0.85(y - 65.55)$

$\Rightarrow y = 0.99x + 1.02$ and $x = 0.85y + 9.46$

Example 24 The regression equations calculated from a given set of observations for two variables x and y are: $x = 9y + 5$ and $y = kx + 9$

Show that $0 < k < \frac{1}{9}$. Also if $k = \frac{1}{10}$, find $\bar{x}, \bar{y}$ and r

Solution: The two equations of regression are:

$x = 9y + 5$ ...①

$y = kx + 9$ ...②

$\Rightarrow b_{xy} = 9$ and $b_{yx} = k$

$\therefore r^2 = b_{xy} b_{yx} = 9k$

$\Rightarrow r = 3\sqrt{k}$ $\because b_{xy} = 9$ is positive, therefore k and r are also positive

Now $0 < r < 1$ or $0 < 3\sqrt{k} < 1$

$\Rightarrow 0 < 9k < 1$ or $0 < k < \frac{1}{9}$

Now if $k = \frac{1}{10}$, equation ② becomes $10y = x + 90$ ...③

Solving ① and ③, the point of intersection of two regression lines is

$\bar{x} = 860$, $\bar{y} = 95$, also $r = 3\sqrt{k} = 3\sqrt{\frac{1}{10}} = 0.949$

Exercise 4

1. Fit a straight line $y = ax + b$ to the following data

x	0	1	3	6	8
y	1	3	2	5	4

2. Fit a straight line $y = a + bx$ to the following data

x	25	19	50	36	40	45	30
y	77	76	85	80	82	83	79

3. Fit a second degree parabola $y = ax^2 + bx + c$ to the following data

x	0	1	2	3	4
y	1	1.8	1.3	2.5	6.3

4. Fit a second degree parabola $y = a + bx + cx^2$ to the following data

x	2	4	6	8	10
y	3.07	12.85	31.47	57.38	91.29

5. Find the coefficient of correlation between x and y from the given data. Also find the two lines of regression.

x	1	2	3	4	5	6	7	8	9	10
y	10	12	16	28	25	36	41	49	40	50

6. Find the rank correlation for the following data:

x	56	42	72	36	63	47	55	49	38	42	68	60
y	147	125	160	118	149	128	150	145	115	140	152	155

7. Following table shows ages of husband and wife of 53 married couples.

Wife→ Husband↓	15-25	25-35	35-45	45-55	55-65	65-75
15-25	1	1	-	-	-	-
25-35	2	12	1	-	-	-
35-45	-	4	10	1	-	-
45-55	-	-	3	6	1	-
55-65	-	-	-	2	4	2
65-75	-	-	-	-	1	2

Calculate the coefficient of correlation between the age of the husband and that of wife.

8. The regression equations of two variables x and y are $x = 0.7y + 5.2$, $y = 0.3x + 2.8$. Find the means of the two variables and the coefficient of correlation between them.

9. If the coefficient of correlation between two variables x and y is 0.5 and the acute angle between their lines of regression is $\tan^{-1}\frac{3}{8}$, show that $\sigma_x = \frac{\sigma_y}{2+\sqrt{3}}$

10. From a partially destroyed lab data, following results were retrieved:
    Lines of regression are:
    $$8x = 10y - 66 \text{ and } 18y = 40x - 214 \text{ , } \sigma_x = 3$$
    Find $\bar{x}, \bar{y}, \sigma_y$ and r for the existing data.

Answers

1. $y = 0.38x + 1.6$
2. $y = 70.052 + 0.292x$
3. $y = 0.55x^2 - 1.07x + 1.42$
4. $y = 0.34 - 0.78x + 0.99x^2$
5. $r = 0.96, x = 0.2y - 0.64, y = 4.69x + 4.9$
6. 0.932
7. 0.91
8. $\bar{x} = 9.06, \bar{y} = 5.52, r = 0.46$
10. $\bar{x} = 13, \bar{y} = 17, \sigma_y = 4, r = 0.6$

Chapter 5: Sampling Distributions & Hypothesis Testing

Sampling Distributions And Hypothesis Testing

5.1 Introduction

Sampling is a statistical method of obtaining representative data (observations) from a group. We have been using sampling concepts in our day to day lives knowingly or unknowingly; for instance we take a handful of rice to check the rice quality of the full lot. This is an example of random sampling from a large population.

Population (Universe): The group of objects (individuals) under study is called population or universe. Universe may be finite or infinite.

Sample: A part containing objects (individuals), selected from the population is called a sample.

Random Sampling: The selection of objects (individuals) from the universe in such a way that each object (individual) of the universe has the same chance of being selected is called random sampling. Lottery system is the most common example of random sampling.

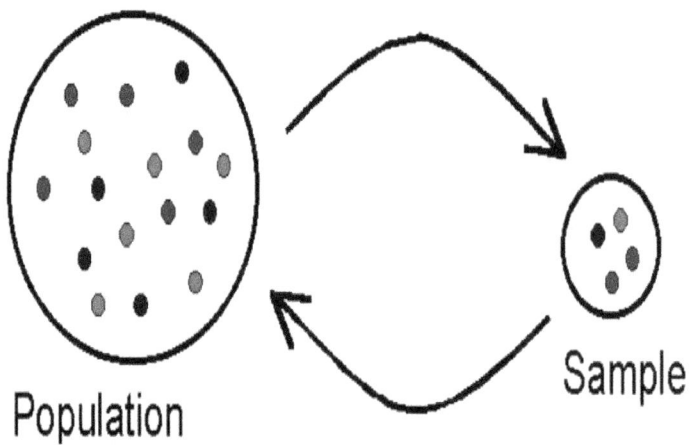

Simple Sampling: Simple sampling is a special case of random sampling in which each event has same probability of success or failure.

Note: Every random sampling need not be simple. For example if balls are drawn without replacement from a bag of balls containing different balls; the probability of success changes in every trial. Thus the sampling though random is not simple.

Hypothesis: A hypothesis is an assumption based on insubstantial evidences that lends itself to further testing and experimentation. For example a farmer claims significant increase in crop production after using a particular fertilizer and after a season of experimenting, his hypothesis may be proved true or false. Any hypothesis may be accepted or rejected as per specific confidence levels and must be admissible to refutation.

Null Hypothesis: A hypothesis which is tested for possible rejection under the assumption of being true is known as null hypothesis. Usually the null hypothesis is stated as 'There is no relationship between two quantities'. It is denoted by H_0.

Alternative Hypothesis: It is the opposite statement of null hypothesis and denoted by H_1.

Significance levels(α): The probability levels below which we reject a hypothesis H_0 are called levels of significance. Most common significance levels employed in hypothesis testing are $\alpha = 5\%$, $\alpha = 1\%$, $\alpha = 0.27\%$ in which critical (rejection) regions occupy 5%, 1% and 0.27% areas of normal curve respectively.

One Tailed and Two Tailed Tests: While testing statistical significance levels; one-tailed test and a two-tailed test are used for accepting or rejecting a hypothesis. One-tailed tests are used for asymmetric distributions (reference value is unidirectional) which have a single tail; such as the chi-square distribution.

A two-tailed test is appropriate if the estimated value may lie on both sides of reference value. Two-tailed tests are only applicable when the probability curve has two tails; such as normal distribution.

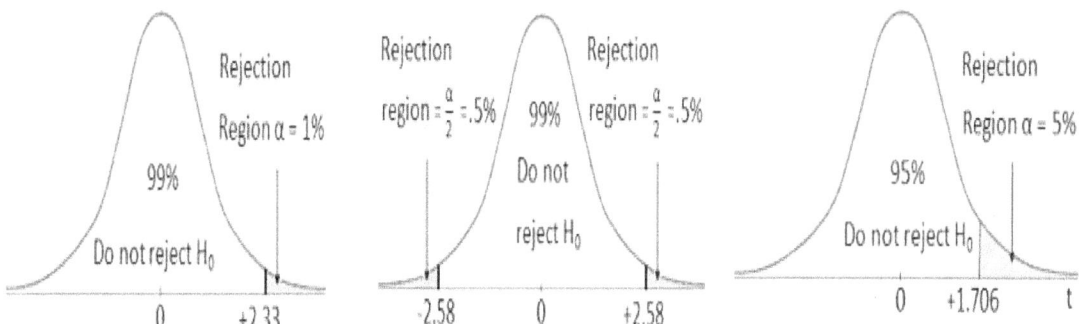

One-tailed test; $\alpha = 1\%$ Two-tailed test; $\alpha = 1\%$ One-tailed test; $\alpha = 5\%$

Reference z values (two-tailed) for various levels are given in the table below:

Significance level	Confidence level	z value
0.27 %	99.73%	3.00
1.00%	99.00%	2.58
4.55%	95.45%	2.00
5.00%	95.00%	1.96
9.89%	90.11%	1.65

Testing at 1% level of confidence means that assumption has 99% chance of being true as rejection zone is just 1% of the total area under the normal curve. Similarly testing at 5% would provide 95% confidence level of acceptance.

Type I and Type II Errors: When we test a statistic at specified confidence level, there are chances of taking wrong decisions due to small sample size or sampling fluctuations etc.

Type I error is the incorrect rejection of a true null hypothesis, i.e. we reject H_0, when it is true.

Type II error is the incorrect acceptance of a false null hypothesis, i.e. we accept H_0, when it is false.

Testing of Hypothesis:

Testing of statistical hypothesis is a procedure designed for accepting or rejecting a hypothesis on the basis of some preset values.

Step1: Plant Null Hypothesis H_0 and alternate hypothesis H_1 (optional) where H_0 is the hypothesis of no difference, i.e. H_0 presumes that there is no significant difference between observed value and expected value.

Step2: Find the most befitting test statistic for the analysis.

Step3: Take a random sample and compute the test statistic.

Step 4: H_0 is accepted if the value of test statistic lies in acceptance zone and rejected if it falls in critical (rejection) region at the desired significance level.

5.2 Sampling Distributions

A sampling distribution is a distribution of all of the possible values of a statistic; computed from randomly drawn samples of the same size from a population.

Some commonly used notations in sampling distributions are given below:

	Population	Sample
Size	N	n
Mean	μ	$\bar{x}$
Variance	σ^2	s^2
Standard Deviation	σ	s

Suppose we take various samples each of size n from a population. If p and q be the probabilities of success and failure of each member of the sample, then the binomial distribution given by $(p+q)^n$ provides the sampling distribution of the number of successes in the sample with mean np and variance npq.

- Mean (expected value) of number of successes $= np$
- Standard deviation $= \sqrt{npq}$.
- Probable occurrence range at 99.73% confidence level i.e. 0.27% significance level is given by: $np \pm 3\sqrt{npq}$
- Probable occurrence range at 99% confidence level i.e. 1% significance level is given by: $np \pm 2.58\sqrt{npq}$
- Probable occurrence range at 95% confidence level i.e. 5% significance level is given by: $np \pm 1.96\sqrt{npq}$

In case of proportion of successes, mean and standard deviation of proportion of successes are obtained by dividing each statistic by n.

- Mean (expected value) of proportion of successes $= \frac{np}{n} = p$
- Standard deviation $= \frac{\sqrt{npq}}{n} = \sqrt{\frac{pq}{n}}$
- Probable occurrence range of the proportion at 99.73% confidence level i.e. 0.27% significance level is given by: $p \pm 3\sqrt{\frac{pq}{n}}$
- Probable occurrence range of the proportion at 99% confidence level i.e. 1% significance level is given by: $p \pm 2.58\sqrt{\frac{pq}{n}}$
- Probable occurrence range of the proportion at 95% confidence level i.e. 5% significance level is given by: $p \pm 1.96\sqrt{\frac{pq}{n}}$

Standard Error: The standard deviation of the sampling distribution of a statistic is known as Standard Error (S.E.).

Precision: Reciprocal of standard error is known as precision.

Probable Error: It is taken as 0.67449 times the standard error and is used sometimes to explain the concept of sampling errors to layman or unprofessional people.

5.3 Sampling of Attributes for large samples ($n > 30$)

Characteristics like language, religion, habits (traits) etc. cannot be measured in numbers as they are attributes. Sampling of attributes means testing how many in a population possess a particular attribute (trait) or whether the two populations share an attribute (trait) in common and to how much confidence level.

When sample size (n) is very large i.e. greater than 30 and neither p nor q are very small, the binomial distribution tends to normal distribution and therefore we choose the variate z as test statistic.

Following procedure is adopted for testing the significance of large samples in terms of attributes.

Step1: Postulate the null hypothesis (H_0); if required.

Step2: If x is the observed number of successes in a sample and z is the standard normal variate, then $z = \frac{x-\mu}{\sigma}$, i.e. $z = \frac{x-np}{\sqrt{npq}}$

Step3: Accept or reject H_0 as per given values of parameter z at different significance levels:

Significance Level	H_0 accepted	H_0 rejected
0.27%	$|z| < 3$	$|z| > 3$
1%	$|z| < 2.58$	$|z| > 2.58$
5%	$|z| < 1.96$	$|z| > 1.96$

In case of absence of any specified significance level, we may consider 0.27% level, i.e. take acceptance range as $-3 < z < 3$.

Example 1 A coin is tossed 400 times and turns up head 216 times. Discuss whether the coin may be unbiased one.

Solution: Let H_0: coin is unbiased

Here $n = 400$, if p denotes probability of success; i.e. getting a head,

then $p = \frac{1}{2}$, $q = \frac{1}{2}$, $\therefore$ Expected number of heads is $np = 400 \times \frac{1}{2} = 200$

Again if x denotes observed number of heads; then the test statistic

$$z = \frac{x-np}{\sqrt{npq}}; \text{ i.e. } z = \frac{216-200}{\sqrt{400 \times \frac{1}{2} \times \frac{1}{2}}} = 1.6 < 1.96 < 2.58$$

$\therefore$ H_0 is acceptable at both 5% and 1% levels of significance.

Hence our assumption that coin is unbiased is true

Example2 A die is rolled 900 times and turns up five or six 338 times. Discuss whether the die may be unbiased one.

Solution: Let H_0: die is unbiased

Here $n = 900$, if p denotes probability of success; i.e. getting 5 or 6, then $p = \frac{1}{3}$, $q = \frac{2}{3}$, $\therefore$ Expected number of successes is $np = 900 \times \frac{1}{3} = 300$

Again if x denotes observed number of successes; then the test statistic

$$z = \frac{x-np}{\sqrt{npq}}; \text{ i.e. } z = \frac{338-300}{\sqrt{900 \times \frac{1}{3} \times \frac{2}{3}}} = 2.69$$

$\therefore$ $2.69 < 3$ but $2.69 > 1.96$ and also $2.69 > 2.58$

Hence H_0 is acceptable at 0.27% level of significance but fails at both 1% and 5% significance levels.

Example3 A random sample of 500 oranges was taken from a large consignment and 65 were found to be bad. Show that the standard error of the proportion of bad ones in a sample of this size is 0.015 and deduce that the percentage of bad oranges in the consignment lies between 8.5 and 17.5.

Solution: Let p denote the proportion of bad oranges in the given sample

Then $p = \frac{65}{500} = 0.13$, $q = 1 - p = 0.87$

Standard Deviation (S.D.) $= \sqrt{\frac{pq}{n}} = \sqrt{\frac{0.13 \times 0.87}{500}} = 0.015$

$\therefore$ Standard Error (S.E.) = S.D. = 0.015

Probable limit of bad oranges in the consignment is given by:

$$p \pm 3\sqrt{\frac{pq}{n}} = 0.13 \pm 3(0.015)$$

$$= 0.085 \text{ to } 0.175$$

$\therefore$ Probable percentage of bad oranges in the consignment is 8.5% to 17.5%

Example4 A random sample of 100 bolts was taken from the lot manufactured by a machine and 10 were found to be defective. Find the 95% confidence limits for the proportion of defective bolts produced by the machine.

Solution: Let p denote the proportion of defective bolts in the given sample

Then $p = \frac{10}{100} = 0.1$, $q = 1 - p = 0.9$

Standard Deviation (S.D.) $= \sqrt{\frac{pq}{n}} = \sqrt{\frac{0.1 \times 0.9}{100}} = 0.03$

Probable limit of defective bolts in the lot at 95% confidence level is given by:

$$p \pm 1.96\sqrt{\frac{pq}{n}} = 0.1 \pm 1.96(0.03) = 0.1 \pm 0.0588 = 0.0412 \text{ to } 0.1588$$

$\therefore$ Probable percentage of proportion of defective bolts in the lot at 95% confidence level is 4.12% to 15.88%

Example5 A sample of 900 days is taken from metrological records of a district and 100 of them are found to be foggy. What is the probable percentage of foggy days in the district?

Solution: Let p denote the probability of a foggy day in the district, then

$$p = \frac{100}{900} = \frac{1}{9}, q = 1 - p = \frac{8}{9}$$

Standard Deviation (S.D.) $= \sqrt{npq} = \sqrt{900 \times \frac{1}{9} \times \frac{8}{9}} = 9.43$

Probable limit of foggy days at 99.73% confidence level is given by:

$np \pm 3\sqrt{npq} = 100 \pm 3(9.43) = 100 \pm 28.29$; i.e. 71.71 to 128.29

Probable percentage is $\frac{71.71}{900} \times 100$ to $\frac{128.29}{900} \times 100$ i.e. 7.97% to 14.25%

5.3.1 Comparing Proportions of Large Samples from Two Different Populations in Terms of Attributes

For an attribute (trait), let proportions p_1 and p_2 be given from two large samples of sizes n_1 and n_2 respectively from two different populations. We may want to test whether two populations are similar regarding the specified attribute (trait).

Working methodology:

1. Set up the hypothesis H_0: The two populations are similar regarding the specified attribute (trait)
2. Find common proportion of two populations for the specified attribute as: $p = \frac{n_1 p_1 + n_2 p_2}{n_1 + n_2}$, $q = 1 - p$
3. Compute combined standard error of two populations as: $e = \sqrt{pq\left(\frac{1}{n_1} + \frac{1}{n_2}\right)}$
4. Compute the statistic parameter $z = \frac{p_1 - p_2}{e}$
5. Accept or reject H_0 as per given values of parameter z at different significance levels:

Significance Level	H_0 accepted	H_0 rejected
0.27%	$|z| < 3$	$|z| > 3$
1%	$|z| < 2.58$	$|z| > 2.58$
5%	$|z| < 1.96$	$|z| > 1.96$

Example 6 In a sample of 600 men from a certain city, 450 are found to be smokers. In another sample of 700 men from another city, 450 are found smokers. Do the data indicate that the cities are significantly different with respect to the habit of smoking among men?

Solution: Let H_0: The two populations are similar regarding smoking habits among men.

Here $n_1 = 600$, $p_1 = \frac{450}{600} = 0.75$

$n_2 = 900$, $p_2 = \frac{450}{700} = 0.64$

$p = \frac{n_1 p_1 + n_2 p_2}{n_1 + n_2} = \frac{600 \times 0.75 + 700 \times 0.64}{600 + 700} = \frac{449}{650} = 0.69$

$q = 1 - p = 1 - 0.69 = 0.31$

$\therefore e = \sqrt{pq\left(\frac{1}{n_1} + \frac{1}{n_2}\right)} = \sqrt{0.69 \times 0.31 \left(\frac{1}{600} + \frac{1}{700}\right)} = 0.026$

$z = \frac{p_1 - p_2}{e} = \frac{0.75 - 0.64}{0.026} = 4.23$

$|z| > 3$, $\therefore H_0$ is rejected, i.e. the difference between two populations is highly significant in terms of smoking habits among men both at 1% and 5% levels of significance.

Example 7 In a large city A, 20% of a random sample of 900 school boys had defective eye-sight. In another large city B, 15.5% of a random sample of 1600 school boys had defective eye-sight. Is the difference between two proportions significant?

Solution: Let H_0: The two populations are similar regarding school boys having defective eye sight.

Here $n_1 = 900$, $p_1 = \frac{20}{100} = \frac{1}{5} = 0.2$

$n_2 = 1600$, $p_2 = \frac{15.5}{100} = \frac{31}{200} = 0.155$

$p = \frac{n_1 p_1 + n_2 p_2}{n_1 + n_2} = \frac{900 \times 0.2 + 1600 \times 0.155}{900 + 1600} = 0.1712$

$q = 1 - p = 1 - 0.1712 = 0.8288$

$\therefore e = \sqrt{pq\left(\frac{1}{n_1} + \frac{1}{n_2}\right)} = \sqrt{0.1712 \times 0.8288 \left(\frac{1}{900} + \frac{1}{1600}\right)} = 0.01568$

$z = \frac{p_1 - p_2}{e} = \frac{0.2 - 0.155}{0.01568} = 2.8699$

$|z| < 3$, $\therefore H_0$ is accepted at 0.27% level of significance, i.e. The difference between two populations is not significant in terms of defective eye-sights.

Remark: H_0 should be rejected at 1% or 5% significance levels.

Example 8 Following data gives proportion of dark coloured people in two cities.

City	Sample size	Percentage of dark coloured people
A	250	42
B	450	48

Can the difference between two percentages taken as sampling fluctuations?

Solution: Let H_0: The two populations are similar regarding smoking habits among men.

Here $n_1 = 250$, $p_1 = 42\%$

$n_2 = 450$, $p_2 = 48\%$

$p = \frac{n_1 p_1 + n_2 p_2}{n_1 + n_2} = \frac{250 \times 42 + 450 \times 48}{250 + 450} = 45.86\%$

$q = 1 - p = 54.14\%$

$\therefore e = \sqrt{pq\left(\frac{1}{n_1} + \frac{1}{n_2}\right)} = \sqrt{45.86 \times 54.14 \left(\frac{1}{250} + \frac{1}{450}\right)} = 3.93\%$

$z = \frac{p_1 - p_2}{e} = \frac{42 - 48}{3.93} = -1.53$

$|z| < 1.96$, $\therefore H_0$ is accepted, i.e. The difference between two populations is not significant in terms of dark coloured people both at 1% and 5% levels of significance.

5.4 Sampling Distributions of Sample means

The sampling distribution of the mean refers to the pattern of sample means, observed by different samples drawn from the population at large.

Result I: If all possible samples of size n are drawn without replacement from a finite population of size N and if μ and σ denote population mean and standard deviation respectively;

$\mu_{\bar{x}}$ and $\sigma_{\bar{x}}$ denote mean and standard deviation respectively of sampling distribution, then $\mu_{\bar{x}} = \mu$ and $\sigma_{\bar{x}} = \frac{\sigma}{\sqrt{n}}\sqrt{\frac{N-n}{N-1}}$

Result II : If all possible samples of size n are drawn with replacement from a finite population of size N and if μ and σ denote population mean and standard deviation respectively; $\mu_{\bar{x}}$ and $\sigma_{\bar{x}}$ denote mean and standard deviation respectively of sampling distribution, then $\mu_{\bar{x}} = \mu$ and $\sigma_{\bar{x}} = \frac{\sigma}{\sqrt{n}}$

Example 9 Suppose a population consists of five numbers: 1,3,5,7,9. Find sampling distribution of sample means. Also calculate mean and standard deviation of the sample means, if random samples of two numbers are drawn

i. without replacement *ii.* with replacement

Verify Results *I* and *II* by comparing these statistics with population mean and population standard deviation.

Solution: Population mean $(\mu) = \frac{1}{N}\sum_1^N x_i$

$\mu = \frac{1}{5}[1 + 3 + 5 + 7 + 9] = 5$

Population variance $(\sigma^2) = \frac{1}{N}\sum_1^N(x_i - \mu)^2$

$\sigma^2 = \frac{1}{5}[(1-5)^2 + (3-5)^2 + 0 + (7-5)^2 + (9-5)^2] = \frac{40}{5} = 8$, $\therefore \sigma = \sqrt{8}$

i. Total number of possible samples each of size two without replacement is $^5C_2 = 10$

Sample	Sample mean ($\bar{x}$)
1,3	2
1,5	3
1,7	4
1,9	5
3,5	4
3,7	5
3,9	6
5,7	6
5,9	7
7,9	8

Sampling distribution of Sample mean ($\bar{x}$)

$\bar{x}$	2	3	4	5	6	7	8
Frequency	1	1	2	2	2	1	1

Mean of sample means $(\mu_{\bar{x}}) = \frac{1}{n}\sum_1^n \bar{x}$

$= \frac{1}{10}[2 + 3 + 2(4) + 2(5) + 2(6) + 7 + 8] = 5$

Variance of sample means

$\sigma_{\bar{x}}^2 = \frac{1}{10}[(2-5)^2 + (3-5)^2 + 2(4-5)^2 + 0 + 2(6-5)^2 + (7-5)^2 + (8-5)^2] = \frac{30}{10} = 3$

Standard deviation of sample means $(\sigma_{\bar{x}}) = \sqrt{3}$

∴ $\mu_{\bar{x}} = 5 = \mu$, also $\frac{\sigma}{\sqrt{n}}\sqrt{\frac{N-n}{N-1}} = \frac{\sqrt{8}}{\sqrt{2}}\sqrt{\frac{5-2}{5-1}} = \sqrt{3} = \sigma_{\bar{x}}$

Hence Result *I* is verified.

ii. Total number of possible samples each of size two with replacement is $5 \times 5 = 25$

Samples of 2 numbers with replacement are shown below; individual sample mean is given under each sample.

(1,1)	(1,3)	(1,5)	(1,7)	(1,9)
1	2	3	4	5
(3,1)	(3,3)	(3,5)	(3,7)	(3,9)
2	3	4	5	6
(5,1)	(5,3)	(5,5)	(5,7)	(5,9)
3	4	5	6	7
(7,1)	(7,3)	(7,5)	(7,7)	(7,9)
4	5	6	7	8
(9,1)	(9,3)	(9,5)	(9,7)	(9,9)
5	6	7	8	9

Mean of sample means is given by:

$$\mu_{\bar{x}} = \frac{1}{25}[1 + 2(2) + 3(3) + 4(4) + 5(5) + 4(6) + 3(7) + 2(8) + 9] = 5$$

Variance of sample means is given by:

$$\sigma_{\bar{x}}^2 = \frac{1}{25}[(1-5)^2 + 2(2-5)^2 + 3(3-5)^2 + 4(4-5)^2 + 5(5-5)^2 +$$
$$4(6-5)^2 + 3(7-5)^2 + 2(8-5)^2 + (9-5)^2]$$
$$= \frac{1}{25}[16 + 18 + 12 + 4 + 0 + 4 + 12 + 18 + 16] = \frac{100}{25} = 4$$

∴ Standard deviation of sample means ($\sigma_{\bar{x}}$) = 2

$\mu_{\bar{x}} = 5 = \mu$ and $\sigma_{\bar{x}} = 2 = \frac{8}{\sqrt{2}} = \frac{\sigma}{\sqrt{n}}$, hence Result *II* is verified.

Result III: If a population is normally distributed with mean and variance μ and σ^2 respectively; also $\bar{x}$ denotes sample mean for a random sample $(x_1, x_2, \cdots, x_n)$; then sampling distribution of $\bar{x}$ is also normally distributed with mean μ and variance $\frac{\sigma^2}{n}$, i.e. If $x \sim N(\mu, \sigma^2)$ then $\bar{x} \sim N\left(\mu, \frac{\sigma^2}{n}\right)$

Proof: Since the population is normally distributed, for any object x_i of the population;
$$E(x_i) = \mu \text{ and } Var(x_i) = \sigma^2$$

∴ $E(\bar{x}) = E\left(\frac{x_1 + x_2 + \cdots + x_n}{n}\right)$

$= E\left(\frac{x_1}{n}\right) + E\left(\frac{x_2}{n}\right) + \cdots + E\left(\frac{x_n}{n}\right)$

$= \frac{1}{n}\mu + \frac{1}{n}\mu + \cdots + \frac{1}{n}\mu = \frac{n\mu}{n} = \mu$

Also $Var(\bar{x}) = Var\left(\frac{x_1 + x_2 + \cdots + x_n}{n}\right)$

$= Var\left(\frac{x_1}{n}\right) + Var\left(\frac{x_2}{n}\right) + \cdots + Var\left(\frac{x_n}{n}\right)$

$$= \frac{1}{n^2}\sigma^2 + \frac{1}{n^2}\sigma^2 + \cdots + \frac{1}{n^2}\sigma^2 = \frac{n\sigma^2}{n^2} = \frac{\sigma^2}{n}$$

$$\because Var(ax_i) = a^2 Var(x_i)$$

∴ Standard deviation of sampling distribution ($\sigma_{\bar{x}}$) = $\frac{\sigma}{\sqrt{n}}$

Standard Error: The standard deviation of the sampling distribution is called the standard error

∴ Standard error of sampling distribution is $\sigma_{\bar{x}} = \frac{\sigma}{\sqrt{n}}$

Remark: We have assumed here that σ is known. However if σ is not known, we take σ to be equal to the standard deviation of the sample.

Properties of the sampling distribution of sample means:

➢ A sample drawn from a normally distributed population follows normal distribution,

∴ Z-value for the distribution of $\bar{x}$ is given by $\frac{\bar{x}-\mu}{\sigma_{\bar{x}}} = \frac{\bar{x}-\mu}{\frac{\sigma}{\sqrt{n}}}$

➢ The mean $\mu_{\bar{x}}$ of the sample means will be the same as population mean from which the samples were drawn, i.e. $\mu_{\bar{x}} = \mu$

➢ The variance $\sigma_{\bar{x}}^2$ of the sampling distribution of $\bar{x}$ will be equal to the variance of the population divided by the sample size i.e. $\sigma_{\bar{x}}^2 = \frac{\sigma^2}{n}$

Result IV : Central Limit Theorem: As the sample size gets large enough (30 or higher); the sampling distribution becomes approximately normal regardless of shape of population.

Remark: For large samples ($n \geq 30$), probability distribution is taken as normal for computational purposes.

Example 10 A population has mean 0.1 and standard deviation 2.1. Find the probability that the mean of a random sample of size 900 will be negative.

Solution: Given that population mean (μ) = 0.1 and standard deviation (σ) = 2.1

Since the sample size is large enough, sampling distribution is approximately normal with mean 0.1 and standard deviation = $\frac{\sigma}{\sqrt{n}} = \frac{2.1}{\sqrt{900}} = 0.07$

i.e. $\bar{x} \sim N\left(0.1, \frac{(2.1)^2}{900}\right)$ and $z = \frac{\bar{x}-\mu}{\frac{\sigma}{\sqrt{n}}} = \frac{\bar{x}-0.1}{\frac{2.1}{30}}$ ⇒ $\bar{x} = 0.1 + 0.07z$

∴ $P(\bar{x} < 0) = P(0.1 + 0.07z < 0)$
$= P\left(z < \frac{-0.1}{0.07}\right)$
$= P(z < -1.43)$
$= P(z > 1.43)$
$= 0.5 - 0.4236 = 0.0764$

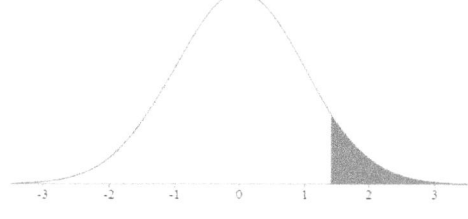

Example 11 Suppose a population has mean 10 and variance 4. What is the probability that the sample of size 36 has mean lying between 9.8 and 10.2?

Solution: Given that population mean (μ) = 10 and standard deviation (σ) = 2. Since the sample size is large enough, sampling distribution is approximately normal with mean 10 and standard deviation = $\frac{\sigma}{\sqrt{n}} = \frac{2}{\sqrt{36}} = 0.33$

i.e. $\bar{x} \sim N\left(10, \frac{4}{36}\right)$ and $z = \frac{\bar{x}-\mu}{\frac{\sigma}{\sqrt{n}}} = \frac{\bar{x}-10}{\frac{2}{6}} \Rightarrow \bar{x} = 10 + 0.33z$

$\therefore P(9.8 < \bar{x} < 10.2)$
$= P(9.8 < 10 + 0.33z < 10.2)$
$= P\left(\frac{9.8-10}{0.33} < z < \frac{10.2-10}{0.33}\right)$
$= P(-0.61 < z < 0.61)$
$= 0.2291 + 0.2291 = 0.4582$

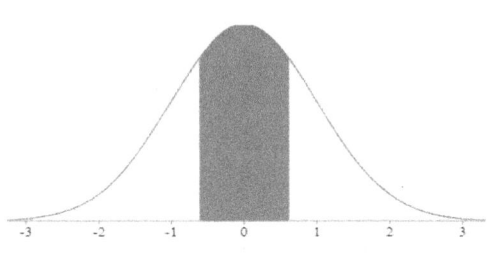

Example 12 A firm produces electric bulbs that have normally distributed mean burning life 800 hours with a standard deviation of 40 hours. Find the probability that a random sample of 16 bulbs will have average burning life of less than 775 hours.

Solution: Given distribution is normal with population mean $(\mu) = 800$ and standard deviation $(\sigma) = 40$, sample size $(n) = 16$

$z = \frac{\bar{x}-\mu}{\frac{\sigma}{\sqrt{n}}} = \frac{\bar{x}-800}{10}$

$\Rightarrow \bar{x} = 800 + 10z$

$\therefore P(\bar{x} < 775) = P(800 + 10z < 775)$
$= P\left(z < \frac{775-800}{10}\right)$
$= P(z < -2.5)$
$= P(z > 2.5)$ By symmetry of standard normal curve
$= 0.5 - P(0 < z < 2.5) = 0.5 - 0.4938 = 0.0062$

Result V : Test of significance of sample mean for a large sample

Here we test the hypotheses about the sample mean of a population in the case when sample size is at least 30 (for large samples, distribution is approximately normal as per central limit theorem). Let μ and σ be mean and variance of a population. If $\bar{x}$ be the mean of a simple large sample of size n from an unknown population, we may want to test whether the sample belongs to given population.

Working rule:
1. Set up the hypothesis H_0: the sample belongs to given population
2. Compute the statistic parameter $z = \frac{\bar{x}-\mu}{\frac{\sigma}{\sqrt{n}}}$
3. Accept or reject H_0 as per given values of parameter z at different significance levels:

Significance Level	H_0 accepted	H_0 rejected
0.27%	$|z| < 3$	$|z| > 3$
1%	$|z| < 2.58$	$|z| > 2.58$
5%	$|z| < 1.96$	$|z| > 1.96$

Remark: The statistic parameter $z = \frac{\bar{x}-\mu}{\frac{\sigma}{\sqrt{n}}}$ can also be used to check whether the sample taken from the given population is random or not.

Example 13 A sample of size 900 is having mean 3.6 mm; could it be reasonably regarded as a random sample from large population whose mean is 3.35 mm and standard deviation 2.6 mm at 1% significance level.

Solution: Let H_0: sample belongs to the given population

Here population mean $\mu = 3.35$mm, population standard deviation $\sigma = 2.6$ mm,

also sample size $n = 900$ and sample mean $\bar{x} = 3.6$ mm

Now $z = \frac{\bar{x}-\mu}{\frac{\sigma}{\sqrt{n}}} = \frac{3.6-3.35}{\frac{2.6}{\sqrt{900}}} = \frac{3.6-3.35}{2.6}(30) = 2.88 > 2.58$

As $|z| > 2.58$, H_0 is rejected, i.e. difference between sample mean ($\bar{x}$) and population mean (μ) is significant at 1% level and the sample may not be considered from the same population.

Example 14 A random sample of 400 canes of vegetable oil with labeled net weight as 5kg has a mean net weight of 4.98 kg with a standard deviation of 0.22 kg. Can we accept the hypothesis of net weight 5 kg per cane on the basis of given sample, at 5 % level of significance?

Solution: Let H_0: Mean net weight of each tin is 5 kg.

Here population mean $\mu = 5$kg, sample size $n = 400$ and sample mean $\bar{x} = 4.98$ kg,

sample standard deviation $s = 0.22$kg

Now $z = \frac{\bar{x}-\mu}{\frac{\sigma}{\sqrt{n}}} = \frac{4.98-5}{\frac{0.22}{\sqrt{400}}} = \frac{4.98-5}{0.22}(20) = -1.81$

As $|z| < 1.96$, H_0 is accepted at 5% significance level, i.e. Mean net weight of each tin is 5 kg.

Note: Here population standard deviation σ is not known, $\therefore$ sample standard deviation is taken as population standard deviation.

Example 15 A pharmaceutical company fills its best-selling 8 ounce jars of liquid medicine by an automatic dispensing machine. The machine is set to dispense a mean of 8.1 ounces per jar with a standard deviation of 0.22 ounce. Uncontrollable factors in the process can shift the mean away from 8.1 and cause either under-fill or overfill, both of which are undesirable. In such a case the dispensing machine is stopped and recalibrated. A quality control engineer routinely selects 30 jars from the assembly line to check the amounts filled. On one occasion, the sample mean is 8.2 ounces with the standard deviation of 0.25 ounce. Determine if there is sufficient evidence in the sample to indicate, that the machine should be recalibrated, at the 5% level of significance.

Solution: Let H_0: machine need not be recalibrated

Here population mean $\mu = 8.1$ ounce, population standard deviation

$\sigma = 0.22$ ounce.

Also sample size $n = 30$ and sample mean $\bar{x} = 8.2$ ounce

$\therefore z = \frac{\bar{x}-\mu}{\frac{\sigma}{\sqrt{n}}} = \frac{8.2-8.1}{\frac{0.22}{\sqrt{30}}} = \frac{8.2-8.1}{0.22}(\sqrt{30}) = 0.45 \times 5.48 = 2.47$

As $|z| > 1.96$, H_0 is rejected, i.e. difference between sample mean ($\bar{x}$) and population mean (μ) is significant at 5% level and the machine needs to be recalibrated.

Result *VI* : Test of significance of difference between means of two large samples

   (a) **If samples are from two different populations**

Let $\bar{x}_1$ and $\bar{x}_2$ be sample means of two populations of sizes n_1 and n_2 from two different normally distributed populations having means μ_1; μ_2 and variances σ_1^2; σ_2^2 respectively;

i.e. $\bar{x}_1 \sim N\left(\mu_1, \frac{\sigma_1^2}{n_1}\right)$ and $\bar{x}_2 \sim N\left(\mu_2, \frac{\sigma_2^2}{n_2}\right)$

Then $E(\bar{x}_1 - \bar{x}_2) = E(\bar{x}_1) - E(\bar{x}_2) = \mu_1 - \mu_2$

Also $Var(\bar{x}_1 - \bar{x}_2) = Var(\bar{x}_1) + Var(\bar{x}_2) = \frac{\sigma_1^2}{n_1} + \frac{\sigma_2^2}{n_2}$

$\therefore$ Standard Error $(e) = \sqrt{\frac{\sigma_1^2}{n_1} + \frac{\sigma_2^2}{n_2}}$

z-value for the distribution of $\bar{x}_1 - \bar{x}_2$ is given by $z = \frac{(\bar{x}_1 - \mu_1) - (\bar{x}_2 - \mu_2)}{e}$

(b) If samples are from same population

If $\bar{x}_1$ and $\bar{x}_2$ be two sample means of sizes n_1 and n_2 taken from a normally distributed population having mean μ and variance σ^2

Then $E(\bar{x}_1 - \bar{x}_2) = E(\bar{x}_1) - E(\bar{x}_2) = \mu - \mu = 0$

Also $Var(\bar{x}_1 - \bar{x}_2) = Var(\bar{x}_1) + Var(\bar{x}_2) = \frac{\sigma^2}{n_1} + \frac{\sigma^2}{n_2}$

$\therefore$ Standard Error $(e) = \sqrt{\frac{\sigma^2}{n_1} + \frac{\sigma^2}{n_2}}$

z-value for the distribution of $\bar{x}_1 - \bar{x}_2$ is given by $z = \frac{\bar{x}_1 - \bar{x}_2}{e}$

Example 16 A random sample of 150 villages was taken from a district A having standard deviation 32 and average population per village was found to be 440. Another random sample of 250 villages from district B with a standard deviation of 56 gave an average population of 480 per village. Is the difference between the averages of two populations significant? Give reasons.

Solution: Let H_0: The differences between averages of two populations is not significant,

i.e. $\mu_1 = \mu_2$

Here $n_1 = 150$, $\bar{x}_1 = 440$, $\sigma_1 = 32$

$n_2 = 250$, $\bar{x}_2 = 480$, $\sigma_2 = 56$

$e = \sqrt{\frac{\sigma_1^2}{n_1} + \frac{\sigma_2^2}{n_2}} = \sqrt{\frac{32^2}{150} + \frac{56^2}{250}} = 4.4$

$\therefore z = \frac{(\bar{x}_1 - \mu_1) - (\bar{x}_2 - \mu_2)}{e} = \frac{(\bar{x}_1 - \bar{x}_2) - (\mu_1 - \mu_2)}{e}$

$= \frac{440 - 480}{4.4} = -9.09$, under the assumption $\mu_1 = \mu_2$

$|z| \gg 3$, thus differences between two averages is highly significant.

$\therefore H_0$ is rejected and the averages of two populations cannot be taken as same.

Example 17 Random samples of 500 and 400 are having means 11.5 and 10.9 respectively. Can the two samples be regarded as drawn from the population of standard deviation 5?

Solution: Let H_0: The two samples are drawn from the same population with standard deviation $(\sigma) = 5$

Here $n_1 = 500$, $\bar{x}_1 = 11.5$
$n_2 = 400$, $\bar{x}_2 = 10.9$

$$e = \sqrt{\frac{\sigma^2}{n_1} + \frac{\sigma^2}{n_2}} = \sqrt{\frac{25}{500} + \frac{25}{400}} = 0.335$$

$$\therefore z = \frac{\bar{x}_1 - \bar{x}_2}{e} = \frac{11.5 - 10.9}{0.335} = 1.79$$

$|z| < 1.96$, $\therefore H_0$ is accepted at 5% level of significance, i.e. the two samples can be considered to be drawn from same population.

Example 18 50 new entrants in a class are found to have an average height of 135cm and 30 old one have an average height of 140cm with a class standard deviation of 8cm. Does this indicate that mean height of old students is greater than that of new entrants?

Solution: Let H_0: Average height of new entrants is same as old students

Here standard deviation of the class is $\sigma = 8$

$n_1 = 50$, $\bar{x}_1 = 135$
$n_2 = 30$, $\bar{x}_2 = 140$,

$$e = \sqrt{\frac{\sigma^2}{n_1} + \frac{\sigma^2}{n_2}} = \sqrt{\frac{64}{50} + \frac{64}{30}} = 1.85$$

$$\therefore z = \frac{\bar{x}_1 - \bar{x}_2}{e} = \frac{135 - 140}{1.85} = -2.7$$

$|z| < 3$, $\therefore H_0$ is accepted at 0.27% significance level, i.e. average height of new entrants is same as old students.

Remark: H_0 will have to be rejected at 1% and 5% levels of significance.

Exercise 5

1. A coin is tossed 400 times and head turns up 225 times. Discuss whether the coin is biased or unbiased at 5% level of significance.
2. A random sample of 600 oranges was taken from a large consignment and 60 were found to be rotten. Show that the standard error of the proportion of bad ones in a sample of this size is 0.1 and deduce that the percentage of bad oranges in the consignment almost lies between 6.3 and 13.7
3. In a city A 20% of a random sample of 900 school children wore spectacles and in another city B 18.5% of a random sample of 1600 school children used to wear spectacles. Is the difference between the proportions significant?
4. In a sample of 500 people from a state 280 take tea and rest take coffee. Can we assume that tea and coffee are equally popular in the state?
5. A sample of 900 members is found to have a mean of 3.4cm. Can it be reasonably regarded as truly random sample from a large population with mean 3.25cm and S.D. 1.61cm.

6. A sample of 100 electric bulbs produced by a manufacturer M_1 showed a mean life time 1190 hours with a standard deviation of 90 hours. Another sample of 75 bulbs produced by manufacturer M_2 showed a mean life time 1230 hours with a standard deviation of 120 hours. Is there a difference between the mean life times of two brands at 5% level of significance?
7. The means of two large samples of 1000 and 2000 members are 168.75 cm and 170cm respectively. Can these be regarded as drawn from the same population of standard deviation 6.25 cm.
8. A stenographer states that he can take dictation at the rate of 120 words per minute. Can we accept his claim on the basis of 100 trials in which he showed a mean of 116 words with standard deviation of 15 words ?
9. A sample of height of 6400 soldiers has a mean of 67.85 inches and a standard deviation of 2.56 inches, while a random sample of heights of 1600 sailors has a mean of 68.55 inches and a standard deviation of 2.52 inches. Does this indicate that the sailors are on average taller than the soldiers?
10. A random sample of 400 students has an average weight of 55 kg. Can we say that the sample comes from a population with mean 58 kg. with a variance of 9 kg. ?
11. In a big city two samples of people are drawn. First sample of size 100, the average daily income of people is 210$ with a standard deviation 10$ and in the second sample of size 150 persons, average daily income is 220$ with a standard deviation of 11$. Test if there is any significant difference in average incomes.

Answers

1. Biased
3. $z = 0.37$ ∴ the difference is not significant
4. $z = 6.56$ ∴ the difference is highly significant
5. $z = 2.8$ ∴ it cannot be regarded as a random sample.
6. Yes
7. No
8. $z = 2.67$ ∴ The claim is not acceptable
9. Yes
10. No
11. $|z| = 7.14$ ∴ the difference is highly significant.

Chapter 6: Tests of Significance for Small Samples

Tests of Significance For Small Samples

6.1 Introduction

For small samples (size < 30); tests proposed for large samples do not hold good as sampling distribution cannot be assumed to be normal for small samples. This led to search of new approaches to deal with small samples. It should be made rational that the methods and theory applicable to small samples can be used for large samples; but the converse is not true

After hypothesis formulation; choice of test may be sometimes baffling unless specified which test to use. Following suggestions should be kept in mind while choosing test of significance for any hypothesis.

- Size of sample: If size of sample is greater than thirty, use any of the applicable large sample tests or Chi-Square test depending upon the applicability.
- Variance of population: If the population variance is known and the underlying distribution is normal, z-test should be used. Also with known variance; if the distribution is not normal, yet for large sample size, z test can be used. But in case of unknown variance, t-test should be used.

Degree of freedom (v): The degrees of freedom are the number of independent quantities that can be assigned to any statistical distribution arbitrarily. Suppose we are required to choose 5 numbers whose sum is 30, then we have choice of only 4 numbers. This suggests that a data of size n has $(n-1)$ degrees of freedom in general. In case of two restrictions, degrees of freedom will be $(n-2)$.

6.2 Student's *t*-Distribution

Gosset; who wrote under the pen-name 'Student', derived a theoretical distribution known as Student's t distribution; which is used to test a hypothesis when the sample size is small and the population variance is not known. As n (size of the sample) increases; t distribution tends to normal distribution. It is evident from the graph below that t-distribution has proportionally larger area at its tails than the normal distribution.

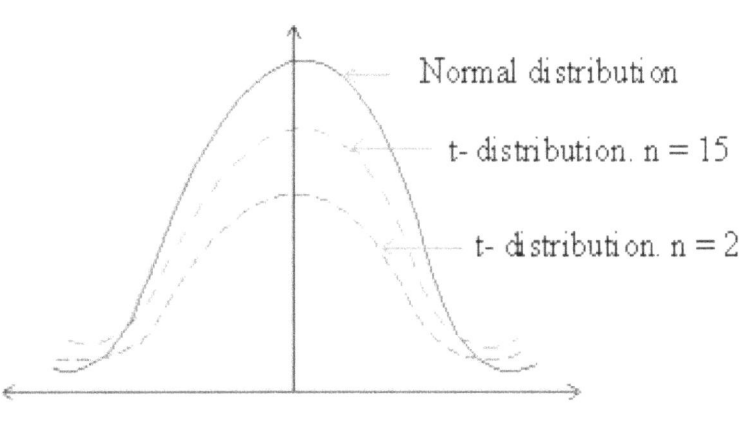

If $x_1, x_2, \cdots, x_n$ be a random sample of size $n \leq 30$, from a normal distribution with mean μ and variance σ^2 (not known); also $\bar{x}$ be the sample mean and s standard deviation of the sample then $t = \dfrac{\bar{x}-\mu}{\frac{s}{\sqrt{n}}}$, where $\bar{x} = \dfrac{\sum x}{n}$, $s = \sqrt{\dfrac{\sum(x-\bar{x})^2}{n-1}}$

Degree of freedom is given by $v = n - 1$

Statistics t is used to test the significance of:

1. Sample mean
2. Difference between two sample means
3. Sample coefficient of correlation
4. Sample Regression Coefficient

Result I Testing significance of sample mean

Set up the hypothesis H_0: sample is drawn from the given population.

To calculate significance of sample mean at 5% level, calculate the statistics

$t = \dfrac{\bar{x}-\mu}{\frac{s}{\sqrt{n}}}$ and compare with the table value of t at $(n-1)$ degrees of freedom.

Let the tabulated value of t be t_1, if $|t| < t_1$; H_0, is accepted, otherwise value of t is considered significant and we reject the hypothesis H_0.

Fiducial (Confidence) limits of population mean

At 5% significance level (95% confidence level):

$\bar{x} \pm \dfrac{s}{\sqrt{n}} t_{0.05}$, where $t_{0.05}$ is table value of t at given degrees of freedom

At 1% significance level (99% confidence level):

$\bar{x} \pm \dfrac{s}{\sqrt{n}} t_{0.01}$, where $t_{0.01}$ is table value of t at given degrees of freedom.

Example1 The mean life time of sample of 16 fluorescent light bulbs produced by a company is computed to be 1550 hours with a standard deviation of 100 hours. The company claims that the average life of the each bulb is 1600 hours. Using the level of significance of 0.05, is the claim accepted? Also find the confidence limits for μ.

Solution: Let H_0: Average life of the each bulb is 1600 hours.

Population mean $(\mu) = 1600$ hours, sample size $(n) = 16$

Sample mean $\bar{x} = 1550$, sample standard deviation $(s) = 100$ hours

$$\text{Statistic } t = \dfrac{\bar{x}-\mu}{s}\sqrt{n} = \dfrac{1550-1600}{100}(4) = -2$$

Table value of t at 5% level of significance for 15 degrees of freedom is 2.13.

∴ Calculated value of $|t| <$ table value of t, hence the hypothesis (H_0) is accepted that the average life of the each bulb is 1600 hours.

Also confidence limits for $\mu = \bar{x} \pm \dfrac{s}{\sqrt{n}} t_{0.05} = 1550 \pm \dfrac{100}{\sqrt{16}}(2.13)$

$=1496.75$ to 1603.25

Example2 Ten individuals are chosen at random from a population and their heights are found in inches as: 63, 63, 64, 65, 66, 69, 69, 70, 70, and 71. Discuss the suggestion that the mean height of universe is 65. Value of t at 5% level of significance for 9 degrees of freedom is 2.262.

Solution: Let H_0: The sample is drawn from the given population.

Population mean $(\mu) = 65$, sample size $(n) = 10$

Calculating sample mean and standard deviation:

x	$x - \bar{x}, \bar{x} = 67$	$(x - \bar{x})^2$
63	-4	16
63	-4	16
64	-3	9
65	-2	4
66	-1	1
69	+2	4
69	+2	4
70	+3	9
70	+3	9
71	+4	16
$\sum x = 670$ $\bar{x} = \frac{670}{10} = 67$		$\sum(x - \bar{x})^2 = 88$

Sample standard deviation: $s = \sqrt{\frac{(x-\bar{x})^2}{n-1}} = \sqrt{\frac{88}{9}} = 3.13$

Statistic $t = \frac{\bar{x}-\mu}{s}\sqrt{n} = \frac{(67-65)\sqrt{10}}{3.13} = \frac{2\sqrt{10}}{3.13} = 2.02$

∴ Calculated value of t is less than tabulated value at 9 degrees of freedom.

Hence the hypothesis (H_0) is accepted that sample is drawn from the given population i.e. mean height of the universe is 65 inches.

Result II Testing difference between means of two small samples

Let two independent samples ($x_1, x_2, \cdots, x_{n_1}$); ($y_1, y_2, \cdots, y_{n_2}$) of sizes n_1; n_2 with means $\bar{x}$; $\bar{y}$ and standard deviations $s_x = \sqrt{\frac{\sum(x-\bar{x})^2}{n_1-1}}$; $s_y = \sqrt{\frac{\sum(y-\bar{y})^2}{n_2-1}}$ be drawn from two normal populations with means μ_1; μ_2 and having same variance σ^2, then to test whether the two population means μ_1 and μ_2 are same:

Calculate $t = \frac{\bar{x}-\bar{y}}{s\sqrt{\frac{1}{n_1}+\frac{1}{n_2}}}$; where $s = \sqrt{\frac{\sum(x-\bar{x})^2+\sum(y-\bar{y})^2}{n_1+n_2-2}}$ or $\sqrt{\frac{(n_1-1)s_x^2+(n_2-1)s_y^2}{n_1+n_2-2}}$

and compare with table value of t (say t_1) at $(n_1 + n_2 - 2)$ degrees of freedom. Hypothesis is accepted if calculated value of $|t| < t_1$.

Example 3 Two independent samples showing weights in ounces of eight and seven items are given below:

Sample I:	10	12	13	11	15	9	12	14
Sample II:	9	11	10	13	9	8	10	

Is the difference between the means of the samples significant?

Value of t at 5% level of significance for 15 degrees of freedom is 2.16.

Solution: Let the null hypothesis (H_0) be: $\mu_1 \approx \mu_2$

Here $n_1 = 8$ and $n_2 = 7$, calculating sample means and standard deviations

x	$(x - \bar{x})$	$(x - \bar{x})^2$	y	$(y - \bar{y})$	$(y - \bar{y})^2$
10	-2	4	9	-1	1
12	0	0	11	1	1
13	1	1	10	0	0
11	-1	1	13	3	9
15	3	9	9	-1	1
9	-3	9	8	-2	4
12	0	0	10	0	0
14	2	4			
$\Sigma x = 96$ $\bar{x} = \frac{96}{8} = 12$		28	$\Sigma y = 70$ $\bar{y} = \frac{70}{7} = 10$		16

$$s = \sqrt{\frac{\Sigma(x-\bar{x})^2 + \Sigma(y-\bar{y})^2}{n_1 + n_2 - 2}} = \sqrt{\frac{\Sigma(x-12)^2 + \Sigma(y-10)^2}{8+7-2}} = \sqrt{\frac{28+16}{8+7-2}} = 1.8$$

$$\therefore t = \frac{\bar{x} - \bar{y}}{s\sqrt{\frac{1}{n_1} + \frac{1}{n_2}}} = \frac{12 - 10}{(1.8)\sqrt{\frac{1}{8} + \frac{1}{7}}} = 2.15$$

Table value of t at 5% level of significance for 13 degrees of freedom is 2.16

$\therefore$ Calculated value of $|t|$ < table value of t

Hence the hypothesis (H_0) is accepted that $\mu_1 \approx \mu_2$, i.e. difference between the means of the two samples is not significant.

Example 4 Two types of batteries were tested for their mean life length and following results are obtained. Is there a significant difference in the two batteries?

	Sample size	Mean life	Variance
Battery A	10	500 hours	100
Battery B	8	540 hours	81

Solution: Let the null hypothesis (H_0) be: $\mu_1 \approx \mu_2$ i.e. there is no significant difference in the two batteries. Here $n_1 = 10$ and $n_2 = 8$, $\bar{x} = 500$ and $\bar{y} = 540$, $s_x = 10$, $s_y = 9$

$$\Rightarrow s = \sqrt{\frac{(n_1-1)s_x^2 + (n_2-1)s_y^2}{n_1+n_2-2}} = \sqrt{\frac{9(100)+7(81)}{10+8-2}} = 9.58$$

$$\therefore t = \frac{\bar{x} - \bar{y}}{s\sqrt{\frac{1}{n_1} + \frac{1}{n_2}}} = \frac{500 - 540}{(9.58)\sqrt{\frac{1}{10} + \frac{1}{8}}} = -8.8$$

Table value of t at 5% level of significance for 16 degrees of freedom is 2.12.

$\therefore$ Calculated value of $|t| \gg$ table value of t, hence the hypothesis (H_0) is rejected that $\mu_1 \approx \mu_2$, i.e. difference between two batteries is highly significant.

Example5: A set of 15 observations has mean 68.57; standard deviation 2.4, another set of 7 observations gives mean as 64.14; standard deviation 2.7.

Use t-test to find whether two sets of data are drawn from populations with same mean, if standard deviations of two populations are assumed to be equal.

Solution: Let the null hypothesis (H_0) be: $\mu_1 \approx \mu_2$

i.e. there is no significant difference between the two population means.

Here $n_1 = 15$ and $n_2 = 7$, $\bar{x} = 68.57$ and $\bar{y} = 64.14$

$s_x = 2.4$, $s_y = 2.7$

$$\Rightarrow s = \sqrt{\frac{(n_1-1)s_x^2 + (n_2-1)s_y^2}{n_1+n_2-2}} = \sqrt{\frac{14(2.4)^2 + 6(2.7)^2}{15+7-2}} = 2.49$$

$$\therefore t = \frac{\bar{x}-\bar{y}}{s\sqrt{\frac{1}{n_1}+\frac{1}{n_2}}} = \frac{68.57-64.14}{(2.49)\sqrt{\frac{1}{15}+\frac{1}{7}}} = 3.89$$

Table value of t at 5% level of significance for 20 degrees of freedom is 2.086

$\therefore$ Calculated value of $|t| >$ table value of t, hence the hypothesis (H_0) is rejected that $\mu_1 \approx \mu_2$, i.e. difference between two means is significant.

Result III Testing difference between two dependent samples or paired observations

A 't' test can be efficiently used to compare two samples of the same population for some treatment effects; for instance to compare efficacy of two drugs on a population or to study the effect of coaching on some students etc.

Compute the statistic $t = \frac{\bar{x}-\Delta}{\frac{s}{\sqrt{n}}}$; where Δ denotes targeted value and is zero for testing equal means and $s = \sqrt{\frac{\sum d^2}{n-1}}$, where d is the deviation from the mean difference.

Example6 A dietitian opts to try out a new type of diet program on ten overweight girls for 2 months. He targets to make them loose 6 kgs on average, and records their weights before and after the diet program. Use 0.05 significance level to test whether this special diet program helped or not.

Before weights	65	77	99	86	84	93	59	72	69	103
After weights	63	72	92	80	80	87	57	67	64	95

Table value of t at 5% level of significance for 9 degrees of freedom is 2.26.

Solution: Let H_0: Average weight loss caused by diet program is 6 kg.

Calculating deviations from the mean weight loss:

Girls	Weight difference (m)	$d = m - \bar{x}$	d^2
1	-2	+3	9
2	-5	0	0
3	-7	-2	4
4	-6	-1	1
5	-4	+1	1
6	-6	-1	1
7	-2	+3	9
8	-5	0	0
9	-5	0	0
10	-8	-3	9
$n = 10$	$\sum m = -50$, $\bar{x} = \frac{\sum m}{n} = \frac{-50}{10} = -5$		$\sum d^2 = 34$

Standard deviation of the differences $s = \sqrt{\frac{\sum d^2}{n-1}} = \sqrt{\frac{34}{9}} = 1.94$

Statistic $t = \frac{\bar{x} - \Delta}{\frac{s}{\sqrt{n}}}$, where targeted weight loss (Δ) = -6 kg

$\Rightarrow t = \frac{-5 - (-6)\sqrt{10}}{1.94} = 1.63$ ∴ Calculated value of $|t|$ < table value of t

Hence the hypothesis (H_0) is accepted that the average weight loss caused by diet program is 6 kg.

Example 7 Two laboratories carried out independent estimates of lead content (in mg) in noodles of a certain brand. A sample is taken from each batch, halved and the separate halves were tested in the two laboratories to obtain the following results:

Batch Number	1	2	3	4	5	6	7	8	9	10
Lab A	9	8	8	4	7	7	9	6	6	6
Lab B	7	8	7	3	8	6	9	4	7	8

Does the testing suggest same average lead content in the brand?

Solution: H_0: Average lead content of two samples are equal.

Calculating deviations from the mean difference:

Batch	Difference (m)	$d = m - \bar{x}$	d^2
1	2	1.7	2.89
2	0	-0.3	0.09
3	1	0.7	0.49
4	1	0.7	0.49
5	-1	-1.3	1.69
6	1	0.7	0.49
7	0	-0.3	0.09
8	2	1.7	2.89
9	-1	-1.3	1.69
10	-2	-2.3	5.29
$n = 10$	$\sum m = 3$, $\bar{x} = \frac{\sum m}{n} = \frac{3}{10} = 0.3$		$\sum d^2 = 16.10$

Standard deviation of the differences $s = \sqrt{\frac{\Sigma d^2}{n-1}} = \sqrt{\frac{16.1}{9}} = 1.34$

$t = \frac{\bar{x} - \Delta}{\frac{s}{\sqrt{n}}} = \frac{0.3 - (0)\sqrt{10}}{1.34} = 0.71$ $\because \Delta = 0$ as we are testing for equal means of two lab tests

Table value of t at 5% level of significance for 9 degrees of freedom is 2.26.

$\therefore$ Calculated value of $|t|$ < table value of t at 9 degrees of freedom at 5% level of significance, hence the hypothesis (H_0) is accepted that the average lead content of two samples are equal.

Result IV Testing significance of population correlation coefficient from sample coefficient of correlation

Let $(x_1, y_1), (x_2, y_2), \ldots, (x_n, y_n)$ having coefficient of correlation r, be a random sample from a bivariate frequency distribution having individual means $\mu_1; \mu_2$ and standard deviations $\sigma_1; \sigma_2$. Then to test the hypothesis that population correlation coefficient (ρ) is zero: Compute the statistic $t = \frac{r\sqrt{n-2}}{\sqrt{1-r^2}}$, where n is the sample size

Hypothesis is accepted if calculated value of $|t|$ is less than tabulated value of t at $(n-2)$ degrees of freedom for the specified significance level.

Example 8 A random sample of size 18 from a bivariate population gave correlation coefficient $r = 0.4$. Does this indicate the existence of correlation in the population?

Solution: Let H_0: Population correlation coefficient (ρ) is zero, i.e. there is no correlation between the population variables.

$t = \frac{r\sqrt{n-2}}{\sqrt{1-r^2}} = \frac{(0.4)\sqrt{18-2}}{\sqrt{1-(0.4)^2}} = 1.7$

Table value of t at 16 degrees of freedom is 2.12

$\therefore$ Calculated value of $|t|$ is less than tabulated value of t at 16 degrees of freedom, hence the hypothesis H_0 is accepted that there is no correlation between the population variables.

6.3 Snedecor's F – test for Testing Equality of Two Population Variances

An F-test is used to test if the variances of two populations are equal. It can be a one-tailed or two-tailed test. The one-tailed version tests in only one direction, i.e. variance of the first population is either greater or less than the second population but not both ways. The two-tailed version tests for the hypothesis that the variances are not equal but one can be greater or less than the other.

Let two independent random samples $(x_1, x_2, \ldots, x_{n_1})$; $(y_1, y_2, \ldots, y_{n_2})$ having standard deviations $s_x = \sqrt{\frac{\Sigma(x-\bar{x})^2}{n_1 - 1}}$; $s_y = \sqrt{\frac{\Sigma(y-\bar{y})^2}{n_2 - 1}}$ be drawn from two normal populations. Snedecor defined the statistic $F = \frac{s_x^2}{s_y^2}$, $s_x^2 > s_y^2$ for testing equality of two population variances. Here greater of the two variances s_x^2 and s_y^2 is to be taken in the numerator and if n_1 corresponds to the greater variance, then degree of freedom is $(n_1 - 1, n_2 - 1)$.

If calculated value of F is less than the table value of F with $v = (n_1 - 1, n_2 - 1)$ at given level of significance, the null hypothesis that 'the two samples might have been drawn from two normal population with the same variance' is accepted.

Example 9 Two samples of size 9 and 8 give the sum of squares of deviations from their respective means equal to 160 inches square and 91 inches square respectively. Can they be regarded as drawn from two normal populations with same variances? (Value of F for 8 and 7 degrees of freedom is 3.73)

Solution: Let H_0: Two samples have been drawn from two normal populations with same variance.

Given $s_x^2 = \frac{\Sigma(x-\bar{x})^2}{n_1-1} = \frac{160}{9-1} = 20$, $s_y^2 = \frac{\Sigma(y-\bar{y})^2}{n_2-1} = \frac{91}{8-1} = 13$

$\therefore F = \frac{s_x^2}{s_y^2} = \frac{20}{13} = 1.54 \quad \because s_x^2 > s_y^2$

Table value of F for (8, 7) degrees of freedom is 3.73, $\therefore$ Calculated value of F is much less than the table value of F at (8,7) degrees of freedom. Hence H_0 is accepted, i.e. two samples may be regarded as drawn from two normal populations with same variances.

Example 10 Show how we can use Student's-t test and Snedecor's F test to decide whether the following two samples have been drawn from the same normal population. Which of the two tests would you apply first and why?

	Size	Mean	Sum of squares of deviations from the mean
Sample I	9	68	36
Sample II	10	69	42

Given that $t_{0.5}(17) = 2.11$, $F_{0.5}(9,8) = 3.39$

Solution: H_0: Two samples have been drawn from the same normal population

To test H_0 using Student's-t test, population variance σ^2 should be same, we can test that two samples have been drawn from two normal populations with same variance using Snedecor's F test. $\therefore$ Snedecor's F test should be applied first.

For the two samples I and II say $(x_1, x_2, \cdots, x_9)$; $(y_1, y_2, \cdots, y_{10})$

$s_x^2 = \frac{\Sigma(x-\bar{x})^2}{n_1-1} = \frac{36}{9-1} = 4.5$, $s_y^2 = \frac{\Sigma(y-\bar{y})^2}{n_2-1} = \frac{42}{10-1} = 4.67$

$\therefore F = \frac{s_y^2}{s_x^2} = \frac{4.67}{4.5} = 1.04 \quad \because s_y^2 > s_x^2$

Table value of F for (9, 8) degrees of freedom is 3.39

$\therefore$ Calculated value of F is much less than the table value of F at (9, 8) degrees of freedom.

Hence the two samples may be regarded as drawn from two normal populations with same variances.

Again to test whether the two population means are same using t-test

Calculate $t = \dfrac{\bar{x}-\bar{y}}{s\sqrt{\frac{1}{n_1}+\frac{1}{n_2}}}$; where $s = \sqrt{\dfrac{(n_1-1)s_x^2 + (n_2-1)s_y^2}{n_1+n_2-2}}$

$$\Rightarrow s = \sqrt{\dfrac{(9-1)(4.5)+(10-1)(4.67)}{9+10-2}} = 2.14$$

$$\therefore t = \dfrac{68-69}{(2.14)\sqrt{\frac{1}{9}+\frac{1}{10}}} = -1.02$$

Table value of t at 5% level of significance for 17 degrees of freedom is 2.11.

$\therefore$ Calculated value of $|t|$ < table value of t, hence the hypothesis that two population means are same is accepted.

Thus using Snedecor's F test, we can say that the two samples have been drawn from two normal populations with same variance and also the two population means are same using t-test, hence we can conclude that the hypothesis H_0 that the two samples have been drawn from the same normal population may be accepted.

6.4 Fisher's Z Test for Testing Significance of Correlation Coefficient for Small Samples

If r is the correlation co-efficient of a sample and ρ be the population co-efficient of correlation, then to test the hypothesis that the given sample has been drawn from the population whose coefficient of correlation is ρ:

Compute the statistic $\dfrac{z-\xi}{\frac{1}{\sqrt{n-3}}}$, where $z = \dfrac{1}{2}\log_e\left(\dfrac{1+r}{1-r}\right)$ and $\xi = \dfrac{1}{2}\log_e\left(\dfrac{1+\rho}{1-\rho}\right)$

If value of $\left|\dfrac{z-\xi}{\frac{1}{\sqrt{n-3}}}\right|$ < 1.96, hypothesis is accepted at 5% level of significance.

Example 11 Test the significance of the correlation $r = 0.6$ for a sample of size 20 against hypothetical population coefficient of correlation $\rho = 0.8$

Solution: Let H_0: correlation coefficient of population is 0.8

$$\text{Here } z = \dfrac{1}{2}\log_e\left(\dfrac{1+r}{1-r}\right) = \dfrac{1}{2}\log_e\left(\dfrac{1+0.6}{1-0.6}\right) = 0.693$$

$$\xi = \dfrac{1}{2}\log_e\left(\dfrac{1+\rho}{1-\rho}\right) = \dfrac{1}{2}\log_e\left(\dfrac{1+0.8}{1-0.8}\right) = 1.099$$

$$\text{Now } \dfrac{z-\xi}{\frac{1}{\sqrt{n-3}}} = \dfrac{0.693-1.099}{\frac{1}{\sqrt{20-3}}} = (0.693-1.099)\sqrt{17} = -1.67$$

$$\therefore \left|\dfrac{z-\xi}{\frac{1}{\sqrt{n-3}}}\right| = 1.67 < 1.96$$

Hence the sample may be regarded as coming from population with coefficient of correlation $(\rho) = 0.8$

Exercise 6

1. A factory makes a machine part with axle diameter of 0.7 inch. A random sample of 10 parts shows a mean diameter of 0.742 inch with a standard deviation of 0.04 inch. On the basis of this sample would you say that the work is inferior?
   Value of t at 5% level of significance for 9 degrees of freedom is 2.262.

2. A random sample of 10 boys had the I.Q. levels: 70, 120, 110, 101, 88, 83, 95, 98, 107 and 100. Does this data support the assumption of population mean I.Q. of 100 at 5% level of significance?

3. In a school the heights of six randomly chosen girls are: 63, 65, 68, 69, 71 and 72 inches and those of nine randomly chosen boys are 61, 62, 65, 66, 69, 70, 71, 72 and 73 inches. Discuss the hypothesis that the girls are taller than boys.
   Value of t at 5% level of significance for 13 degrees of freedom is 1.77.

4. A random sample of size 16 has mean 53. The sum of squares of deviations from the mean is 135. Can the sample be regarded as taken from a population having mean as 56?

5. A new medicine is given to 12 patients whose B.P. increases by 5, 2, 8, -1, 3, 0, -2, 1, 5, 0, 4, 6 units. Can we conclude that the medicine results in increased blood pressure?

6. Two samples of different brands were tested for average life; a sample from first brand of size 7 shows a mean life of 1036 hours with a standard deviation of 40 hours and a sample of size 8 shows a mean life of 1234 hours with a standard deviation of 36 hours. Is the difference in the two sample means significant to conclude that the second brand has more life than first brand?

7. A researcher hypothesizes that people who are allowed to sleep for only four hours will score significantly lower in an objective skills test than people who are allowed to sleep for eight hours. He selects sixteen participants and randomly assigns them to one of two groups. In one group he makes participants sleep for eight hours and in the other group he allows them to sleep only for four hours. The next morning he administers the skill test to all participants. Scores range from 1-9 with high scores representing better performance.

Test scores→								
8 hours sleep group	5	7	5	3	5	3	3	9
4 hours sleep group	8	1	4	6	6	4	1	2

   Test the hypothesis, given that $t_{0.5}(14) = 2.145$

8. A group of 10 rats fed on the diet A and another group of 8 rats fed on the diet B recorded the following increase in weights in a week:

Weight gains (grams)→									
Diet A	5	6	8	1	12	4	3	9	6	10
Diet B	2	3	6	8	10	1	2	8	-	-

   Does it show superiority of Diet A over that of Diet B?

9. Test runs with 6 models of an experimental engine showed that they operated for 24, 28, 21, 23, 32 and 22 minutes with a gallon of fuel. If the probability of a Type I error is at most 0.01, is this an evidence against the hypothesis that on average this kind of engine will operate for at least 27 minutes per gallon on the same fuel?

10. Test whether the following two samples have been drawn from the same normal population.

	Size	Mean	Sum of squares of deviations from the mean
Sample I	10	15	90
Sample II	12	14	108

Given that $t_{0.5}(20) = 2.086$, $F_{0.5}(9,11) = 2.9$

Answers

1. $t = 3.16$ which is greater than the table value at 5% significance level, ∴ work can be considered to be inferior.
2. $t = 0.62$ ∴ given data supports the mean I.Q. as 100.
3. $t = 0.3031$ ∴ there is no significant difference between the sample means
4. No
5. No
6. $t = 18.15$ ∴ significantly different to conclude that the second brand has more life than first brand
7. $t = 0.847$ ∴ there is no significant difference between the performances of two groups.
8. No
9. No
10. Yes, the two samples can be considered to be drawn from the same normal population.

Chapter 7: Tests of significance based on Chi-Square

Tests of Significance Based on Chi-Square (χ^2)

7.1 Chi-Square (χ^2) Test

Tests like t, F and Z are based on the assumption that the samples are drawn from a normally distributed population. As these tests require assumptions about the type of population or parameters, these tests are called 'parametric tests'.

Sometimes it is unrealistic to make any rigid assumption about the distribution of the population from which samples are drawn. Studies came out with Chi-Square (read as Ki-square) tests which provide non-parametric approach for testing of goodness of fit and independence of attributes. Although it assumes population to be normal while assessing population variance with sample parameters; thus it is parametric in this case.

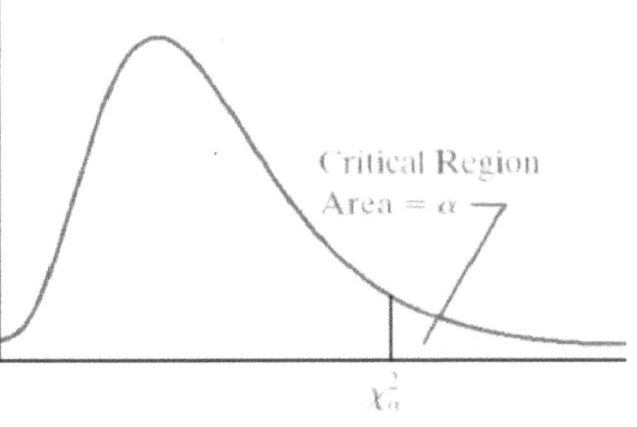

Chi-square (χ^2) is a right tailed test and describes the magnitude of discrepancy between theory and observation. If $\chi^2 = 0$; the observed and expected frequencies completely coincide, $\therefore \chi^2$ provides a measure of correspondence between theory and observations. It is one of the simplest and most general tests and can be used to perform

1. test of significance of sample variance
2. test of goodness of fit
3. test of independence of attributes in a contingency table

7.2 Chi-Square (χ^2) Test for Testing Significance of Sample Variance

Chi-Square test can be used to check the population variance for a specified value and also if it needs a revision (if population variance is already given), looking into sample parameters.

If $x_1, x_2, \cdots, x_n$ be a random sample of size $n \leq 30$; with variance s^2, from a normal population with given variance σ^2, then under null hypothesis that population variance is unchanged,

the statistic Chi-Square is defined by: $\chi^2 = \sum \frac{(x-\bar{x})^2}{\sigma^2} = \frac{1}{\sigma^2}\sum(x-\bar{x})^2$ or $\chi^2 = \frac{(n-1)s^2}{\sigma^2}$

$$\text{where } \bar{x} = \frac{\sum x}{n}, \quad s = \sqrt{\frac{\sum(x-\bar{x})^2}{n-1}} \quad \Rightarrow \sum(x-\bar{x})^2 = (n-1)s^2$$

$\therefore \chi^2$ is unbiased estimator of testing population variance with degrees of freedom $(v) = n - 1$.

Example1 The standard deviation of a certain dimension of articles produced by a machine is 7.5 over a long period. A random sample of 25 articles gave a standard deviation of 10.0. Is it justified to conclude that variability has increased?

Value of χ^2 for 24 degrees of freedom at 5% significance level is 36.415

Solution: Let H_0: Population variance $(\sigma^2) = 7.5$ has not changed.

H_1: Variance has increased

$$\chi^2 = \frac{(n-1)s^2}{\sigma^2} = \frac{24(10)^2}{(7.5)^2} = 42.67$$

Table value of χ^2 for 24 degrees of freedom at 5% level is 36.415

Calculated value of χ^2 is greater than table value. $\therefore$ H_0 is rejected and it is justified to conclude that variability has increased.

Example2 Eleven measurements of the same object on an instrument; at different times are given by: 2.5, 2.3, 2.4, 2.5, 2.7, 2.5, 2.6, 2.6, 2.7, 2.5 and 2.3.

Test at 1% level of significance that variance of the instrument is not more than 0.16.

Solution: Let H_0: Population variance $(\sigma^2) = 0.16$

$H_1: \sigma^2 > 0.16$

$$\chi^2 = \frac{1}{\sigma^2}\Sigma(x-\bar{x})^2, \text{ where } \bar{x} = \frac{\Sigma x}{n}$$

Here $\bar{x} = \frac{2.5+2.3+2.4+2.5+2.7+2.5+2.6+2.6+2.7+2.5+2.3}{11} = 2.51$

$\therefore \Sigma(x-\bar{x})^2 = 2(2.3-2.51)^2 + (2.4-2.51)^2 + 4(2.5-2.51)^2 +$
$\qquad 2(2.6-2.51)^2 + 2(2.7-2.51)^2 = 0.1891$

$\therefore \chi^2 = \frac{1}{\sigma^2}\Sigma(x-\bar{x})^2 = \frac{0.1891}{0.16} = 1.182$

Table value of χ^2 for 10 degrees of freedom at 1% level is 23.209

Calculated value of χ^2 is greater than table value. $\therefore$ H_0 is rejected and it is justified to conclude that variability has increased.

7.3 Chi-Square (χ^2) Test for Testing Goodness of Fit

If O_i $(i = 1,2,\ldots,n)$ be a set of observed (experimental) frequencies and E_i $(i = 1,2,\ldots,n)$ be the corresponding set of expected (theoretical) frequencies,

then $\chi^2 = \sum_{i=1}^{n}\frac{(O_i-E_i)^2}{E_i}$; with degrees of freedom $(\nu) = n - 1$.

There are some underlying conditions for applying χ^2 test such as:

1. Sum of frequencies should be large (at least 50)
2. No theoretical cell-frequencies should be small, if small (less than 5) theoretical frequencies occur, regrouping of two or more cells must be done before calculating $(O_i - E_i)$. Degree of freedom is determined by the number of classes after regrouping.

Example3 Apply χ^2 test of goodness to fit for the following data:

Observed frequency:	1	5	20	28	42	22	15	5	2
Theoretical Frequency:	1	6	18	25	40	25	18	6	1

Value of χ^2 for 6, 7, 8 degrees of freedom at 5% significance level are 12.592, 14.067 and 15.507 respectively.

Solution: Since first two and the last two cells frequencies are smaller than prescribed, regrouping the data to 7 cells:

O_i	6	20	28	42	22	15	7
E_i:	7	18	25	40	25	18	7

Degrees of freedom $(\nu) = 7 - 1 = 6$

$\chi^2 = \sum_{i=1}^{n} \frac{(O_i - E_i)^2}{E_i}$

$= \frac{(6-7)^2}{7} + \frac{(20-18)^2}{18} + \frac{(28-25)^2}{25} + \frac{(42-40)^2}{40} + \frac{(22-25)^2}{25} + \frac{(15-18)^2}{18} + \frac{(7-7)^2}{7} = 1.685$

Table value of χ^2 for 6 degrees of freedom at 5% level = 12.592

Calculated value of χ^2 is much less than table value ∴ the fit is very good.

Example 4 In experimental breeding, Mendal got the following frequencies of seeds:

315 round and yellow, 101 wrinkled and yellow, 108 round and green, 32 wrinkled and green, total 556. Theory predicts that the frequencies should be proportional in the order 9:3:3:1. Examine the correspondence between theory and experiments.

(Value of χ^2 for 3 degrees of freedom at 5% significance level = 7.815)

Solution: Expected (theoretical) frequencies are:

$\frac{9}{16} \times 556 \approx 313, \frac{3}{16} \times 556 \approx 104, \frac{3}{16} \times 556 \approx 104, \frac{1}{16} \times 556 \approx 35$

Comparing observed and expected frequencies:

O_i:	315	101	108	32
E_i:	313	104	104	35

Degrees of freedom $(\nu) = 4 - 1 = 3$

$\chi^2 = \sum_{i=1}^{n} \frac{(O_i - E_i)^2}{E_i}$

$= \frac{(315-313)^2}{313} + \frac{(101-104)^2}{104} + \frac{(108-104)^2}{104} + \frac{(32-35)^2}{35} = 0.51$

Table value of χ^2 for 3 degrees of freedom at 5% level = 7.815

Calculated value of χ^2 is much less than table value, ∴ there is much correspondence between theory and experiment.

Example 5 A die is thrown 246 times and results of these throws are given as:

Number on the die	1	2	3	4	5	6
Frequency	32	35	59	57	39	24

Find the value of χ^2 on the hypothesis that the die is unbiased.

Solution: Let H_0: die is unbiased.

Expected frequency of each number for an unbiased die is $\frac{246}{6} = 41$

Comparing observed and expected frequencies:

O_i	32	35	59	57	39	24
E_i:	41	41	41	41	41	41

Degrees of freedom $(\nu) = 6 - 1 = 5$

$$\chi^2 = \sum_{i=1}^{n} \frac{(O_i - E_i)^2}{E_i}$$

$$= \frac{(32-41)^2}{41} + \frac{(35-41)^2}{41} + \frac{(59-41)^2}{41} + \frac{(57-41)^2}{41} + \frac{(39-41)^2}{41} + \frac{(24-41)^2}{41} = 24.15$$

Table value of χ^2 for 5 degrees of freedom at 5% level = 11.07

Calculated value of χ^2 is greater than table value, $\therefore$ the die can be concluded to be biased one.

7.4 Chi-Square (χ^2) Test for Testing Independence of Attributes

The Chi-Square test of independence is used to determine if there is a significant relationship between two nominal (categorical) variables.

Consider two attributes A and B to be tested for independence with categories A_i, $i = 1, 2, \ldots, m$ and B_j, $j = 1, 2, \ldots, n$ respectively.

Attribute		$B \rightarrow$				Total
		B_1	B_2		B_n	
$A \downarrow$	A_1	O_{11}	O_{12}	...	O_{1n}	$\sum_{j=1}^{n} O_{1j}$
	A_2	O_{21}	O_{22}		O_{2n}	$\sum_{j=1}^{n} O_{2j}$
	$\vdots$					
	A_m	O_{m1}	O_{m2}		O_{mn}	$\sum_{j=1}^{n} O_{mj}$
Total		$\sum_{i=1}^{m} O_{i1}$	$\sum_{i=1}^{m} O_{i2}$		$\sum_{i=1}^{m} O_{in}$	$N = \sum_{i,j} O_{ij}$

Here O_{ij} is observed frequency for attributes $A_i B_j$ in the given contingency table.

Expected frequency E_{ij} is calculated as shown:

$$E_{11} = \frac{(\sum_{j=1}^{n} O_{1j})(\sum_{i=1}^{m} O_{i1})}{N}, \quad E_{12} = \frac{(\sum_{j=1}^{n} O_{1j})(\sum_{i=1}^{m} O_{i2})}{N}, \ldots, E_{mn} = \frac{(\sum_{j=1}^{n} O_{mj})(\sum_{i=1}^{m} O_{in})}{N}$$

Chi-Square value is given by: $\chi^2 = \sum \frac{(O_{ij} - E_{ij})^2}{E_{ij}}$

For a $m \times n$ contingency table (m rows and n columns), $\nu = (m-1)(n-1)$

Example 6 Test the hypothesis that flower colour is independent of leaf pattern using the following information:

Leaves ↓	Flower colour→		
	White	Red	Total
Flat	68	32	100
Curled	52	58	110
Total	120	90	210

Value of χ^2 for 2 and 1 degree of freedom at 5% level is 5.991 and 3.841 respectively.

Solution: Let H_0: Flower colour and leaf pattern are independent

On the basis of null hypothesis, expected (theoretical) frequencies are given as:

Leaves ↓	Flower colour→		
	White	Red	Total
Flat	$\frac{(100)(120)}{210} = 57.14$	$\frac{(100)(90)}{210} = 42.86$	100
Curled	$\frac{(110)(120)}{210} = 62.86$	$\frac{(110)(90)}{210} = 47.14$	110
Total	120	90	210

Degrees of freedom $(v) = (2-1)(2-1) = 1$

$$\chi^2 = \sum \frac{(O_{ij} - E_{ij})^2}{E_{ij}} = \frac{(68-57.14)^2}{57.14} + \frac{(32-42.86)^2}{42.86} + \frac{(52-62.86)^2}{62.86} + \frac{(58-47.14)^2}{47.14} = 9.19$$

Table value of χ^2 for 1 degrees of freedom at 5% level = 3.841

Calculated value of χ^2 is greater than table value, $\therefore$ H_0 is rejected, i.e. flower colour and leaf pattern have some connection.

Example 7 For the attributes A and B, in a contingency table $\begin{matrix} a & b \\ c & d \end{matrix}$

show that $\chi^2 = \frac{(a+b+c+d)(ad-bc)^2}{(a+b)(c+d)(b+d)(a+c)}$

Solution: The 2 × 2 contingency table for the attributes A and B is given below

Attribute	B →		Total
A	a	b	a + b
↓	c	d	c + d
Total	a + c	b + d	N = a + b + c + d

Here a, b, c, d are observed frequencies and under the assumption that the two attributes A and B are independent, expected frequencies are given as $\frac{(a+b)(a+c)}{N}, \frac{(a+b)(b+d)}{N}, \frac{(c+d)(a+c)}{N}, \frac{(c+d)(b+d)}{N}$ respectively.

Now $\chi^2 = \sum \frac{(O_{ij} - E_{ij})^2}{E_{ij}} = \frac{\left[a - \left(\frac{(a+b)(a+c)}{N}\right)\right]^2}{\frac{(a+b)(a+c)}{N}} + \cdots + \frac{\left[d - \left(\frac{(c+d)(b+d)}{N}\right)\right]^2}{\frac{(c+d)(b+d)}{N}}$

$= \frac{[a(a+b+c+d) - (a+b)(a+c)]^2}{(a+b+c+d)(a+b)(a+c)} + \cdots + \frac{[d(a+b+c+d) - (c+d)(b+d)]^2}{(a+b+c+d)(c+d)(b+d)}$

$= \frac{(ad-bc)^2}{(a+b+c+d)} \left[\frac{1}{(a+b)(a+c)}\right] + \cdots + \frac{(ad-bc)^2}{(a+b+c+d)} \left[\frac{1}{(c+d)(b+d)}\right]$

$= \frac{(ad-bc)^2}{(a+b+c+d)} \left[\frac{1}{(a+b)(a+c)} + \frac{1}{(a+b)(b+d)} + \frac{1}{(c+d)(a+c)} + \frac{1}{(c+d)(b+d)}\right]$

$= \frac{(ad-bc)^2}{(a+b+c+d)} \left[\frac{b+d+a+c}{(a+b)(a+c)(b+d)} + \frac{b+d+a+c}{(c+d)(a+c)(b+d)}\right]$

$= \frac{(ad-bc)^2(a+b+c+d)}{(a+b+c+d)} \left[\frac{c+d+a+b}{(a+b)(a+c)(b+d)(c+d)}\right]$

$= \frac{(a+b+c+d)(ad-bc)^2}{(a+b)(c+d)(b+d)(a+c)}$

Example 8 A public opinion poll surveyed a simple random sample of 1000 voters. Respondents were classified by voting preferences to parties A, B and C and by gender (male or female). Results are shown in the contingency table below.

	Voting Preferences			Total
	Party A	Party B	Party C	
Male	200	150	50	400
Female	250	300	50	600
Total	450	450	100	1000

Do the men's voting preferences differ significantly from the women's preferences indicating a gender bias? Use a 0.05 level of significance.

Solution: Let H_0: Voting preferences and gender are independent.

	Voting Preferences			Total
	Party A	Party B	Party C	
Male	$\frac{(400)(450)}{1000} = 180$	$\frac{(400)(450)}{1000} = 180$	$\frac{(400)(100)}{1000} = 40$	400
Female	$\frac{(600)(450)}{1000} = 270$	$\frac{(600)(450)}{1000} = 270$	$\frac{(600)(100)}{1000} = 60$	600
Total	450	450	100	1000

On the basis of null hypothesis, expected frequencies are given in the above table.

Degrees of freedom $(\nu) = (2-1)(3-1) = 2$

$$\chi^2 = \sum \frac{(O_{ij}-E_{ij})^2}{E_{ij}} = \frac{(200-180)^2}{180} + \frac{(150-180)^2}{180} + \frac{(50-40)^2}{40} + \frac{(250-270)^2}{270} + \frac{(300-270)^2}{270} + \frac{(50-60)^2}{60}$$

= 16.20, table value of χ^2 for 2 degrees of freedom at 5% level = 5.991

Calculated value of χ^2 is greater than table value, $\therefore H_0$ is rejected, i.e. voting preferences and gender cannot be considered independent.

Exercise 7

1. Following tables shows number of male and female births in 800 families having four children:

Number of male births	0	1	2	3	4
Number of female births	4	3	2	1	0
Number of families	32	178	290	236	94

   Test if the data is consistent with the hypothesis that the binomial distribution holds and the probability of mail birth is same as that of a female birth

2. A random sample of 395 people was surveyed and each person was asked to report the highest education level they obtained. The data that resulted from the survey is summarized in the following table:

	Senior Secondary School	Bachelors	Masters	Ph.d.	Total
Female	60	54	46	41	201
Male	40	44	53	57	194
Total	100	98	99	98	395

   Test whether gender and education level independent at 5% level of significance? Value of χ^2 at 3 degrees of freedom is 7.815.

3. Following data shows number of good and bad parts produced by each of the three shifts in a factory.

	Good parts	Bad parts	Total
Day shift	960	40	1000
Evening shift	940	50	990
Night shift	950	45	995
Total	2850	135	2985

   Test whether the production of bad parts is independent of the shift on which they were produced. Value of χ^2 at 2 degrees of freedom is 5.991.

4. Out of a sample of 120 persons in a village, 76 persons were administered a new drug for controlling influenza and out of them, 24 persons were attacked by influenza. Out of those who were not administered the new drug, 12 persons were not affected by influenza. Use Chi-square test for finding whether the new drug is effective or not. Value of χ^2 for 1 degree of freedom is 3.84.

5. In a survey of 200 boys of which 75 are intelligent, 40 have educated fathers, while 85 of the unintelligent boys have uneducated fathers. Do these figures support the hypothesis that educated fathers have intelligent boys? Value of χ^2 for 1 degree of freedom is 3.84.

6. The following table shows the number of people having T.B. out of number of enquired people in different age- groups.

Age-group	Number of people enquired	T.B. cases
15-25	199	1
25-35	300	8
35-45	1128	38
45-55	1375	96
55-65	1089	105
65-75	625	56
75-85	155	12
Total	4871	316

Do these figures support the hypothesis that T. B. is equally spread in all age-groups?

Value of χ^2 at 6 degrees of freedom is 12.59

7. Two hundred digits were chosen at random from a set of tables and their frequencies are given below:

Digits	0	1	2	3	4	5	6	7	8	9
Frequency	18	19	23	21	16	25	21	20	21	15

Test the hypothesis that the digits were distributed in an equal manner in the tables from which they are chosen. Value of χ^2 at 9 degrees of freedom is 16.919.

Following data gives number of accidents in a city from years 2005 to 2014. Use χ^2 test of goodness of fit to prove the hypothesis that the number of accidents reported for each year from 2005 to 2014 does not differ significantly from an equal number of accidents in each year.

Year	2005	2006	2007	2008	2009	2010	2011	2012	2013	2014
Number of Accidents	164	142	153	171	155	148	136	133	138	140

Answers

1. $\chi^2 = 54.43$ ∴ probability of mail birth is not same as that of a female birth
2. $\chi^2 = 8.006$, which is greater than table value ∴ gender and education levels seem dependent on each other.
3. $\chi^2 = 1.28$, which is less than table value ∴ production of bad parts is independent of the shifts
4. $\chi^2 = 18.97$, which is much greater than table value ∴ null hypothesis is not supported, i.e. the new drug is definitely effective for controlling influenza.
5. $\chi^2 = 8.88$, ∴ education of fathers has a significant effect on intelligence of boys.
6. $\chi^2 = 57.6$, which is much greater than table value ∴ hypothesis is not supported.
7. $\chi^2 = 4.3$, which is less than table value ∴ hypothesis is accepted.

Chapter 8: Analysis of Variance

Analysis of Variance (ANOVA)

8.1 Introduction

Introduced by Ronald Fisher; Analysis of Variance (ANOVA) has found wide range of applications in number of statistical measures and it's most useful application is in analyzing the differences among three or more group means. ANOVA is conceptually similar to multiple two-sample *t*-tests, but it is more meticulous as it fits a wide range of practical problems.

8.2 One-Way ANOVA

One way ANOVA is a hypothesis test; having only one independent variable and is used to test the equality of three of more population means simultaneously.

Suppose n measurements (or observations) are obtained for each of the m groups being tested for a treatment (or attribute) as shown in table below:

Groups↓	Measurements →					
X_1	x_{11}	x_{12}	...	x_{1n}	$\sum X_1 = \sum_j x_{1j}$	$\bar{X}_1 = \frac{\sum X_1}{n} = \frac{1}{n}\sum_j x_{1j}$
X_2	x_{21}	x_{22}	...	x_{2n}	$\sum X_2 = \sum_j x_{2j}$	$\bar{X}_2 = \frac{\sum X_2}{n} = \frac{1}{n}\sum_j x_{2j}$
	⋮	⋮	⋮	⋮	⋮	
X_m	x_{m1}	x_{m2}	...	x_{mn}	$\sum X_m = \sum_j x_{mj}$	$\bar{X}_m = \frac{\sum X_m}{n} = \frac{1}{n}\sum_j x_{mj}$

If $\bar{X}_1, \bar{X}_2, ..., \bar{X}_m$ represent mean of each of the m groups respectively,

then total mean of all measurements $(\bar{X}) = \frac{1}{m}(\bar{X}_1 + \bar{X}_2 + \cdots + \bar{X}_m)$ ... ①

$$= \frac{1}{mn}\left(\sum_j x_{1j} + \sum_j x_{2j} + \cdots + \sum_j x_{mj}\right)$$

$$= \frac{1}{mn}\left(\sum_{ij} x_{ij}\right)$$

If $N = mn$ are total number of observations, then total mean of all measurements is given by

$$\bar{X} = \frac{1}{N}\sum_{ij} x_{ij} \qquad \text{... ②}$$

Now total variation of all measurements $(V_{total}) = \sum_{ij}(x_{ij} - \bar{X})^2$

$\Rightarrow V_{total} = \sum_{ij}\left[x_{ij}^2 - 2x_{ij}\bar{X} + \bar{X}^2\right]$

$\qquad = \sum_{ij} x_{ij}^2 - 2\bar{X}\sum_{ij} x_{ij} + N\bar{X}^2 \qquad \because \sum_{ij}\bar{X}^2 = N\bar{X}^2$

$$= \sum_{ij} x_{ij}^2 - 2N\bar{X}^2 + N\bar{X}^2 \qquad \because \sum_{ij} x_{ij} = N\bar{X} \qquad \text{from } ②$$

$$= \sum_{ij} x_{ij}^2 - N\bar{X}^2$$

$$= \sum_{ij} x_{ij}^2 - N\left(\frac{1}{N}\sum_{ij} x_{ij}\right)^2 \qquad \text{from } ②$$

$$= \sum_{ij} x_{ij}^2 - \frac{1}{N}(\sum_i X_i)^2$$

By adding squares of measurements group-wise, we get

$$\therefore V_{total} = \sum X_1^2 + \sum X_2^2 + \cdots + \sum X_m^2 - \frac{(\sum X_1 + \sum X_2 + \cdots + \sum X_m)^2}{N}$$

Again variation between groups $(V_{between}) = n \sum_i (\bar{X}_i - \bar{X})^2$

$$\Rightarrow V_{between} = n \sum_i (\bar{X}_i^2 - 2\bar{X}_i \bar{X} + \bar{X}^2)$$

$$= n[\sum_i \bar{X}_i^2 - 2\bar{X} \sum_i \bar{X}_i + m\bar{X}^2] \qquad \because \sum_i \bar{X}^2 = m\bar{X}^2$$

$$= n[\sum_i \bar{X}_i^2 - 2m\bar{X}^2 + m\bar{X}^2] \qquad \because \sum_i \bar{X}_i = m\bar{X} \quad \text{from } ①$$

$$= n[\sum_i \bar{X}_i^2 - m\bar{X}^2]$$

$$= n \sum_i \left(\frac{1}{n}\sum_j x_{ij}\right)^2 - mn\bar{X}^2 \qquad \because \bar{X}_i = \frac{1}{n}\sum_j x_{ij}$$

$$= \frac{1}{n} \sum_i (\sum_j x_{ij})^2 - N\left(\frac{1}{N}\sum_{ij} x_{ij}\right)^2 \qquad \because mn = N$$

$$= \frac{1}{n} \sum_i (\sum X_i)^2 - \frac{1}{N}(\sum_i X_i)^2 \qquad \because \sum_j x_{ij} = \sum X_i$$

By adding squares of measurements group-wise, we get

$$V_{between} = \frac{(\sum X_1)^2}{n} + \frac{(\sum X_2)^2}{n} + \cdots + \frac{(\sum X_m)^2}{n} - \frac{(\sum X_1 + \sum X_2 + \cdots + \sum X_m)^2}{N}$$

Also $V_{total} = V_{within} + V_{between}$ (where V_{within} gives variation within groups)

$\therefore V_{within} = V_{total} - V_{between}$

Working methodology to compute one-way ANOVA

Let there be n measurements (or observations) for each of the m groups being tested for a treatment (or attribute). Thus in total there are $N = mn$ measurements.

Step1: Assume null hypothesis (H_0) that there is no significant difference among the groups being tested for the treatment (or attribute). Then the alternative hypothesis would be that there is at least one significant difference among the groups.

Step2: Calculate sum of measurements (observations) for each group as $\sum X_1, \sum X_2, \ldots, \sum X_m$ and compute their squares as $(\sum X_1)^2, (\sum X_2)^2, \ldots, (\sum X_m)^2$

Step3: Find square of each element in all the groups and record the sum of squares as $\sum X_1^2, \sum X_2^2, \ldots, \sum X_m^2$.

Step4: Now total variation of all measurements is given by:

$$V_{total} = \sum X_1^2 + \sum X_2^2 + \cdots + \sum X_m^2 - \frac{(\sum X_1 + \sum X_2 + \cdots + \sum X_m)^2}{N}$$

Step5: Also variation between groups is given by:

$$V_{between} = \frac{(\sum X_1)^2}{n} + \frac{(\sum X_2)^2}{n} + \cdots + \frac{(\sum X_m)^2}{n} - \frac{(\sum X_1 + \sum X_2 + \cdots + \sum X_m)^2}{N}$$

Step6: Now compute variations within groups as:

$$V_{within} = V_{total} - V_{between}$$

Step7: Compute mean variations as $M_{between} = \frac{V_{between}}{v_1}$ and $M_{within} = \frac{V_{within}}{v_2}$,

Where $v_1 = m - 1$ are the degrees of freedom between groups

$v_2 = N - m$ are the degrees of freedom within groups

Step8: Calculate $F = \frac{M_{between}}{M_{within}}$ and compare with table value of F statistic.

Hypothesis (H_0) is accepted if calculated value of $|F|$ is less than table value of F at (v_1, v_2) degrees of freedom at the specified level of significance.

➢ If sizes of each group differ say $n_1, n_2, \ldots, n_m$ then $N = n_1 + n_2 + \cdots + n_m$ and also variation between groups is given by:

$$V_{between} = \frac{(\sum X_1)^2}{n_1} + \frac{(\sum X_2)^2}{n_2} + \cdots + \frac{(\sum X_m)^2}{n_m} - \frac{(\sum X_1 + \sum X_2 + \cdots + \sum X_m)^2}{N}$$

Example1 Three varieties A, B, C of rice are sown in eight plots of equal area and following quantities per acre were recorded:

Varieties↓	Plots →							
A	7	4	6	8	6	6	2	9
B	5	5	3	4	4	7	2	2
C	2	4	7	1	2	1	5	5

Test the significance of difference between the yields of varieties.

Solution: Let H_0: Rice varieties do not significantly differ from each other.

Number of varieties (m) = 3, Plots for each variety (n) = 8, ∴ $N = mn = 24$

Varieties↓		Plots →								
A	X_1	7	4	6	8	6	6	2	9	$\sum X_1 = 48$, $(\sum X_1)^2 = 2304$
	X_1^2	49	16	36	64	36	36	4	81	$\sum X_1^2 = 322$
B	X_2	5	5	3	4	4	7	2	2	$\sum X_2 = 32$, $(\sum X_2)^2 = 1024$
	X_2^2	25	25	9	16	16	49	4	4	$\sum x_2^2 = 148$
C	X_3	2	4	7	1	2	1	5	5	$\sum X_3 = 27$, $(\sum X_3)^2 = 729$
	X_3^2	4	16	49	1	4	1	25	25	$\sum X_3^2 = 125$

Calculating total variation (V_{total})

$$V_{total} = \sum X_1^2 + \sum X_2^2 + \sum X_3^2 - \frac{(\sum X_1 + \sum X_2 + \sum X_3)^2}{N}$$

$$\Rightarrow V_{total} = 322 + 148 + 125 - \frac{(48+32+27)^2}{24} = 117.96$$

Now calculating variation between groups ($V_{between}$)

$$V_{between} = \frac{(\sum X_1)^2}{n} + \frac{(\sum X_2)^2}{n} + \frac{(\sum X_3)^2}{n} - \frac{(\sum X_1 + \sum X_2 + \sum X_3)^2}{N}$$

$$\Rightarrow V_{between} = \frac{2304}{8} + \frac{1024}{8} + \frac{729}{8} - \frac{(48+32+27)^2}{24} = 30.08$$

$$\therefore V_{within} = V_{total} - V_{between} = 117.96 - 30.08 = 87.88$$

Source of Variation	Variation	DF	Mean variations	F
Between groups	30.8	$v_1 = 2$	$M_{between} = \frac{30.08}{2} = 15.04$	$\frac{15.04}{4.18} = 3.59$
Within groups	87.88	$v_2 = 21$	$M_{within} = \frac{87.88}{21} = 4.18$	

Table value of F for (2, 21) degrees of freedom at 5% level is 3.47 and at 1% level it is 5.78. $\therefore$ Calculated value of F is less than table value at 1% level but greater than one at 5% level. Thus the hypothesis H_0 saying 'Rice varieties do not significantly differ from each other' may be accepted at 1% level of significance but has to be rejected at 5% level.

Example 2 Following table shows monthly income in million dollars of top four firms in each of the three cities.

Cities↓	Annual income (million$)→			
A	50	49	49	48
B	49	48	48	47
C	51	50	50	49

Calculate combined mean monthly trade income of top firms in three cities. Also test the significance of difference in the three cities in terms of monthly trade income of top firms, given that tabulated value of F at (2, 9) degrees of freedom is 4.26.

Solution: H_0: Three cities do not significantly differ from each other in terms of monthly trade income of top firms.

Subtracting 45 from each income figure to simplify the calculations, we get

Cities↓	Annual income (million$)→			
A	5	4	4	3
B	4	3	3	2
C	6	5	5	4

Number of groups (m) = 3, measures for each group (n) = 4, $\therefore N = mn = 12$

Cities↓	Annual income (million$)→					
A	X_1	5	4	4	3	$\sum X_1 = 16$, $(\sum X_1)^2 = 256$
	X_1^2	25	16	16	9	$\sum X_1^2 = 66$
B	X_2	4	3	3	2	$\sum X_2 = 12$, $(\sum X_2)^2 = 144$
	X_2^2	16	9	9	4	$\sum X_2^2 = 38$
C	X_3	6	5	5	4	$\sum X_3 = 20$, $(\sum X_3)^2 = 400$
	X_3^2	36	25	25	16	$\sum X_3^2 = 102$

Combined mean for all the income figures $(\bar{X}) = \frac{1}{N}(\sum X_1 + \sum X_2 + \sum X_3)$

$$= \frac{1}{12}(16 + 12 + 20) = 4$$

∴ Actual mean combined monthly trade income of top firms in three cities

$$= 45 + 4 = 49 \text{ million dollars}$$

Calculating total variation (V_{total})

$$V_{total} = \sum X_1^2 + \sum X_2^2 + \sum X_3^2 - \frac{(\sum X_1 + \sum X_2 + \sum X_3)^2}{N}$$

$$\Rightarrow V_{total} = 66 + 38 + 102 - \frac{(16+12+20)^2}{12} = 14$$

Now calculating variation between groups $(V_{between})$

$$V_{between} = \frac{(\sum X_1)^2}{n} + \frac{(\sum X_2)^2}{n} + \frac{(\sum X_3)^2}{n} - \frac{(\sum X_1 + \sum X_2 + \sum X_3)^2}{N}$$

$$\Rightarrow V_{between} = \frac{256}{4} + \frac{144}{4} + \frac{400}{4} - \frac{(16+12+20)^2}{12} = 8$$

∴ $V_{within} = V_{total} - V_{between} = 14 - 8 = 6$

Source of Variation	Variation	DF	Mean variations	F
Between groups	8	$v_1 = 2$	$M_{between} = \frac{8}{2} = 4$	$\frac{4}{0.67} = 5.97$
Within groups	6	$v_2 = 9$	$M_{within} = \frac{6}{9} = 0.67$	

Table value of **F** for (2, 9) degrees of freedom at 5% level is 4.26

∴ Calculated value of F is more than table value and thus the hypothesis H_0 is rejected at 5% level of significance and hence we can say that the three cities significantly differ from each other in terms of annual trade income of top firms.

Example3 A music teacher assumes that students will learn most effectively with constant background music, compared to an unpredictable sound or no sound at all. He randomly divides twelve students into three groups of four each. All students study a passage of text for half an hour. Those in group I study with a sound at a constant volume in the background. Those in group II study with noise that changes volume periodically. Those in group III study with no sound in the background at all. After studying, all students take a multiple choice test of 10 marks over the material and their scores are given in the following table:

Groups↓	Test scores→			
I	8	4	6	7
II	7	5	5	3
III	2	5	4	4

Test the significance of difference of scores in the three groups, given that tabulated value of F at $(2, 9)$ degrees of freedom is 4.26.

Solution: H_0: test scores of three groups do not significantly differ from each other.

Number of groups $(m) = 3$, Test scores for each group $(n) = 4$, $\therefore N = mn = 12$

Groups↓		Test scores→				
I	X_1	8	4	6	7	$\sum X_1 = 25$, $(\sum X_1)^2 = 625$
	X_1^2	64	16	36	49	$\sum X_1^2 = 165$
II	X_2	7	5	5	3	$\sum X_2 = 20$, $(\sum X_2)^2 = 400$
	X_2^2	49	25	25	9	$\sum X_2^2 = 108$
III	X_3	2	5	4	4	$\sum X_3 = 15$, $(\sum X_3)^2 = 225$
	X_3^2	4	25	16	16	$\sum X_3^2 = 61$

Calculating total variation (V_{total})

$$V_{total} = \sum X_1^2 + \sum X_2^2 + \sum X_3^2 - \frac{(\sum X_1 + \sum X_2 + \sum X_3)^2}{N}$$

$$\Rightarrow V_{total} = 165 + 108 + 61 - \frac{(25+20+15)^2}{12} = 34$$

Now calculating variation between groups $(V_{between})$

$$V_{between} = \frac{(\sum X_1)^2}{n} + \frac{(\sum X_2)^2}{n} + \frac{(\sum X_3)^2}{n} - \frac{(\sum X_1 + \sum X_2 + \sum X_3)^2}{N}$$

$$\Rightarrow V_{between} = \frac{625}{4} + \frac{400}{4} + \frac{225}{4} - \frac{(25+20+15)^2}{12} = 12.5$$

$\therefore V_{within} = V_{total} - V_{between} = 34 - 12.5 = 21.5$

Source of Variation	Variation	DF	Mean variations	F
Between groups	12.5	$v_1 = 2$	$M_{between} = \frac{12.5}{2} = 6.25$	$\frac{6.25}{2.39} = 2.62$
Within groups	21.5	$v_2 = 9$	$M_{within} = \frac{21.5}{9} = 2.39$	

Table value of F for (2, 9) degrees of freedom at 5% level is 4.26 ∴ Calculated value of F is less than table value and hence the hypothesis H_0 saying 'test scores of three groups do not significantly differ from each other' may be accepted at 5% level of significance.

Example 4 Three methods are used in a production procedure and following outputs were obtained:

Methods↓	Output →									
I	20	15	18	19	20	22	16	18	29	17
II	18	16	15	22	21	20	18			
III	22	20	18	17	18	22				

Test at 5% level whether the three methods can be considered to be equivalent as far as outputs are concerned.

Solution: Let H_0: There is no significant difference between the three methods.

Number of methods $(m) = 3$, $n_1 = 10$, $n_2 = 7$, $n_3 = 6$ ∴ $N = n_1 + n_2 + n_3 = 23$

Variety↓		Plots →										
I	X_1	20	15	18	19	20	22	16	18	29	17	$\sum X_1 = 194$ $(\sum X_1)^2 = 37636$
	X_1^2	400	225	324	361	400	484	256	324	841	289	$\sum X_1^2 = 3904$
II	X_2	18	16	15	22	21	20	18				$\sum X_2 = 130$ $(\sum X_2)^2 = 16900$
	X_2^2	324	256	225	484	441	400	324				$\sum X_2^2 = 2454$
III	X_3	22	20	18	17	18	22					$\sum X_3 = 117$ $(\sum X_3)^2 = 13689$
	X_3^2	484	400	324	289	324	484					$\sum X_3^2 = 2305$

Calculating total variation (V_{total})

$$V_{total} = \sum X_1^2 + \sum X_2^2 + \sum X_3^2 - \frac{(\sum X_1 + \sum X_2 + \sum X_3)^2}{N}$$

$$\Rightarrow V_{total} = 3904 + 2454 + 2305 - \frac{(194 + 130 + 117)^2}{23} = 207.3$$

Now calculating variation between groups $(V_{between})$

$$V_{between} = \frac{(\sum X_1)^2}{n_1} + \frac{(\sum X_2)^2}{n_2} + \frac{(\sum X_3)^2}{n_3} - \frac{(\sum X_1 + \sum X_2 + \sum X_3)^2}{N}$$

$$\Rightarrow V_{between} = \frac{37636}{10} + \frac{16900}{7} + \frac{13689}{6} - \frac{(194+130+117)^2}{23} = 3.69$$

$$\therefore V_{within} = V_{total} - V_{between} = 207.3 - 3.69 = 203.61$$

Source of Variation	Variation	DF	Mean variations	F
Between groups	3.69	$v_1 = 2$	$M_{between} = \frac{3.69}{2} = 1.845$	$\frac{1.845}{10.18} = 0.18$
Within groups	203.61	$v_2 = 20$	$M_{within} = \frac{203.61}{20} = 10.18$	

Table value of **F** for (2, 20) degrees of freedom at 5% level is 3.49 ∴ Calculated value of *F* is much less than table value at 5% level Thus the hypothesis H_0 saying 'There is no significant difference between the three methods' may be accepted at 5% level of significance.

8.3 Two-Way ANOVA

Two ways ANOVA comprising of two independent variables (called factors); is a statistical technique wherein the interaction between factors and influencing variables can be studied.

Two-Way ANOVA is useful when we need to compare the effect of multiple levels of two factors and there are multiple observations at each level.

Working methodology to compute two-way ANOVA

Let there be m rows of combination of a given factor; each row subject to n columns for another specified factor (treatment). Thus in total there are $N = mn$ measurements such that each measurement corresponds to two factors (say factor I and factor II) at a time.

Then to test two simultaneous null hypotheses:

H_{01}: There is no significant difference in factor I.

H_{02}: There is no significant difference in application of factor II.

Step1: Calculate sum of measurements (observations) for factor I as $\sum X_1, \sum X_2, ..., \sum X_m$ and compute their squares as $(\sum X_1)^2, (\sum X_2)^2, ..., (\sum X_m)^2$.

Step2: Find square of each element row-wise and record the sum of squares as $\sum X_1^2, \sum X_2^2, ..., \sum X_m^2$.

Step3: Calculate sum of measurements (observations) for factor II as $\sum Y_1, \sum Y_2, ..., \sum Y_n$ and compute their squares as $(\sum Y_1)^2, (\sum Y_2)^2, ..., (\sum Y_n)^2$.

Step4: Find square of each element column-wise and record the sum of squares as $\sum Y_1^2, \sum Y_2^2, ..., \sum Y_n^2$.

Step5: Calculate total variation of all measurements given by:

$$V_{total} = \sum X_1^2 + \sum X_2^2 + \cdots + \sum X_m^2 - \frac{(\sum X_1 + \sum X_2 + \cdots + \sum X_m)^2}{N}$$

$$\text{Or } V_{total} = \sum Y_1^2 + \sum Y_2^2 + \cdots + \sum Y_n^2 - \frac{(\sum Y_1 + \sum Y_2 + \cdots + \sum Y_n)^2}{N}$$

Step6: Compute variations between factor I given by:

$$V_{between\ (X)} = \frac{(\sum X_1)^2}{n} + \frac{(\sum X_2)^2}{n} + \cdots + \frac{(\sum X_m)^2}{n} - \frac{(\sum X_1 + \sum X_2 + \cdots + \sum X_m)^2}{N}$$

Step7: Compute variations between factor II given by:

$$V_{between\ (Y)} = \frac{(\sum Y_1)^2}{m} + \frac{(\sum Y_2)^2}{m} + \cdots + \frac{(\sum Y_n)^2}{m} - \frac{(\sum Y_1 + \sum Y_2 + \cdots + \sum Y_n)^2}{N}$$

Step8: Now compute variations within two factors I and II as:

$$V_{within(XY)} = V_{total} - V_{between\ (X)} - V_{between\ (Y)}$$

Step9: Taking $v_1 = m - 1$ as the degrees of freedom for factor I,

$v_2 = n - 1$ as the degrees of freedom for factor II

$v_3 = N - m - n$ as degrees of freedom within two factors I and II,

compute mean variations as $M_{between\ (X)} = \dfrac{V_{between\ (X)}}{v_1}$

$$M_{between\ (Y)} = \frac{V_{between\ (Y)}}{v_2}$$

$$M_{within(XY)} = \frac{V_{within(XY)}}{v_3}$$

Step10: Calculate $F_1 = \dfrac{M_{between\ (X)}}{M_{within(XY)}}$ and compare with table value of F statistic.

Hypothesis (H_{01}) is accepted if calculated value of $|F_1|$ is less than table value of F at (v_1, v_3) degrees of freedom at the specified level of significance.

Also compute $F_2 = \dfrac{M_{between\ (Y)}}{M_{within(XY)}}$ and compare with table value of F statistic.

Hypothesis (H_{02}) is accepted if calculated value of $|F_2|$ is less than table value of F at (v_2, v_3) degrees of freedom at the specified level of significance.

Example5 Twenty four dogs were divided into four groups of six each as per their weights (in kilograms) at the beginning of the experiment, so that factor I (initial weight) has 4 levels W_1, W_2, W_3, W_4 and each group member was fed with factor II i.e. different diets $D_1, D_2, \ldots, D_6$. Weight gains at different levels on specific diets are given in the table below:

Groups↓	Diet→					
	D_1	D_2	D_3	D_4	D_5	D_6
W_1	9	12	9	6	9	10
W_2	7	3	7	7	5	5
W_3	6	5	9	11	3	11
W_4	6	8	11	2	2	10

Test the hypothesis that neither the initial weights nor the type of diet affects the weight gain.

Solution: H_{01}: There is no difference in weight gains.

H_{02}: There is no effect of type of diet.

Number of levels of weights $(m) = 4$, and types of diets $(n) = 6$, $\therefore N = mn = 24$

Groups↓	Diet→						Sum $\sum X_i$ ($\sum X_i^2$)	$(\sum X_i)^2$
	D_1 (Y_1)	D_2 (Y_2)	D_3 (Y_3)	D_4 (Y_4)	D_5 (Y_5)	D_6 (Y_6)		
W_1 (X_1)	9 (81)	12 (144)	9 (81)	6 (36)	9 (81)	10 (100)	55 $\sum X_1^2 = 523$	3025
W_2 (X_2)	7 (49)	3 (9)	7 (49)	7 (49)	5 (25)	5 (25)	34 $\sum X_2^2 = 206$	1156
W_3 (X_3)	6 (36)	5 (25)	9 (81)	11 (121)	3 (9)	11 (121)	45 $\sum X_3^2 = 393$	2025
W_4 (X_4)	6 (36)	8 (64)	11 (121)	2 (4)	2 (4)	10 (100)	39 $\sum X_4^2 = 329$	1521
Sum $\sum Y_j$	28	28	36	26	19	36		
$\sum Y_j^2$	202	242	332	210	119	346		
$(\sum Y_j)^2$	784	784	1296	676	361	1296		

Here numbers in brackets in each cell are the squares of each observation.

Calculating total variation (V_{total})

$$V_{total} = \sum X_1^2 + \sum X_2^2 + \sum X_3^2 + \sum X_4^2 - \frac{(\sum X_1 + \sum X_2 + \sum X_3 + \sum X_4)^2}{N}$$

$$\Rightarrow V_{total} = 523 + 206 + 393 + 329 - \frac{(55+34+45+39)^2}{24} = 203.96$$

Now calculating variation between weight gains (factor I)

$$V_{between\ (X)} = \frac{(\sum X_1)^2}{n} + \frac{(\sum X_2)^2}{n} + \frac{(\sum X_3)^2}{n} + \frac{(\sum X_4)^2}{n} - \frac{(\sum X_1 + \sum X_2 + \sum X_3 + \sum X_4)^2}{N}$$

$$\Rightarrow V_{between\ (X)} = \frac{3025}{6} + \frac{1156}{6} + \frac{2025}{6} + \frac{1521}{6} - \frac{(55+34+45+39)^2}{24} = 40.79$$

Now calculating variation between diets (factor II)

$$V_{between\ (Y)} = \frac{(\sum Y_1)^2}{m} + \frac{(\sum Y_2)^2}{m} + \cdots + \frac{(\sum Y_6)^2}{m} - \frac{(\sum Y_1 + \sum Y_2 + \cdots + \sum Y_6)^2}{N}$$

$$\Rightarrow V_{between\ (Y)} = \frac{2(784)}{4} + \frac{2(1296)}{4} + \frac{676}{4} + \frac{361}{4} - \frac{(2(28)+2(36)+26+19)^2}{24} = 52.21$$

$$\therefore V_{within(XY)} = V_{total} - V_{between\ (X)} - V_{between\ (Y)}$$

$$= 203.96 - 40.79 - 52.21 = 110.96$$

Source of Variation	Variation	DF	Mean variations	F
Between groups	40.79	$v_1 = 3$	$M_{between\ (X)} = \frac{40.79}{3} = 13.6$	$F_1 = \frac{13.6}{7.4} = 1.84$
Between diets	52.21	$v_2 = 5$	$M_{between\ (Y)} = \frac{52.21}{5} = 10.44$	$F_2 = \frac{10.44}{7.4} = 1.41$
Within groups and diets	110.96	$v_3 = 15$	$M_{within(XY)} = \frac{110.96}{15} = 7.4$	

Table value of **F** for (2, 15) degrees of freedom at 5% level is 3.29 and for (5, 15) degrees of freedom is 2.90. ∴ Calculated values of F are less than table values for both weight gains and diets.

Hence the hypothesis H_{01} saying 'There is no difference in weight gains' may be accepted at 5% level of significance.

Also hypothesis H_{02} saying 'There is no effect of type of diet' may be accepted at 5% level of significance.

Exercise 8

1. A transport company wishes to test the average life of the four tyre brands it uses. The company uses all the tyre brands on randomly selected trucks. Following data shows tyre lives (in thousand miles):

Brand I	Brand II	Brand III	Brand IV
20	19	21	15
23	15	19	17
18	17	20	16
17	20	17	18
	16	16	

Test the hypothesis that average life of each brand is same at 5% significance level.

2. A professor wishes to check the responses of text books from four authors A, B, C, D in the market. He has 37 students whom he distributes at random into four groups of 9, 10, 11

and 7 students and assigns them with a book at random. After the course is over, all the students are given with a test and their scores are shown in the table given below:

Text books →	Book A	Book B	Book C	Book D
Test Scores	68	41	54	44
	68	47	44	51
	69	54	51	69
	60	65	56	59
	73	32	47	59
	64	73	61	55
	71	44	59	66
	67	48	49	
	75	64	41	
		54	61	
			73	

Test the hypothesis that all the books are equally good.

3. The varieties A, B, C, D, E, F, G of rice are sown in 5 plots each and the following yields in quintals per acre were obtained:

A	B	C	D	E	F	G
13	15	14	14	17	15	16
11	11	10	10	15	9	12
10	13	12	15	14	13	13
16	18	13	17	19	14	15
12	12	11	10	12	10	11

Do the data indicate the significant difference in yields of the varieties?

4. Following are weekly sales records (in thousand Rs.) of three salesmen A, B and C of a company during 13 sale-calls.

Salesmen↓	Sale records→				
A	300	400	300	500	
B	600	300	300	400	
C	700	300	400	600	500

Test whether the three salesmen are different.

5. Data given below represents number of units of production per day turned out by 5 different workmen using different types of machines.

Workmen ↓	Machine type →			
	A	B	C	D
1	46	40	49	38
2	48	42	54	45
3	36	38	46	34
4	35	40	48	35
5	40	44	51	41

   i. Test whether the machine productivity is same for the four brands of machine types.

   ii. Test whether the 5 workmen significantly differ with respect to productivity.

6. An experiment was conducted to determine the effects of different months of planting and different methods of planting on the yield of sugarcane. Following data gives the production output of sugarcane in 100kgs.

Method of planting ↓	Month of planting →			
	October	November	March	April
I	7.10	3.69	4.70	1.90
II	10.29	4.79	4.58	2.64
III	8.30	3.58	4.90	1.80

Carry out analysis of variance of the above data to test the significance of planting months as well as methods of planting.

Answers

1. $F = 1.67$, ∴There is no significant difference in the average lives of four brands.
2. $F = 5.66$, ∴Hypothesis is rejected, thus the books significantly differ from each other.
3. $F = 1.1$, ∴There is no significant difference in yields of different varieties.
4. $F = 1.072$, ∴There is no significant difference in the sales records of three salesmen.
5. F (Workmen) $= 8.20$, significant, F (Machines) $= 19.20$, significant
   ∴ There is significant difference between the productivities of 4 machines and also the 5 workmen in terms of productivity.
6. F (Methods of planting) $= 2.98$, F (Months of planting) $= 19.20$ significant
   ∴ Methods of planting are not significant different but months of planting have significant effect on yield of sugarcane.

Chapter 9: Numerical Solutions of Algebraic and Transcendental Equations

Numerical Solutions of Algebraic
And
Transcendental Equations

9.1 Introduction

An expression of the form $f(x) = a_0 x^n + a_1 x^{n-1} + \cdots + a_{n-1} x + a_n$, $a_0 \neq 0$ is called a polynomial of degree 'n' and the polynomial $f(x) = 0$ is called an algebraic equation of nth degree.

If $f(x)$ contains trigonometric, logarithmic or exponential functions, then $f(x) = 0$ is called a transcendental equation. For example $x^2 + 2\sin x + e^x = 0$ is a transcendental equation.

If $f(x)$ is an algebraic polynomial of degree less than or equal to 4, direct methods for finding the roots of such equation are available, but if $f(x)$ is of higher degree or it involves transcendental functions, direct methods do not exist and we need to apply numerical methods to find the roots of the equation $f(x) = 0$.

Some useful results

- If α is root of the equation $f(x) = 0$, then $f(\alpha) = 0$
- Every equation of nth degree has exactly n roots (real or imaginary)
- **Intermediate Value Theorem**: If $f(x)$ is a continuous function in a closed interval $[a, b]$ and $f(a) \& f(b)$ are having opposite signs, then the equation $f(x) = 0$ has atleast one real root or odd number of roots between a and b.
- If $f(x)$ is a continuous function in the closed interval $[a, b]$ and $f(a) \& f(b)$ are of same signs, then the equation $f(x) = 0$ has no root or even number of roots between a and b.

9.2 Numerical methods to find roots of algebraic and transcendental equations

Most numerical methods use iterative procedures to find an approximate root of an equation $f(x) = 0$. They require an initial guess of the root as starting value and each subsequent iteration leads closer to the actual root.

Order of convergence: For any iterative numerical method, each successive iteration gives an approximation that moves progressively closer to actual solution. This is known as convergence. Any numerical method is said have order of convergence ρ, if ρ is the largest positive number such that $|\epsilon_{n+1}| \leq k|\epsilon_n|^\rho$, where ϵ_n and ϵ_{n+1} are errors in n^{th} and $(n+1)^{th}$ iterations, k is a finite positive constant.

9.2.1 Direct Iteration Method

Direct iteration method is a fairly simple technique which requires the equation $f(x) = 0$ to be rewritten in the form $x = \phi(x)$.

We start with initial guess as x_0 and find first approximation as $x_1 = \phi(x_0)$.

Second and subsequent approximations are obtained using the expression $x_{n+1} = \phi(x_n)$; until desired accuracy is obtained.

- Sequence of approximations $x_0, x_1, x_2, ..., x_n$ converges if $\phi(x)$ is so chosen that $|\phi'(x)| < 1$ for the choice of x_0.
- The method has slow convergence.

Example 1 Find a root of the equation $2x = \cos x + 3$ correct to three decimal places using direct iteration method.

Solution: Rewriting in the form $x = \frac{1}{2}(\cos x + 3)$ ... ①

Here $\phi(x) = \frac{1}{2}(\cos x + 3)$, $|\phi'(x)| = \frac{1}{2}|\sin x| < 1$

Let $x_0 = \frac{\pi}{2} \Rightarrow x_1 = \frac{1}{2}\left(\cos\frac{\pi}{2} + 3\right) = 1.5$ from ①

$x_2 = \frac{1}{2}(\cos 1.5 + 3) = 1.5354$

$x_3 = \frac{1}{2}(\cos 1.5354 + 3) = 1.5179$

$x_4 = \frac{1}{2}(\cos 1.5179 + 3) = 1.5264$

$x_5 = \frac{1}{2}(\cos 1.5264 + 3) = 1.5222$

$x_6 = \frac{1}{2}(\cos 1.5222 + 3) = 1.5243$

$x_7 = \frac{1}{2}(\cos 1.5243 + 3) = 1.5232$

$x_8 = \frac{1}{2}(\cos 1.5232 + 3) = 1.5238$

$x_9 = \frac{1}{2}(\cos 1.5238 + 3) = 1.5235$

$x_{10} = \frac{1}{2}(\cos 1.5235 + 3) = 1.5236$

Hence 1.524 is the real root correct to three decimal places.

Example 2 Find a root of the equation $x^3 - 3x^2 - 3x - 4 = 0$ up to five iterations; taking an initial approximation 3; using direct iteration method.

Solution: Rewriting in the form $x = \frac{1}{3}(x^3 - 3x^2 - 4)$

Let initial approximation $x_0 = 3$

Here $\phi(x) = \frac{1}{3}(x^3 - 3x^2 - 4)$, $|\phi'(x)| = \frac{1}{3}|3x^2 - 6x| = |x^2 - 2x| > 1$ at $x = 3$

∴ Rewriting again in the form $x = (3x^2 + 3x + 4)^{\frac{1}{3}}$ ... ①

Here $\phi(x) = (3x^2 + 3x + 4)^{\frac{1}{3}}$, $|\phi'(x)| = \frac{1}{3}\left|(3x^2 + 3x + 4)^{\frac{-2}{3}}(6x + 3)\right| < 1$ at $x = 3$

∴ $x_0 = 3 \Rightarrow x_1 = (3(3)^2 + 3(3) + 4)^{\frac{1}{3}} = 3.420$ from ①

$$x_2 = (3(3.42)^2 + 3(3.42) + 4)^{\frac{1}{3}} = 3.668$$

$$x_3 = (3(3.668)^2 + 3(3.668) + 4)^{\frac{1}{3}} = 3.8114$$

$$x_4 = (3(3.8114)^2 + 3(3.8114) + 4)^{\frac{1}{3}} = 3.8933$$

$$x_5 = (3(3.8933)^2 + 3(3.8933) + 4)^{\frac{1}{3}} = 3.9398$$

Hence 3.9398 is the real root after five iterations.

9.2.2 Bisection Method (or Bolzano Method)

Bisection method is used to find an approximate root in an interval by repeatedly bisecting into subintervals. It is a very simple and robust method but it is also relatively slow. Because of this it is often used to obtain a rough approximation of a solution which is then used as a starting point for more rapidly converging methods. This method is based on the intermediate value theorem for continuous functions.

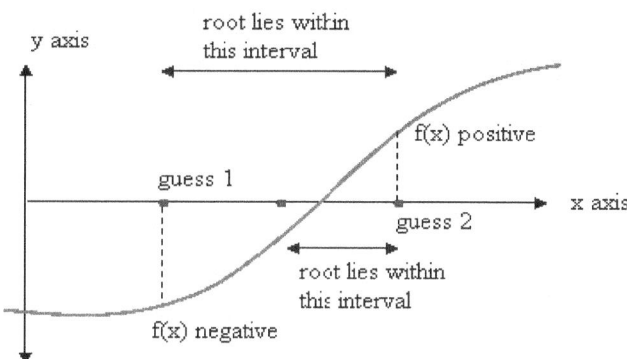

Algorithm:

Let $f(x)$ be a continuous function in the interval $[a, b]$, such that $f(a)$ and $f(b)$ are of opposite signs, i.e. $f(a).f(b) < 0$.

Step1. Take the initial approximation given by $x_0 = \frac{a+b}{2}$, one of the three conditions arises for finding the 1st approximation x_1

i. $f(x_0) = 0$, we have a root at x_0.
ii. If $f(a).f(x_0) < 0$, the root lies between a and x_0 $\therefore$ $x_1 = \frac{a+x_0}{2}$ and repeat the procedure by halving the interval again.
iii. If $f(b).f(x_0) < 0$, the root lies between x_0 and b $\therefore$ $x_1 = \frac{x_0+b}{2}$ and repeat the procedure by halving the interval again.
iv. Continue the process until root is found to be of desired accuracy.

Remarks:

- Convergence is not unidirectional as none of the end points is fixed. As a result convergence of bisection method is very slow.
- Repeating the procedure n times, the new interval will be exactly half the length of the previous one, until the root is found of desired accuracy (error less than ϵ). $\therefore$ and at the end of nth iteration, the interval containing the root will be of length $\frac{|b-a|}{2^n}$, such that $\frac{|b-a|}{2^n} < \epsilon$

$$\Rightarrow \log \frac{|b-a|}{2^n} < \log \epsilon$$

$$\Rightarrow \log|b-a| - \log 2^n < \log \epsilon$$

$$\Rightarrow \log|b-a| - \log \epsilon < n \log 2 \quad \Rightarrow n > \frac{\log|b-a| - \log \epsilon}{\log 2}$$

∴ In bisection method, the minimum number of iterations required to achieve the desired accuracy (error less than ϵ) are $\frac{\log \frac{|b-a|}{\epsilon}}{\log 2}$.

Example 3 Apply bisection method to find a root of the equation $x^4 + 2x^3 - x - 1 = 0$

Solution: $f(x) = x^4 + 2x^3 - x - 1$

Here $f(0) = -1$ and $f(1) = 1 \Rightarrow f(0).f(1) < 0$

Also $f(x)$ is continuous in $[0,1]$, ∴ atleast one root exists in $[0,1]$

Initial approximation: $a = 0, b = 1$

$x_0 = \frac{0+1}{2} = 0.5, \ f(0.5) = -1.1875, \ f(0.5).f(1) < 0$

First approximation: $a = 0.5, b = 1$

$x_1 = \frac{0.5+1}{2} = 0.75, \ f(0.75) = -0.5898, \ f(0.75).f(1) < 0$

Second approximation: $a = 0.75, b = 1$

$x_2 = \frac{0.75+1}{2} = 0.875, \ f(0.875) = 0.051, \ f(0.75).f(0.875) < 0$

Third approximation: $a = 0.75, b = 0.875$

$x_3 = \frac{0.75+0.875}{2} = 0.8125, \ f(0.8125) = -0.30394, \ f(0.8125).f(0.875) < 0$

Fourth approximation: $a = 0.8125, b = 0.875$

$x_4 = \frac{0.8125+0.875}{2} = 0.84375, \ f(0.84375) = -0.135, \ f(0.84375).f(0.875) < 0$

Fifth approximation: $a = 0.84375, b = 0.875$

$x_5 = \frac{0.84375+0.875}{2} = 0.8594, \ f(0.8594) = -0.0445, \ f(0.8594).f(0.875) < 0$

Sixth approximation: $a = 0.8594, b = 0.875$

$x_6 = \frac{0.8594 + 0.875}{2} = 0.8672, \ f(0.8672) = 0.0027, \ f(0.8594).f(0.8672) < 0$

Seventh approximation: $a = 0.8594, b = 0.8672$

$x_7 = \frac{0.8594+0.8672}{2} = 0.8633$

First 2 decimal places have been stabilized; hence 0.8633 is the real root correct to two decimal places.

Example 4 Apply bisection method to find a root of the equation $xe^x = 1$ correct to three decimal places.

Solution: $f(x) = xe^x - 1$

Here $f(0) = -1$ and $f(1) = e - 1 = 1.718 \Rightarrow f(0).f(1) < 0$

Also $f(x)$ is continuous in $[0,1]$, ∴ atleast one root exists in $[0,1]$

Initial approximation: $a = 0, b = 1$

$x_0 = \frac{0+1}{2} = 0.5, \ f(0.5) = -0.1756, \ f(0.5).f(1) < 0$

First approximation: $a = 0.5, b = 1$

$x_1 = \frac{0.5+1}{2} = 0.75, \ f(0.75) = 0.5877, \ f(0.5).f(0.75) < 0$

Second approximation: $a = 0.5, b = 0.625$
$x_2 = \frac{0.5+0.75}{2} = 0.625, f(0.625) = 0.8682, f(0.5).f(0.625) < 0$

Third approximation: $a = 0.5, b = 0.625$
$x_3 = \frac{0.5+0.625}{2} = 0.5625, f(0.5625) = -0.0128, f(0.5625).f(0.625) < 0$

Fourth approximation: $a = 0.5625, b = 0.625$
$x_4 = \frac{0.5625+0.625}{2} = 0.59375, f(0.59375) = 0.0751, f(0.5625).f(0.59375) < 0$

Fifth approximation: $a = 0.5625, b = 0.59375$
$x_5 = \frac{0.5625+0.59375}{2} = 0.5781, f(0.5781) = 0.0305, f(0.5625).f(0.5781) < 0$

Sixth approximation: $a = 0.5625, b = 0.5781$
$x_6 = \frac{0.5625+0.5781}{2} = 0.5703, f(0.5703) = 0.0087, f(0.5625).f(0.5703) < 0$

Seventh approximation: $a = 0.5625, b = 0.5703$
$x_7 = \frac{0.5625+0.5703}{2} = 0.5664, f(0.5664) = -0.002, f(0.5664).f(0.5703) < 0$

Eighth approximation: $a = 0.5664, b = 0.5703$
$x_8 = \frac{0.5664+0.5703}{2} = 0.5684, f(0.5684) = 0.0035, f(0.5664).f(0.5684) < 0$

Ninth approximation: $a = 0.5664, b = 0.5684$
$x_9 = \frac{0.5664+0.5684}{2} = 0.5674, f(0.5674) = .0007, f(0.5664).f(0.5674) < 0$

Tenth approximation: $a = 0.5664, b = 0.5674$
$x_{10} = \frac{0.5664+0.5674}{2} = 0.5669, f(0.5669) = -0.0007, f(0.5669).f(0.5674) < 0$

Eleventh approximation: $a = 0.5669, b = 0.5674$
$x_{11} = \frac{0.5669+0.5674}{2} = 0.56715, f(0.56715) = .00001 \sim 0$

Hence 0.56715 is the real root correct to three decimal places.

Example 5 Using bisection method find an approximate root of the equation $\sin x = \frac{1}{x}$ correct to two decimal places.

Solution: $f(x) = x\sin x - 1$

Here $f(1) = \sin 1 - 1 = -0.1585$ and $f(2) = 2\sin 2 - 1 = 0.8186$

Also $f(x)$ is continuous in $[1,2]$, $\therefore$ atleast one root exists in $[1,2]$

Initial approximation: $a = 1, b = 2$
$x_0 = \frac{1+2}{2} = 1.5, f(1.5) = 0.4963, f(1).f(1.5) < 0$

First approximation: $a = 1, b = 1.5$
$x_1 = \frac{1+1.5}{2} = 1.25, f(1.25) = 0.1862, f(1).f(1.25) < 0$

Second approximation: $a = 1, b = 1.25$
$x_2 = \frac{1+1.25}{2} = 1.125, f(1.125) = 0.0151, f(1).f(1.125) < 0$

Third approximation: $a = 1, b = 1.125$
$x_3 = \frac{1+1.125}{2} = 1.0625$, $f(1.0625) = -0.0718$, $f(1.0625).f(1.125) < 0$

Fourth approximation: $a = 1.0625, b = 1.125$
$x_4 = \frac{1.0625+1.125}{2} = 1.09375$, $f(1.09375) = -0.0284$, $f(1.09375).f(1.125) < 0$

Fifth approximation: $a = 1.09375, b = 1.125$
$x_5 = \frac{1.09375+1.125}{2} = 1.10937$, $f(1.10937) = -0.0066$, $f(1.10937).f(1.125) < 0$

Sixth approximation: $a = 1.10937, b = 1.125$
$x_6 = \frac{1.10937+1.125}{2} = 1.11719$, $f(1.11719) = .0042$, $f(1.10937).f(1.11719) < 0$

Seventh approximation: $a = 1.10937, b = 1.11719$
$x_7 = \frac{1.10937+1.11719}{2} = 1.11328$, $f(1.11328) = -.0012 \sim 0$

Hence 1.11328 is the real root correct to two decimal places.

9.2.3 Regula- Falsi Method (Geometrical Interpretation)

Regula-Falsi method is also known as method of false position as false position of the curve is taken as initial approximation.

Let $y = f(x)$ be represented by the curve AB.

The real root of equation $f(x) = 0$ is α as shown in given figure. The false position of curve AB is taken as chord AB and initial approximation x_0 is the point of intersection of chord AB with x − axis. Successive approximations x_1, x_2, ... are given by point of intersection of chord A^1B, A^2B... with x − axis, until the root is found to be of desired accuracy.

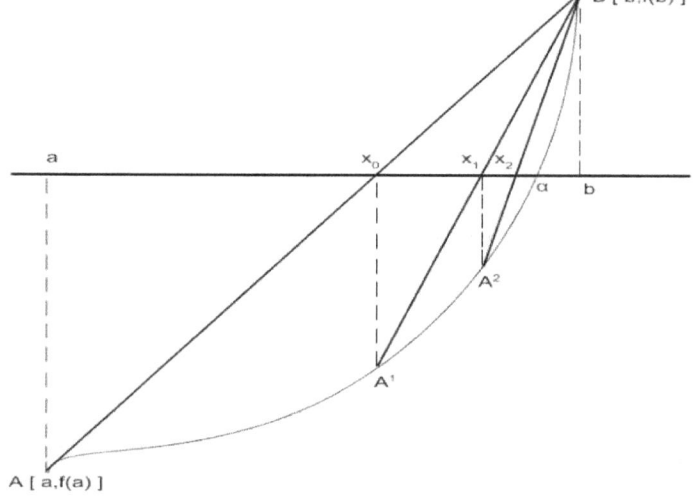

Now equation of chord AB in two-point form is given by:

$y - f(a) = \frac{f(b)-f(a)}{b-a}(x - a)$

To find x_0 (point of intersection of chord AB with x-axis), put $y = 0$

$\Rightarrow -f(a) = \frac{f(b)-f(a)}{b-a}(x_0 - a)$

$\Rightarrow (x_0 - a) = \frac{-(b-a)f(a)}{f(b)-f(a)}$

$\Rightarrow x_0 = a - \frac{(b-a)}{f(b)-f(a)}f(a)$

Repeat the procedure until the root is found to the desired accuracy.

Remarks:
- Rate of convergence is much faster than that of bisection method.

- Unlike bisection method, one end point will converge to the actual root a, whereas the other end point always remains fixed. As a result Regula-Falsi method has linear convergence.

Example6 Apply Regula-Falsi method to find a root of the equation $x^3 + x - 1 = 0$ correct to two decimal places.

Solution: $f(x) = x^3 + x - 1$

Here $f(0) = -1$ and $f(1) = 1 \Rightarrow f(0).f(1) < 0$

Also $f(x)$ is continuous in $[0,1]$, $\therefore$ atleast one root exists in $[0,1]$

Initial approximation: $x_0 = a - \frac{(b-a)}{f(b)-f(a)}f(a); a = 0, b = 1$

$\Rightarrow x_0 = 0 - \frac{(1-0)}{f(1)-f(0)}f(0) = 0 - \frac{1}{1-(-1)}(-1) = 0.5$

$f(0.5) = -0.375, \; f(0.5).f(1) < 0$

First approximation: $a = 0.5, b = 1$

$x_1 = 0.5 - \frac{(1-0.5)}{f(1)-f(0.5)}f(0.5) = 0 - \frac{0.5}{1-(-0.375)}(-0.375) = 0.636$

$f(0.636) = -0.107, \; f(0.636).f(1) < 0$

Second approximation: $a = 0.636, b = 1$

$x_2 = 0.636 - \frac{(1-0.636)}{f(1)-f(0.636)}f(0.636) = 0.636 - \frac{0.364}{1-(-0.107)}(-0.107) = 0.6711$

$f(0.6711) = -0.0267, \; f(0.6711).f(1) < 0$

Third approximation: $a = 0.6711, b = 1$

$x_3 = .6711 - \frac{(1-0.6711)}{f(1)-f(0.6711)}f(.6711) = .6711 - \frac{0.3289}{1-(-0.0267)}(-.0267) = 0.6796$

First two decimal places have been stabilized; hence 0.6796 is the real root correct to two decimal places.

Example7 Use Regula-Falsi method to find a root of the equation $x \log_{10} x - 1.2 = 0$ correct to two decimal places.

Solution: $f(x) = x \log_{10} x - 1.2$

Here $f(2) = -0.5979$ and $f(3) = 0.2314 \Rightarrow f(2).f(3) < 0$

Also $f(x)$ is continuous in $[2,3]$, $\therefore$ atleast one root exists in $[2,3]$

Initial approximation: $x_0 = a - \frac{(b-a)}{f(b)-f(a)}f(a)\,; a = 2, b = 3$

$\Rightarrow x_0 = 2 - \frac{(3-2)}{f(3)-f(2)}f(2) = 2 - \frac{1}{0.2314-(-0.5979)}(-0.5979) = 2.721$

$f(2.721) = -0.0171, \; f(2.721).f(3) < 0$

First approximation: $a = 2.721, b = 3$

$x_1 = 2.721 - \frac{(3-2.721)}{f(3)-f(2.721)} f(2.721) = 2.721 - \frac{0.279}{0.2314-(-0.0171)}(-0.0171) = 2.7402$

$f(2.7402) = -0.0004, \; f(2.7402).f(3) < 0$

Second approximation: $a = 2.7402, b = 3$

$x_2 = 2.7402 - \frac{(3-2.7402)}{f(3)-f(2.7402)} f(2.7402) = 2.7402 - \frac{0.2598}{0.2314-(-.0004)}(-.0004) = 2.7407$

Hence 2.7407 is the real root correct to two decimal places.

Example8 Use Regula-Falsi method to find a root of the equation $\tan x + \tanh x = 0$ upto three iterations only.

Solution: $f(x) = \tan x + \tanh x$

Here $f(2) = -1.2210$ and $f(3) = 0.8525 \Rightarrow f(2).f(3) < 0$

Also $f(x)$ is continuous in [2,3], $\therefore$ atleast one root exists in [2,3]

Initial approximation: $x_0 = a - \frac{(b-a)}{f(b)-f(a)} f(a) \; ; a = 2, b = 3$

$\Rightarrow x_0 = 2 - \frac{(3-2)}{f(3)-f(2)} f(2) = 2 - \frac{1}{0.8525-(-1.221)}(-1.221) = 2.5889$

$f(2.5889) = 0.3720, \; f(2).f(2.5889) < 0$

First approximation: $a = 2, b = 2.5889$

$x_1 = 2 - \frac{(2.5889-2)}{f(2.5889)-f(2)} f(2) = 2 - \frac{0.5889}{0.3720-(-1.2210)}(-1.2210) = 2.4514$

$f(2.4514) = 0.1596, \; f(2).f(2.4514) < 0$

Second approximation: $a = 2, b = 2.4514$

$x_2 = 2 - \frac{(2.4514-2)}{f(2.4514)-f(2)} f(2) = 2 - \frac{0.4514}{0.1596-(-1.2210)}(-1.2210) = 2.3992$

$f(2.3992) = 0.0662, \; f(2).f(2.3992) < 0$

Third approximation: $a = 2, b = 2.3992$

$x_3 = 2 - \frac{(2.3992-2)}{f(2.3992)-f(2)} f(2) = 2 - \frac{0.3992}{0.0662-(-1.2210)}(-1.2210) = 2.3787$

$\therefore$ Real root of the equation $\tan x + \tanh x = 0$ after three iterations is 2.3787

Example9 Use Regula-Falsi method to find a root of the equation $xe^x - 2 = 0$ correct to three decimal places.

Solution: $f(x) = xe^x - 2$

Here $f(0) = -2$ and $f(1) = 0.7183 \Rightarrow f(0).f(1) < 0$

Also $f(x)$ is continuous in [0,1], $\therefore$ atleast one root exists in [0,1]

Initial approximation: $x_0 = a - \frac{(b-a)}{f(b)-f(a)} f(a) \; ; a = 0, b = 1$

$\Rightarrow x_0 = 0 - \frac{(1-0)}{f(1)-f(0)} f(0) = 0 - \frac{1}{0.7183-(-2)}(-2) = 0.7358$

$f(0.7358) = -0.4643, \ f(0.7358).f(1) < 0$

First approximation: $a = 0.7358, b = 1$

$x_1 = 0.7358 - \frac{(1-0.7358)}{f(1)-f(0.7358)} f(0.7358) = 0.7358 - \frac{0.2642}{0.7183-(-0.4643)}(-0.4643) = 0.8395$

$f(0.8395) = -0.0564, \ f(0.8395).f(1) < 0$

Second approximation: $a = 0.8395, b = 1$

$x_2 = 0.8395 - \frac{(1-0.8395)}{f(1)-f(0.8395)} f(0.8395)$

$f(0.8512) = -0.006, \ f(0.8512).f(1) < 0$

Third approximation: $a = 0.8512, b = 1$

$x_3 = 0.8512 - \frac{(1-0.8512)}{f(1)-f(0.8512)} f(0.8512) = 0.8512 - \frac{0.1488}{0.7183-(-0.006)}(-0.006) = 0.8524$

$f(0.8524) = -0.009, \ f(0.8524).f(1) < 0$

Fourth approximation: $a = 0.8474 \ b = 1$

$x_4 = 0.8524 - \frac{(1-0.8524)}{f(1)-f(0.8524)} f(0.8524) = 0.8524 - \frac{0.1476}{0.7183-(-0.0009)}(-0.0009) = 0.8526$

$f(0.8526) = -0.00002 \sim 0$

$\therefore$ Real root of the equation $xe^x - 2 = 0$ correct to three decimal places is 0.8526

9.2.4 Newton-Raphson Method (Geometrical Interpretation)

Newton-Raphson method named after Isaac Newton and Joseph Raphson is a powerful technique for solving equations numerically. The Newton-Raphson method in one variable is implemented as follows:

Let α be an exact root and x_0 be the initial approximate root of the equation $f(x) = 0$.

First approximation x_1 is taken by drawing a tangent to curve $y = f(x)$ at the point $(x_0, f(x_0))$. If θ is the angle which tangent through the point $(x_0, f(x_0))$ makes with x- axis, then slope of the tangent is given by: $\tan \theta = \frac{f(x_0)}{x_0 - x_1} = f'(x_0)$

$\Rightarrow x_1 = x_0 - \frac{f(x_0)}{f'(x_0)}$

Similarly $x_2 = x_1 - \frac{f(x_1)}{f'(x_1)}$ and so on

The required root to desired accuracy is obtained by drawing tangents to the curve at points $(x_n, f(x_n))$ successively. $\therefore x_{n+1} = x_n - \frac{f(x_n)}{f'(x_n)}$

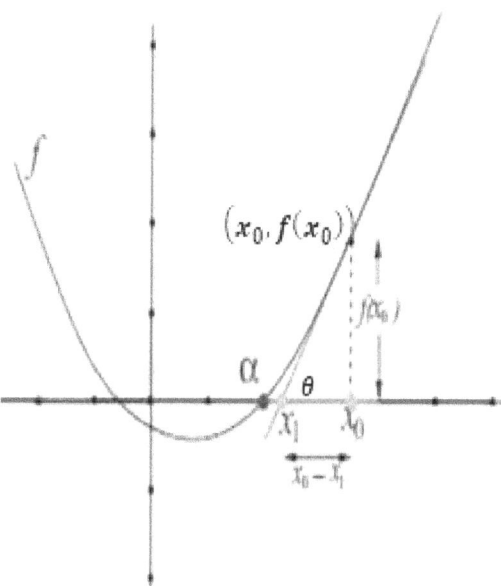

Newton-Raphson method works very fast but sometimes it fails to converge as shown below:

Case I:
If any of the approximations encounters a zero derivative (extreme point), then the tangent at that point goes parallel to x-axis, resulting in no further approximations as shown in given figure where third approximation tends to infinity.

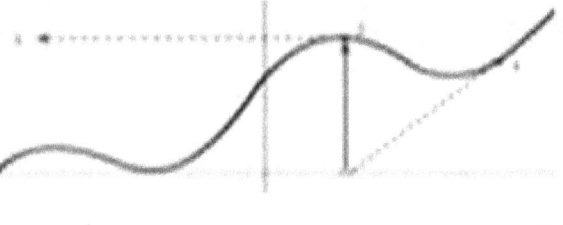

Case II:
Sometimes Newton-Raphson method may run into an infinite cycle or loop as shown in adjoining figure. Change in initial approximation may untangle the problem.

Case III:
In case of a point of discontinuity, as shown in given figure, subsequent roots may diverge instead of converging.

Remarks:
- Newton-Raphson method can be used for solving both algebraic and transcendental equations and it can also be used when roots are complex.
- Initial approximation x_0 can be taken randomly in the interval $[a,b]$, such that $f(a).f(b) < 0$
- Newton-Raphson method has quadratic convergence, but in case of bad choice of x_0 (the initial guess), Newton- Raphson method may fail to converge
- This method is useful in case of large value of $f'(x_n)$ i.e. when graph of $f(x)$ while crossing x-axis is nearly vertical

Example 10 Use Newton-Raphson method to find a root of the equation $x^3 - 5x + 3 = 0$ correct to three decimal places.

Solution: $f(x) = x^3 - 5x + 3 \Rightarrow f'(x) = 3x^2 - 5$

Here $f(0) = 3$ and $f(1) = -1 \Rightarrow f(0).f(1) < 0$

Also $f(x)$ is continuous in $[0,1]$, $\therefore$ atleast one root exists in $[0,1]$

Initial approximation: Let initial approximation (x_0) in the interval $[0,1]$ be 0.8

By Newton-Raphson method $x_{n+1} = x_n - \frac{f(x_n)}{f'(x_n)}$

First approximation:

$x_1 = x_0 - \frac{f(x_0)}{f'(x_0)}$, where $x_0 = 0.8$, $f(0.8) = -0.488$, $f'(0.8) = -3.08$

$\Rightarrow x_1 = 0.8 - \frac{-0.488}{-3.08} = 0.6416$

Second approximation:

$x_2 = x_1 - \frac{f(x_1)}{f'(x_1)}$, where $x_1 = 0.6415, f(0.6416) = 0.0561, f'(0.6416) = -3.7650$

$$\Rightarrow x_2 = 0.6416 - \frac{0.05611}{-3.7650} = 0.6565$$

Third approximation:

$x_3 = x_2 - \frac{f(x_2)}{f'(x_2)}$, where $x_2 = 0.6565, f(0.6565) = 0.0004, f'(0.6565) = -3.7070$

$$\Rightarrow x_3 = 0.6565 - \frac{0.0004}{-3.7070} = 0.6566$$

Hence a root of the equation $x^3 - 5x + 3 = 0$ correct to three decimal places is 0.6566

Example 11 Find the approximate value of $\sqrt{28}$ correct to 3 decimal places using Newton Raphson method.

Solution: $x = \sqrt{28} \Rightarrow x^2 - 28 = 0$

$\therefore f(x) = x^2 - 28$

$\Rightarrow f'(x) = 2x$

Here $f(5) = -3$ and $f(6) = 8 \Rightarrow f(5).f(6) < 0$

Also $f(x)$ is continuous in $[5,6]$, $\therefore$ atleast one root exists in $[5,6]$

Initial approximation: Let initial approximation (x_0) in the interval $[5,6]$ be 5.5

By Newton-Raphson method $x_{n+1} = x_n - \frac{f(x_n)}{f'(x_n)}$

First approximation:

$x_1 = x_0 - \frac{f(x_0)}{f'(x_0)}$, where $x_0 = 5.5, f(5.5) = 2.25, f'(5.5) = 11$

$$\Rightarrow x_1 = 5.5 - \frac{2.25}{11} = 5.2955$$

Second approximation:

$x_2 = x_1 - \frac{f(x_1)}{f'(x_1)}$, where $x_1 = 5.2955, f(5.2955) = 0.0423, f'(5.2955) = 10.591$

$$\Rightarrow x_2 = 5.2955 - \frac{0.0423}{10.591} = 5.2915$$

Third approximation:

$x_3 = x_2 - \frac{f(x_2)}{f'(x_2)}$, where $x_2 = 5.2915, f(5.2915) = -0.00003, f'(5.2915) = 10.583$

$$\Rightarrow x_3 = 5.2915 - \frac{-0.00003}{10.583} = 5.2915$$

Hence value of $\sqrt{28}$ correct to three decimal places is 5.2915

Example 12 Use Newton-Raphson method to find a root of the equation $x \sin x + \cos x = 0$ correct to three decimal places.

Solution: $f(x) = x \sin x + \cos x$

$\Rightarrow f'(x) = x \cos x + \sin x - \sin x = x \cos x$

Here $f\left(\frac{\pi}{2}\right) = 1.5708$ and $f(\pi) = -1 \Rightarrow f\left(\frac{\pi}{2}\right).f(\pi) < 0$

Also $f(x)$ is continuous in $\left[\frac{\pi}{2}, \pi\right]$ $\therefore$ atleast one root exists in $\left[\frac{\pi}{2}, \pi\right]$

Initial approximation: Let initial approximation (x_0) in the interval $\left[\frac{\pi}{2}, \pi\right]$ be π

By Newton-Raphson method $x_{n+1} = x_n - \dfrac{f(x_n)}{f'(x_n)}$

First approximation:

$x_1 = x_0 - \dfrac{f(x_0)}{f'(x_0)}$, where $x_0 = \pi$, $f(\pi) = -1$, $f'(\pi) = -3.1416$

$\Rightarrow x_1 = 3.1416 - \dfrac{-1}{-3.1416} = 2.8233$

Second approximation:

$x_2 = x_1 - \dfrac{f(x_1)}{f'(x_1)}$, where $x_1 = 2.8233$, $f(2.8233) = -0.0662$, $f'(2.8233) = -2.6815$

$\Rightarrow x_2 = 2.8233 - \dfrac{-0.0662}{-2.6815} = 2.7986$

Third approximation:

$x_3 = x_2 - \dfrac{f(x_2)}{f'(x_2)}$, where $x_2 = 2.798$, $f(2.7986) = -0.0006$, $f'(2.7986) = -2.6356$

$\Rightarrow x_3 = 2.7986 - \dfrac{-0.0006}{-2.6356} = 2.7984$

Hence a root of the equation $x \sin x + \cos x = 0$ correct to three decimal places is 2.7984

Example 13 Use Newton Raphson method to derive a formula to find $\sqrt[5]{N}$, $N \in R$.
Hence evaluate $\sqrt[5]{43}$ correct to 3 decimal places.

Solution: $x = \sqrt[5]{N} \Rightarrow x^5 - N = 0$

Let $f(x) = x^5 - N \Rightarrow f'(x) = 5x^4$

By Newton-Raphson method $x_{n+1} = x_n - \dfrac{f(x_n)}{f'(x_n)}$

$\Rightarrow x_{n+1} = x_n - \dfrac{x_n^5 - N}{5x_n^4} = \dfrac{4}{5}x_n + \dfrac{N}{5x_n^4}$

To evaluate $\sqrt[5]{43}$, putting $N = 43$, $\therefore$ Newton-Raphson formula is given by

$x_{n+1} = \dfrac{4}{5}x_n + \dfrac{43}{5x_n^4}$

Let initial approximation x_0 be 2

First approximation:

$x_1 = \dfrac{4}{5}x_0 + \dfrac{43}{5x_0^4}$, where $x_0 = 2$

$\Rightarrow x_1 = \dfrac{8}{5} + \dfrac{43}{80} = 2.1375$

Second approximation:

$x_2 = \dfrac{4}{5}x_1 + \dfrac{43}{5x_1^4}$, where $x_1 = 2.1375$

$\Rightarrow x_2 = \dfrac{4(2.1375)}{5} + \dfrac{43}{5(2.1375)^4} = 2.1220$

Third approximation:

$x_3 = \dfrac{4}{5}x_2 + \dfrac{43}{5x_2^4}$, where $x_2 = 2.1220$

$\Rightarrow x_3 = \dfrac{4(2.1220)}{5} + \dfrac{43}{5(2.1220)^4} = 2.1217$

Fourth approximation:

$$x_4 = \frac{4}{5}x_3 + \frac{43}{5x_3^4}, \text{ where } x_3 = 2.1217$$

$$\Rightarrow x_4 = \frac{4(2.1217)}{5} + \frac{43}{5(2.1217)^4} = 2.1217$$

Hence value of $\sqrt[5]{43}$ correct to four decimal places is 2.1217

9.2.4.1 Generalized Newton's Method for Multiple Roots

Result: If α is a root of equation $f(x) = 0$ with multiplicity m, then it is also a root of equation $f'(x) = 0$ with multiplicity $(m-1)$ and also of the equation $f''(x) = 0$ with multiplicity $(m-2)$ and so on.

For example $(x-1)^3 = 0$ has '1' as a root with multiplicity 3

$3(x-1)^2 = 0$ has '1' as the root with multiplicity 2

$6(x-1) = 0$ has '1' as the root with multiplicity 1

$\therefore$ The expressions $x_n - m\frac{f(x_n)}{f'(x_n)}$, $x_n - (m-1)\frac{f'(x_n)}{f''(x_n)}$, $x_n - (m-2)\frac{f''(x_n)}{f'''(x_n)}$ are equivalent

Generalized Newton's method is used to find repeated roots of an equation as is given as:

If α be a root of equation $f(x) = 0$ which is repeated m times,

Then $x_{n+1} = x_n - m\frac{f(x_n)}{f'(x_n)} \sim x_n - (m-1)\frac{f'(x_n)}{f''(x_n)}$

Example 14 Use Newton-Raphson method to find a double root of the equation

$$x^3 - x^2 - x + 1 = 0 \text{ upto three iterations.}$$

Solution: $f(x) = x^3 - x^2 - x + 1$

$f'(x) = 3x^2 - 2x - 1$

$f''(x) = 6x - 2$

Let the initial approximation $x_0 = 0.7$

First approximation:

$x_1 = x_0 - \frac{2f(x_0)}{f'(x_0)}$ Also $x_1 = x_0 - \frac{f'(x_0)}{f''(x_0)}$

$\Rightarrow x_1 = 0.7 - \frac{0.306}{-0.93} = 1.0290$ And $x_1 = 0.7 - \frac{-0.93}{2.2} = 1.1227$

$\therefore x_1 = \frac{1.029 + 1.1227}{2} = 1.0759, f(x_1) = .012$

Second approximation:

$x_2 = x_1 - \frac{2f(x_1)}{f'(x_1)}$ Also $x_2 = x_1 - \frac{f'(x_1)}{f''(x_1)}$

$\Rightarrow x_2 = 1.0759 - \frac{0.0239}{0.3209} = 1.001$ And $x_2 = 1.0759 - \frac{0.3209}{4.4554} = 1.004$

$\therefore x_2 = \frac{1.001 + 1.004}{2} = 1.0025, f(x_2) = .00001$

Third approximation:

$x_3 = x_2 - \frac{2f(x_2)}{f'(x_2)}$ Also $x_3 = x_2 - \frac{f'(x_2)}{f''(x_2)}$

$\Rightarrow x_3 = 1.0025 - \frac{0.00003}{0.0100} = 0.995$ And $x_3 = 1.0025 - \frac{0.0100}{4.015} = 1.0000$

$\therefore x_3 = \frac{0.995 + 1.000}{2} = 0.9975$, $f(x_3) = .00001$

The double root of the equation $x^3 - x^2 - x + 1 = 0$ after three iterations is 0.9975.

9.2.4.2 Convergence of Newton Raphson Method

Let α be an exact root of the equation $f(x) = 0 \Rightarrow f(\alpha) = 0$

Also let x_n and x_{n+1} be two successive approximations to the root α.

If ϵ_n and ϵ_{n+1} are the corresponding errors in the approximations x_n and x_{n+1}

Then $x_n = \alpha + \epsilon_n$ ... ①

and $x_{n+1} = \alpha + \epsilon_{n+1}$ ... ②

Now by Newton Raphson method

$x_{n+1} = x_n - \frac{f(x_n)}{f'(x_n)}$ ... ③

Using ① and ② in ③

$\Rightarrow \alpha + \epsilon_{n+1} = \alpha + \epsilon_n - \frac{f(\alpha + \epsilon_n)}{f'(\alpha + \epsilon_n)}$

$\Rightarrow \epsilon_{n+1} = \frac{\epsilon_n f'(\alpha + \epsilon_n) - f(\alpha + \epsilon_n)}{f'(\alpha + \epsilon_n)}$

$\Rightarrow \epsilon_{n+1} = \frac{\epsilon_n [f'(\alpha) + \epsilon_n f''(\alpha) + \ldots] - [f(\alpha) + \epsilon_n f'(\alpha) + \frac{\epsilon_n^2}{2!} f''(\alpha) + \cdots]}{f'(\alpha) + \epsilon_n f''(\alpha) + \ldots}$ by Taylor's expansion

$\Rightarrow \epsilon_{n+1} = \frac{\epsilon_n^2 f''(\alpha) - \frac{\epsilon_n^2}{2!} f''(\alpha) + \cdots}{f'(\alpha)\left[1 + \frac{\epsilon_n f''(\alpha)}{f'(\alpha)} + \cdots\right]}$ $\because f(\alpha) = 0$

$\Rightarrow \epsilon_{n+1} = \left[\frac{\epsilon_n^2}{2} \frac{f''(\alpha)}{f'(\alpha)} + \cdots\right]\left[1 + \frac{\epsilon_n f''(\alpha)}{f'(\alpha)} + \cdots\right]^{-1}$

$\Rightarrow \epsilon_{n+1} = \left[\frac{\epsilon_n^2}{2} \frac{f''(\alpha)}{f'(\alpha)} + \cdots\right]\left[1 - \frac{\epsilon_n f''(\alpha)}{f'(\alpha)} + \cdots\right]$

$\Rightarrow \epsilon_{n+1} = \frac{\epsilon_n^2}{2} \frac{f''(\alpha)}{f'(\alpha)}$ Neglecting higher order terms

$\Rightarrow \epsilon_{n+1} = K \epsilon_n^2$ Where $k = \frac{1}{2} \frac{f''(\alpha)}{f'(\alpha)}$

$\therefore$ Newton Raphson method has convergence of order 2 or quadratic convergence.

9.2.5 Secant Method

Secant method can be seen as a combination of Regula-Falsi and Newton Raphson methods. Secant method algorithm requires the selection of two initial approximations x_0 and x_1; such that they are reasonably close to the exact root, which may or may not satisfy intermediate theorem.

The method utilizes two most recent approximations to obtain the next approximation instead of checking through intermediate value theorem every time in between.

Third and higher estimates are obtained using the expression:
$$x_{n+1} = x_n - \frac{x_n - x_{n-1}}{f(x_n) - f(x_{n-1})} f(x_n), \ n \geq 2$$

Remarks:
- Secant method requires two initial approximation x_0 and x_1 such that $f(x_0)$ and $f(x_1)$ are small (less than one). If x_0 and x_1 are far away from the root, the method may fail.
- Its convergence is faster than Regula-Falsi method but slower than Newton Raphson method.

Example15 Use secant method to find a root of the equation $xe^x - 2 = 0$ correct to three decimal places.

Solution: $f(x) = xe^x - 2$

$$x_{n+1} = x_n - \frac{x_n - x_{n-1}}{f(x_n) - f(x_{n-1})} f(x_n)$$

Let $x_0 = 0.9$ and $x_1 = 1$ be two initial approximations.

Here $f(0.9) = 0.2136$ and $f(1) = 0.7183$

First approximation: $x_0 = 0.9, \ x_1 = 1$

$$x_2 = x_1 - \frac{x_1 - x_0}{f(x_1) - f(x_0)} f(x_1) = 1 - \frac{1 - 0.9}{0.7183 - 0.2136}(0.7183) = 0.8577$$

Second approximation: $x_1 = 1, \ x_2 = 0.8577, \ f(x_2) = 0.0222$

$$x_3 = x_2 - \frac{x_2 - x_1}{f(x_2) - f(x_1)} f(x_2) = 0.8577 - \frac{0.8577 - 1}{0.0222 - 0.7183}(0.0222) = 0.8532$$

Third approximation: $x_2 = 0.8577, \ x_3 = 0.8532, \ f(x_3) = 0.0026$

$$x_4 = x_3 - \frac{x_3 - x_2}{f(x_3) - f(x_2)} f(x_3) = 0.8532 - \frac{0.8532 - 0.8577}{0.0026 - 0.0222}(0.0026) = 0.8526$$

Fourth approximation: $x_3 = 0.8532, \ x_4 = 0.8526, \ f(x_4) = -0.00002$

$$x_5 = x_4 - \frac{x_4 - x_3}{f(x_4) - f(x_3)} f(x_4) = 0.8526 - \frac{0.8526 - 0.8532}{-0.00002 - 0.0026}(-0.00002) = 0.8526$$

∴ Real root of the equation $xe^x - 2 = 0$ correct to three decimal places is 0.8526

9.3 Iterative Methods for Solving Simultaneous Linear Equations

Consider a system of linear equations:
$$\left.\begin{array}{l} a_1 x + b_1 y + c_1 z = d_1 \\ a_2 x + b_2 y + c_2 z = d_2 \\ a_3 x + b_3 y + c_3 z = d_3 \end{array}\right\} \ldots ①$$

We have been using direct methods for solving a system of linear equations. Direct methods produce exact solution after a finite number of steps whereas iterative methods give a sequence of approximate solutions until solution is obtained up to desired accuracy. Common iterative methods for solving a system of linear equations are:

1. Gauss-Jacobi's iteration method
2. Gauss-Seidal's iteration method

These methods require partial pivoting before application.

Partial pivoting: It is about changing rows of a system of linear equations given by ① such that $a_1 \geq a_2, a_3$; $b_2 \geq b_3$.

Complete pivoting: It is the process of selecting the largest element in the magnitude as the pivot element, by interchanging row as well as columns of the system. Order of variables is also changed in the procedure. In particular for the system given by ①, complete pivoting would require $a_1 \geq a_2, a_3$; $b_2 \geq b_1, b_3$, if a_1 and b_2 are to be taken as pivots.

9.3.1 Gauss-Jacobi's Iteration Method

The concept of the Gauss- Jacobi's iteration scheme is extremely simple with the assumptions that the system has unique solution and diagonal elements are non-zeros.

Algorithm: Gauss-Jacobi's iteration method

1. Take the system of linear equations given by ① after partial pivoting and solve each equation in the system for the diagonal value of variables such that

$$\left.\begin{aligned} x &= \frac{1}{a_1}(d_1 - b_1 y - c_1 z) \\ y &= \frac{1}{b_2}(d_2 - a_2 x - c_2 z) \\ z &= \frac{1}{c_3}(d_3 - a_3 x - b_3 y) \end{aligned}\right\} \dots ②$$

2. Rewrite ② in generalized form given by:

$$\left.\begin{aligned} x_{n+1} &= \frac{1}{a_1}(d_1 - b_1 y_n - c_1 z_n) \\ y_{n+1} &= \frac{1}{b_2}(d_2 - a_2 x_n - c_2 z_n) \\ z_{n+1} &= \frac{1}{c_3}(d_3 - a_3 x_n - b_3 y_n) \end{aligned}\right\} \dots ③$$

3. Take $x_0 = y_0 = z_0 = 0$ as initial approximation (in general if a better approximation can not be judged) and substitute in the system given by ③

$$\therefore x_1 = \frac{d_1}{a_1},\ y_1 = \frac{d_2}{b_2},\ z_1 = \frac{d_3}{c_3}$$

4. Putting $n = 1$, substitute the values of x_1, y_1 and z_1 in ③ to get next approximations of x_2, y_2 and z_2. Continue the procedure until the difference between two consecutive approximations is negligible.

Example 16 Solve the following system of equations using Gauss Jacobi's method

$$5x - 2y + 3z = -1$$
$$-3x + 9y + z = 2$$
$$2x - y - 7z = 3$$

Solution: The given system of equations is satisfying rules of partial pivoting.

Rewriting in general form as given in ③

$$x_{n+1} = \frac{1}{5}(-1 + 2y_n - 3z_n)$$
$$y_{n+1} = \frac{1}{9}(2 + 3x_n - z_n)$$
$$z_{n+1} = \frac{1}{7}(-3 + 2x_n - y_n)$$

Taking $x_0 = y_0 = z_0 = 0$ as initial approximation

First Approximation:
$$x_1 = -\frac{1}{5} = -0.2, \quad y_1 = \frac{2}{9} = 0.222, \quad z_1 = -\frac{3}{7} = -0.429$$

Second Approximation:
$$x_2 = \frac{1}{5}(-1 + 2y_1 - 3z_1), \quad y_2 = \frac{1}{9}(2 + 3x_1 - z_1), \quad z_2 = \frac{1}{7}(-3 + 2x_1 - y_1)$$
$$\Rightarrow x_2 = \frac{1}{5}(-1 + 2(0.222) - 3(-0.429)) = 0.146$$
$$y_2 = \frac{1}{9}(2 + 3(-0.2) + 0.429) = 0.203$$
$$z_2 = \frac{1}{7}(-3 + 2(-0.2) - 0.222) = -0.517$$

Third Approximation:
$$x_3 = \frac{1}{5}(-1 + 2y_2 - 3z_2), \quad y_3 = \frac{1}{9}(2 + 3x_2 - z_2), \quad z_3 = \frac{1}{7}(-3 + 2x_2 - y_2)$$
$$\Rightarrow x_3 = \frac{1}{5}(-1 + 2(0.203) - 3(-0.517)) = 0.191$$
$$y_3 = \frac{1}{9}(2 + 3(0.146) + 0.517) = 0.328$$
$$z_3 = \frac{1}{7}(-3 + 2(0.146) - 0.203) = -0.416$$

Fourth Approximation:
$$x_4 = \frac{1}{5}(-1 + 2y_3 - 3z_3), \quad y_4 = \frac{1}{9}(2 + 3x_3 - z_3), \quad z_4 = \frac{1}{7}(-3 + 2x_3 - y_3)$$
$$\Rightarrow x_4 = \frac{1}{5}(-1 + 2(0.328) - 3(-0.416)) = 0.181$$
$$y_4 = \frac{1}{9}(2 + 3(0.191) + 0.416) = 0.332$$
$$z_4 = \frac{1}{7}(-3 + 2(0.191) - 0.328) = -0.421$$

Fifth Approximation:
$$x_5 = \frac{1}{5}(-1 + 2y_4 - 3z_4), \quad y_5 = \frac{1}{9}(2 + 3x_4 - z_4), \quad z_5 = \frac{1}{7}(-3 + 2x_4 - y_4)$$
$$\Rightarrow x_5 = \frac{1}{5}(-1 + 2(0.332) - 3(-0.421)) = 0.185$$
$$y_5 = \frac{1}{9}(2 + 3(0.181) + 0.421) = 0.329$$
$$z_5 = \frac{1}{7}(-3 + 2(0.181) - 0.332) = -0.424$$

Sixth Approximation:
$$x_6 = \frac{1}{5}(-1 + 2y_5 - 3z_5), \quad y_6 = \frac{1}{9}(2 + 3x_5 - z_5), \quad z_6 = \frac{1}{7}(-3 + 2x_5 - y_5)$$
$$\Rightarrow x_6 = \frac{1}{5}(-1 + 2(0.329) - 3(-0.424)) = 0.186$$
$$y_6 = \frac{1}{9}(2 + 3(0.185) + 0.424) = 0.331$$
$$z_6 = \frac{1}{7}(-3 + 2(0.185) - 0.329) = -0.423$$

Values of variables have been stabilized, ∴ approximate solution is given by
$$x = 0.186, \quad y = 0.331 \text{ and } z = -0.423$$

Example 17 Compute 4 iterations to find an approximate solution of the given system of equations using Gauss Jacobi's method.
$$x + y + 5z = -1, \quad 5x - y + z = 10, \quad 2x + 4y = 12$$
Solution: Rearranging the given equations by partial pivoting
$$5x - y + z = 10$$
$$2x + 4y = 12$$
$$x + y + 5z = -1$$
Rewriting in general form as given in ③
$$x_{n+1} = \frac{1}{5}(10 + y_n - z_n)$$
$$y_{n+1} = \frac{1}{4}(12 - 2x_n)$$
$$z_{n+1} = \frac{1}{5}(-1 - x_n - y_n)$$
Taking $x_0 = y_0 = z_0 = 0$ as initial approximation

First Approximation:
$$x_1 = \frac{10}{5} = 2, \; y_1 = \frac{12}{4} = 3, \; z_1 = -\frac{1}{5}$$

Second Approximation:
$$x_2 = \frac{1}{5}(10 + y_1 - z_1), \; y_2 = \frac{1}{4}(12 - 2x_1), \; z_2 = \frac{1}{5}(-1 - x_1 - y_1)$$
$$\Rightarrow x_2 = \frac{1}{5}\left(10 + 3 + \frac{1}{5}\right), \; y_2 = \frac{1}{4}(12 - 4), \; z_2 = \frac{1}{5}(-1 - 2 - 3)$$
$$\therefore x_2 = 2.64, \; y_2 = 2, \; z_2 = -1.2$$

Third Approximation:
$$x_3 = \frac{1}{5}(10 + y_2 - z_2), \; y_3 = \frac{1}{4}(12 - 2x_2), \; z_3 = \frac{1}{5}(-1 - x_2 - y_2)$$
$$\Rightarrow x_3 = \frac{1}{5}(10 + 2 + 1.2), \; y_3 = \frac{1}{4}(12 - 2(2.64)), \; z_3 = \frac{1}{5}(-1 - 2.64 - 2)$$
$$\therefore x_3 = 2.64, \; y_3 = 1.68, \; z_3 = -0.928$$

Fourth Approximation:
$$x_4 = \frac{1}{5}(10 + y_3 - z_3), \; y_4 = \frac{1}{4}(12 - 2x_3), \; z_4 = \frac{1}{5}(-1 - x_3 - y_3)$$
$$\Rightarrow x_4 = \frac{1}{5}(10 + 1.68 + 0.928), \; y_4 = \frac{1}{4}(12 - 2(2.64)), \; z_4 = \frac{1}{5}(-1 - 2.64 - 1.68)$$
$$\therefore x_4 = 2.52, \; y_4 = 1.68, \; z_4 = -1.064$$

Approximate solution after 4 iterations is given by $x = 2.52, \; y = 1.68, \; z = -1.064$

9.3.2 Gauss-Seidal's Iteration Method

Gauss-Seidel method is an improvement of the basic Gauss-Jordan method. Here the improved values of variables are utilized as soon as they are obtained.

∴ System of equations given in ③ is improved by taking latest values of the variables as

$$x_{n+1} = \frac{1}{a_1}(d_1 - b_1 y_n - c_1 z_n)$$
$$y_{n+1} = \frac{1}{b_2}(d_2 - a_2 x_{n+1} - c_2 z_n)$$
$$z_{n+1} = \frac{1}{c_3}(d_3 - a_3 x_{n+1} - b_3 y_{n+1})$$

Gauss-Seidel scheme usually converges faster than Jacobi's iteration method.

Example 18 Solve the system of equations given in Example 14 using Gauss Seidal's method
$$5x - 2y + 3z = -1, \quad -3x + 9y + z = 2, \quad 2x - y - 7z = 3$$
Also compare the results obtained in two methods.

Solution: The given system of equations is satisfying rules of partial pivoting.

Using Gauss Seidal's approximations, system can be rewritten as
$$x_{n+1} = \frac{1}{5}(-1 + 2y_n - 3z_n)$$
$$y_{n+1} = \frac{1}{9}(2 + 3x_{n+1} - z_n)$$
$$z_{n+1} = \frac{1}{7}(-3 + 2x_{n+1} - y_{n+1})$$

Taking $x_0 = y_0 = z_0 = 0$ as initial approximation

First Approximation:
$$x_1 = -\frac{1}{5} = -0.2 \quad y_1 = \frac{2}{9} = 0.222, \quad z_1 = -\frac{3}{7} = -0.429$$

Second Approximation:
$$x_2 = \frac{1}{5}(-1 + 2y_1 - 3z_1), \quad y_2 = \frac{1}{9}(2 + 3x_2 - z_1), \quad z_2 = \frac{1}{7}(-3 + 2x_2 - y_2)$$
$$\Rightarrow x_2 = \frac{1}{5}(-1 + 2(0.222) - 3(-0.429)) = 0.146$$
$$y_2 = \frac{1}{9}(2 + 3(0.146) + 0.429) = 0.319$$
$$z_2 = \frac{1}{7}(-3 + 2(0.146) - 0.319) = -0.432$$

Third Approximation:
$$x_3 = \frac{1}{5}(-1 + 2y_2 - 3z_2), \quad y_3 = \frac{1}{9}(2 + 3x_3 - z_2), \quad z_3 = \frac{1}{7}(-3 + 2x_3 - y_3)$$
$$\Rightarrow x_3 = \frac{1}{5}(-1 + 2(0.319) - 3(-0.432)) = 0.187$$
$$y_3 = \frac{1}{9}(2 + 3(0.187) + 0.432) = 0.333$$
$$z_3 = \frac{1}{7}(-3 + 2(0.187) - 0.333) = -0.423$$

Fourth Approximation:
$$x_4 = \frac{1}{5}(-1 + 2y_3 - 3z_3), \quad y_4 = \frac{1}{9}(2 + 3x_4 - z_3), \quad z_4 = \frac{1}{7}(-3 + 2x_4 - y_4)$$
$$\Rightarrow x_4 = \frac{1}{5}(-1 + 2(0.333) - 3(-0.423)) = 0.187$$
$$y_4 = \frac{1}{9}(2 + 3(0.187) + 0.423) = 0.332$$
$$z_4 = \frac{1}{7}(-3 + 2(0.187) - 0.332) = -0.423$$

Values of variables have been stabilized, ∴ approximate solution is given by
$x = 0.187, y = 0.332$ and $z = -0.423$

Clearly numbers of iterations for the solution to converge in Gauss Seidal's method are much less than Gauss Jacobi's method.

Exercise 9

1. Solve $e^{-x} = 10x$ correct to three decimal places using direct iteration method.
2. Find the real roots of the following equations using Bisection method:
   i. $x^3 - 2x - 5 = 0$ (Correct to three decimal places)
   ii. $x^3 - 4x - 9 = 0$ (Up to four iterations)
3. Find a root of the equation $x^3 - 2x - 5 = 0$ correct to three decimal places using Regula-Falsi method.
4. Using method of false position, find the fourth root of 32 correct to three decimal places.
5. Perform three iterations to find root of the equation $xe^x - \cos x = 0$ using Regula-Falsi method.
6. Find the real root of the equation $x^3 - 3x - 5 = 0$ correct to three decimal places using Newton- Raphson method.
7. Find the real root of the equation $\cos x - x^2 = 0$ correct to three decimal places using Newton- Raphson method.
8. Find the real root of the equation $3x = \cos x + 1$ correct to four decimal places using Newton- Raphson method.
9. Solve the system of equations:
$$10x_1 - 2x_2 - x_3 - x_4 = 3$$
$$2x_1 - 10x_2 + x_3 + x_4 = -15$$
$$x_1 + x_2 - 10x_3 + 2x_4 = -27$$
$$x_1 + x_2 + 2x_3 - 10x_4 = 9$$
using Gauss- Jacobi method. Compute results for 2 iterations.
10. Solve the system of equations given in Q4 upto 2 iterations, using Gauss- Seidal method.

Answers

1. 0.09128
2. i. 2.0944   ii. 2.6875
3. 2.094
4. 2.378
5. 0.494015
6. 2.279
7. 0.8241
8. 0.6071
9. $x_1 = 0.78, x_2 = 1.74, x_3 = 2.7, x_4 = -0.18$ taking initial approximations as zero.
10. $x_1 = 0.8869, x_2 = 1.9523, x_3 = 2.9566, x_4 = -0.0248$ taking initial approximations as zero.

Chapter 10: Finite Differences

Finite Differences

10.1 Introduction

For a function $y = f(x)$, finite differences refer to changes in values of y (dependent variable) for any finite (equal or unequal) variation in x (independent variable).

In this chapter, we shall study various differencing techniques for equal deviations in values of x and associated differencing operators; also their applications will be extended for finding missing values of a data and series summation.

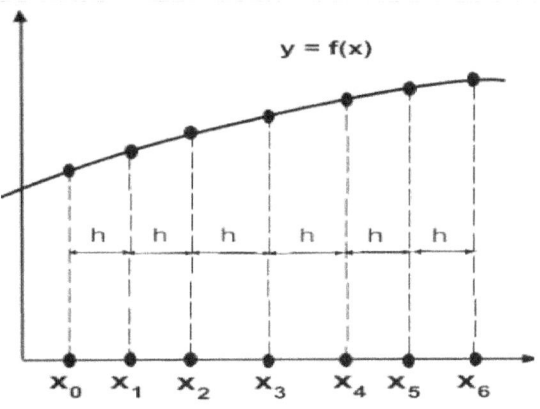

10.2 Shift or Increment Operator (E)

Shift (Increment) operator denoted by 'E' operates on $f(x)$ as $Ef(x) = f(x + h)$

Or $Ey_x = y_{x+h}$, where 'h' is the step height for equi-spaced data points.

Clearly effect of the shift operator E is to shift the function value to the next higher value $f(x + h)$ or y_{x+h}

Also $E^2 f(x) = E(Ef(x)) = Ef(x+h) = f(x+2h)$

$\therefore E^n f(x) = f(x + nh)$

Moreover $E^{-1} f(x) = f(x - h)$, where E^{-1} is the inverse shift operator.

10.3 Differencing Operators

If $y_0, y_1, y_2, \ldots y_n$ be the values of y for corresponding values of $x_0, x_1, x_2, \ldots x_n$, then the differences of y are defined by $(y_1 - y_0), (y_2 - y_1), \ldots, (y_n - y_{n-1})$, and are denoted by different operators discussed in this section.

10.3.1 Forward Difference Operator (Δ)

Forward difference operator 'Δ' operates on y_x as $\Delta y_x = y_{x+1} - y_x$

Or $\Delta f(x) = f(x + h) - f(x)$, where h is the height of differencing.

$\therefore \Delta y_0 = y_1 - y_0$

$\Delta y_1 = y_2 - y_1$

$\vdots$

$\Delta y_n = y_{n+1} - y_n$

Also $\Delta^2 y_0 = \Delta y_1 - \Delta y_0 = (y_2 - y_1) - (y_1 - y_0) = y_2 - 2y_1 + y_0$

$\vdots$

$\Delta^n y_0 = y_n - {}^nC_1 y_{n-1} + {}^nC_2 y_{n-2} - \cdots + (-1)^{n-1} {}^nC_{n-1} y_1 + (-1)^n y_0$

Generalizing $\Delta^n y_r = y_{n+r} - {}^nC_1 y_{n+r-1} + {}^nC_2 y_{n+r-2} - \cdots + (-1)^r y_r$

Here Δ^n is the nth order forward difference; Table 10.1 shows the forward differences of various orders.

Table 10.1 Forward Differences

x	y	Δ	Δ^2	Δ^3	Δ^4	Δ^5
x_0	y_0					
		Δy_0				
x_1	y_1		$\Delta^2 y_0$			
		Δy_1		$\Delta^3 y_0$		
x_2	y_2		$\Delta^2 y_1$		$\Delta^4 y_0$	
		Δy_2		$\Delta^3 y_1$		$\Delta^5 y_0$
x_3	y_3		$\Delta^2 y_2$		$\Delta^4 y_1$	
		Δy_3		$\Delta^3 y_2$		
x_4	y_4		$\Delta^2 y_3$			
		Δy_4				
x_5	y_5					

The arrow indicates the direction of differences from top to bottom. Differences in each column notate difference of two adjoining consecutive entries of the previous column.

Relation between Δ and E

Δ and E are connected by the relation $\Delta \equiv E - 1$

Proof: we know that $\Delta y_n = y_{n+1} - y_n$
$$= E y_n - y_n$$
$$\Rightarrow \Delta y_n = (E - 1) y_n$$
$$\Rightarrow \Delta \equiv E - 1 \text{ or } E \equiv 1 + \Delta$$

Properties of operator 'Δ'

- $\Delta C = 0$, C being a constant
- $\Delta C f(x) = C f(x)$
- $\Delta [a f(x) \pm b g(x)] = a \Delta f(x) \pm b \Delta g(x)$
- $\Delta [f(x) g(x)] = f(x+h) \Delta g(x) + g(x) \Delta f(x)$, f & g may be interchanged
- $\Delta \left[\dfrac{f(x)}{g(x)}\right] = \dfrac{g(x) \Delta f(x) - f(x) \Delta g(x)}{g(x+h) g(x)}$

Result 1: The nth differences of a polynomial of degree 'n' are constant and all higher order differences are zero.

Proof: Consider the polynomial $f(x)$ of nth degree
$$f(x) = a_0 x^n + a_1 x^{n-1} + a_2 x^{n-2} + \cdots + a_{n-1} x + a_n$$
First differences of the polynomial $f(x)$ are calculated as:
$$\Delta f(x) = f(x+h) - f(x)$$
$$= a_0[(x+h)^n - x^n] + a_1[(x+h)^{n-1} - x^{n-1}] + \cdots + a_{n-1}[(x+h) - x]$$
$$= a_0 n h \, x^{n-1} + a_1' x^{n-1} + a_2' x^{n-2} + \cdots + a_{n-1}' h + a_n'$$
where $a_1', a_2', \ldots, a_{n-1}', a_n'$ are new constants

$\Rightarrow$ First difference of a polynomial of degree n is a polynomial of degree $(n-1)$

Similarly $\Delta^2 f(x) = \Delta f(x+h) - \Delta f(x)$
$$= a_0 n(n-1) h^2 \, x^{n-2} + a_1'' x^{n-3} + \ldots + a_n''$$

∴ Second difference of a polynomial of degree n is a polynomial of degree $(n-2)$

Repeating the above process $\Delta^n f(x) = a_0 n(n-1)\ldots 2.1 h^n x^{n-n}$
$$\Rightarrow \Delta^n f(x) = a_0 n! \, h^n \text{ which is a constant}$$
$\therefore n^{th}$ Difference of a polynomial of degree n is a polynomial of degree zero.

Thus $(n+1)^{th}$ and higher order differences of a polynomial of nth degree are all zero.

> The converse of above result is also true, i.e. if the nth difference of a polynomial given at equally spaced points are constant then the function is a polynomial of degree 'n'.

10.3.2 Backward Difference Operator (∇)

Backward difference operator '∇' operates on y_n as $\nabla y_n = y_n - y_{n-1}$

$\therefore$ The differences $(y_1 - y_0), (y_2 - y_1), \ldots, (y_n - y_{n-1})$ when denoted by $\nabla y_1, \nabla y_2, \ldots, \nabla y_n$ are called first backward differences.

Also $\nabla^2 y_n = \nabla y_n - \nabla y_{n-1}$, $\nabla^3 y_n = \nabla^2 y_n - \nabla^2 y_{n-1}$ denote second and third backward differences respectively.

Table 10.2 shows the backward differences of various orders.

Table 10.2 Backward Differences

x	y	∇	∇^2	∇^3	∇^4	∇^5
x_0	y_0					
		∇y_1				
x_1	y_1		$\nabla^2 y_2$			
		∇y_2		$\nabla^3 y_3$		
x_2	y_2		$\nabla^2 y_3$		$\nabla^4 y_4$	
		∇y_3		$\nabla^3 y_4$		$\nabla^5 y_5$
x_3	y_3		$\nabla^2 y_4$		$\nabla^4 y_5$	
		∇y_4		$\nabla^3 y_5$		
x_4	y_4		$\nabla^2 y_5$			
		∇y_5				
x_5	y_5					

The arrow indicates the direction of differences from bottom to top. Differences in each column notate difference of two adjoining consecutive entries of the previous column

i.e. $\nabla y_1 = y_1 - y_0$, $\nabla^2 y_2 = \nabla y_2 - \nabla y_1, \ldots, \nabla^5 y_5 = \nabla^4 y_5 - \nabla^4 y_4$.

Relation between ∇ and E

∇ and E are connected by the relation $\nabla \equiv 1 - E^{-1}$

Proof: we know that $\nabla y_n = y_n - y_{n-1}$
$$= y_n - E^{-1} y_n$$
$$\Rightarrow \nabla y_n = (1 - E^{-1}) y_n$$
$$\Rightarrow \nabla \equiv 1 - E^{-1}$$

10.3.3 Central Difference Operator (δ)

Central difference operator 'δ' operates on y_n as $\delta y_n = y_{n+\frac{1}{2}} - y_{n-\frac{1}{2}}$

∴ The differences $(y_1 - y_0), (y_2 - y_1), \ldots, (y_n - y_{n-1})$ when denoted by $\delta y_{\frac{1}{2}}, \delta y_{\frac{3}{2}}, \ldots, \delta y_{n-\frac{1}{2}}$ are called first central differences.

Also $\delta^2 y_n = \delta y_{n+\frac{1}{2}} - \delta y_{n-\frac{1}{2}}$, $\delta^3 y_n = \delta^2 y_{n+\frac{1}{2}} - \delta^2 y_{n-\frac{1}{2}}$ denote second and third central differences respectively as shown in Table 10.3.

Table 10.3 Central Differences

x	y	Δ	δ^2	δ^3	δ^4	δ^5
x_0	y_0					
		$\delta y_{\frac{1}{2}}$				
x_1	y_1		$\delta^2 y_1$			
		$\delta y_{\frac{3}{2}}$		$\delta^3 y_{\frac{3}{2}}$		
x_2	y_2		$\delta^2 y_2$		$\delta^4 y_2$	
		$\delta y_{\frac{5}{2}}$		$\delta^3 y_{\frac{5}{2}}$		$\delta^5 y_{\frac{5}{2}}$
x_3	y_3		$\delta^2 y_3$		$\delta^4 y_3$	
		$\delta y_{\frac{7}{2}}$		$\delta^3 y_{\frac{7}{2}}$		
x_4	y_4		$\delta^2 y_4$			
		$\delta y_{\frac{9}{2}}$				
x_5	y_5					

Central differences in each column notate difference of two adjoining consecutive entries of the previous column, i.e. $\delta y_{\frac{1}{2}} = y_1 - y_0, \ldots, \delta^5 y_{\frac{5}{2}} = \delta^4 y_3 - \delta^4 y_2$.

Relation between δ and E

δ and E are connected by the relation $\delta \equiv E^{\frac{1}{2}} - E^{-\frac{1}{2}}$

Proof: we know that $\delta y_n = y_{n+\frac{1}{2}} - y_{n-\frac{1}{2}}$

$$= E^{\frac{1}{2}} y_n - E^{-\frac{1}{2}} y_n$$

$$\Rightarrow \delta y_n = \left(E^{\frac{1}{2}} - E^{-\frac{1}{2}}\right) y_n$$

$$\therefore \delta \equiv \left(E^{\frac{1}{2}} - E^{-\frac{1}{2}}\right)$$

Observation: It is only the notation which changes and not the difference.

$$\therefore y_1 - y_0 = \Delta y_0 = \nabla y_1 = \delta y_{\frac{1}{2}}$$

10.3.4 Averaging Operator (μ)

Averaging operator 'μ' operates on y_x as $\mu y_x = \frac{1}{2}\left(y_{x+\frac{h}{2}} + y_{x-\frac{h}{2}}\right)$

Or $\mu f(x) = \frac{1}{2}\left(f\left(x + \frac{h}{2}\right) + f\left(x - \frac{h}{2}\right)\right)$, '$h$' is the height of the interval.

Relation between μ and E

We know that $\mu y_n = \frac{1}{2}\left(y_{n+\frac{h}{2}} + y_{n-\frac{h}{2}}\right)$

$\qquad\qquad = \frac{1}{2}\left(E^{\frac{1}{2}} y_n + E^{-\frac{1}{2}} y_n\right)$

$\Rightarrow \mu y_n = \frac{1}{2}\left(E^{\frac{1}{2}} + E^{-\frac{1}{2}}\right) y_n$

$\therefore \mu \equiv \frac{1}{2}\left(E^{\frac{1}{2}} + E^{-\frac{1}{2}}\right)$

Result 2: Relation between E and D, where $D \equiv \frac{d}{dx}$

We know $y(x+h) = y(x) + h\, y'(x) + \frac{h^2}{2!} y''(x) + \ldots$ By Taylor's theorem

$\qquad\qquad = y(x) + h\, Dy(x) + \frac{h^2}{2!} D^2 y(x) + \ldots$

$\qquad\qquad = \left(1 + hD + \frac{h^2}{2!} D^2 + \cdots\right) y(x)$

$\Rightarrow E\, y(x) = e^{hD} y(x)$

$\therefore E = e^{hD},\; D \equiv \frac{d}{dx}$

Result 3: Relation between Δ and D, where $D \equiv \frac{d}{dx}$

We know that $\Delta \equiv E - 1$

$\Rightarrow \Delta \equiv e^{hD} - 1$ $\because E = e^{hD}$

Result 4: Relation between ∇ and D, where $D \equiv \frac{d}{dx}$

We know that $\nabla \equiv 1 - E^{-1} = 1 - e^{-hD}$ $\because E = e^{hD}$

Result 5: Relation between Δ and ∇

We know that $E \equiv 1 + \Delta$ … ①

Also $E^{-1} \equiv 1 - \nabla$

$\Rightarrow E \equiv \frac{1}{1-\nabla}$ … ②

$\Rightarrow 1 + \Delta \equiv \frac{1}{1-\nabla}$ From ① and ②

$\Rightarrow \Delta \equiv \frac{1}{1-\nabla} - 1$

$\Rightarrow \Delta \equiv \frac{\nabla}{1-\nabla}$

Result 6: Relation between μ, δ and E

We have $\mu \equiv \frac{1}{2}\left(E^{\frac{1}{2}} + E^{-\frac{1}{2}}\right)$

Also $\delta \equiv \left(E^{\frac{1}{2}} - E^{-\frac{1}{2}}\right)$

$\Rightarrow \mu\delta \equiv \frac{1}{2}\left(E^{\frac{1}{2}} + E^{-\frac{1}{2}}\right)\left(E^{\frac{1}{2}} - E^{-\frac{1}{2}}\right)$

$\Rightarrow \mu\delta \equiv \frac{1}{2}(E - E^{-1})$

Result 7: Relation between μ, δ, Δ and ∇

We have $\mu\delta \equiv \frac{1}{2}(E - E^{-1}) = \frac{1}{2}[(1+\Delta) - (1-\nabla)]$

$\Rightarrow \quad \mu\delta \equiv \frac{1}{2}(\Delta + \nabla)$

Result 8: $\quad \Delta^n y_r = \nabla^n y_{n+r}$

We have $\Delta^n y_r = (E-1)^n y_r \qquad \because \Delta = E - 1$

$= y_{n+r} - {}^nC_1 y_{n+r-1} + {}^nC_2 y_{n+r-2} - \cdots + (-1)^r y_r$

$= (E^n - {}^nC_1 E^{n-1} + {}^nC_2 E^{n-2} - \cdots + (-1)^n) y_r$

$= E^n y_r - {}^nC_1 E^{n-1} y_r + {}^nC_2 E^{n-2} y_r - \cdots + (-1)^n y_r$

$= y_{n+r} - {}^nC_1 y_{n+r-1} + {}^nC_2 y_{n+r-2} - \cdots + (-1)^n y_r$

Also $\nabla^n y_{n+r} = (1 - E^{-1})^n y_{n+r} \qquad \because \nabla \equiv 1 - E^{-1}$

$= (1 - {}^nC_1 E^{-1} + {}^nC_2 E^{-2} - \cdots + (-1)^n E^{-n}) y_{n+r}$

$= y_{n+r} - {}^nC_1 y_{n+r-1} + {}^nC_2 y_{n+r-2} - \cdots + (-1)^n y_r$

$\therefore \Delta^n y_r = \nabla^n y_{n+r}$

Example 1 Evaluate the following:

i. Δe^x ii. $\Delta^2 e^x$ iii. $\Delta \tan^{-1} x$ iv. $\Delta\left(\frac{x+1}{x^2-3x+2}\right)$ v. $\Delta f_k^2 = (f_k + f_{k+1})\Delta f_k$

Solution: i. $\Delta e^x = e^{x+h} - e^x = e^x(e^h - 1)$

$\Delta e^x = e^x(e - 1)$, if $h = 1$

ii. $\Delta^2 e^x = \Delta(\Delta e^x)$

$= \Delta[e^x(e^h - 1)]$

$= (e^h - 1) \Delta e^x$

$= (e^h - 1)[e^{x+h} - e^x]$

$= (e^h - 1) e^x(e^h - 1)$

$= e^x(e^h - 1)^2$

iii. $\Delta \tan^{-1} x = \tan^{-1}(x+h) - \tan^{-1} x$

$= \tan^{-1}\left(\frac{x+h-x}{1+(x+h)x}\right)$

$= \tan^{-1}\frac{h}{1+(x+h)x}$

iv. $\Delta\left(\frac{x+1}{x^2-3x+2}\right) = \Delta\left(\frac{x+1}{(x-1)(x-2)}\right)$

$= \Delta\left(\frac{-2}{x-1} + \frac{3}{x-2}\right) = \Delta\left(\frac{-2}{x-1}\right) + \Delta\left(\frac{3}{x-2}\right)$

$= -2\left(\frac{1}{x+1-1} - \frac{1}{x-1}\right) + 3\left(\frac{1}{x+1-2} - \frac{1}{x-2}\right)$

$$= -2\left(\frac{1}{x} - \frac{1}{x-1}\right) + 3\left(\frac{1}{x-1} - \frac{1}{x-2}\right)$$

$$= -\frac{(x+4)}{x(x-1)(x-2)}$$

v. $\Delta f_k^2 = f_{k+1}^2 - f_k^2 = (f_{k+1} + f_k)(f_{k+1} - f_k) = (f_k + f_{k+1})\Delta f_k$

Example 2 Evaluate the following:

i. $\Delta e^x \log 2x$ ii. $\Delta\left(\frac{x^2}{\cos 2x}\right)$

Solution: i. Let $f(x) = e^x$ and $g(x) = \log 2x$

We have $\Delta[f(x)g(x)] = f(x+h)\Delta g(x) + g(x)\Delta f(x)$

$\therefore \Delta e^x \log 2x = e^{x+h}\Delta \log 2x + \log 2x\, \Delta e^x$

$$= e^{x+h}[\log 2(x+h) - \log 2x] + \log 2x\,[e^{x+h} - e^x]$$

$$= e^x e^h \log\left(1 + \frac{h}{x}\right) + e^x \log 2x\,[e^h - 1]$$

$$= e^x \left[e^h \log\left(1 + \frac{h}{x}\right) + \log 2x\,[e^h - 1]\right]$$

ii. Let $f(x) = x^2$ and $g(x) = \cos 2x$

We have $\Delta\left[\frac{f(x)}{g(x)}\right] = \frac{g(x)\Delta f(x) - f(x)\Delta g(x)}{g(x+h)g(x)}$

$$= \frac{\cos 2x\,[(x+h)^2 - x^2] - x^2[\cos 2(x+h) - \cos 2x]}{\cos 2(x+h)\cos 2x}$$

$$= \frac{(h^2 + 2hx)\cos 2x + 2x^2 \sin(2x+h)\sin h}{\cos 2(x+h)\cos 2x}$$

Example 3 Evaluate $\Delta^4[(1-2x)(1-3x)(1-4x)(1-x)]$

where interval of differencing is one.

Solution: $\Delta^4[(1-2x)(1-3x)(1-4x)(1-x)]$

$$= \Delta^4[24x^4 + \cdots + 1] = 24.4!.\,1^4 = 576$$

$\because \Delta^n f(x) = a_0 n!\, h^n$ and $\Delta^4 x^n = 0$ when $n < 4$

Example 4 Prove that $\Delta^3 y_3 = \nabla^3 y_6$

Solution: $\Delta^3 y_3 = (E-1)^3 y_3$ $\qquad \because \Delta = E - 1$

$$= (E^3 - 1 - 3E^2 + 3E)y_3$$

$$= E^3 y_3 - y_3 - 3E^2 y_3 + 3E y_3$$

$$= y_6 - y_3 - 3y_5 + 3y_4$$

Also $\nabla^3 y_6 = (1 - E^{-1})^3 y_6$ $\qquad \because \nabla \equiv 1 - E^{-1}$

$$= (1 - E^{-3} - 3E^{-1} + 3E^{-2})y_6$$

$$= y_6 - y_3 - 3y_5 + 3y_4$$

Example 5 Prove that $\Delta + \nabla = \dfrac{\Delta}{\nabla} - \dfrac{\nabla}{\Delta}$

Solution: L.H.S. $= \Delta + \nabla = (E - 1) + (1 - E^{-1})$

$$= E - E^{-1}$$

R.H.S. $= \dfrac{\Delta}{\nabla} - \dfrac{\nabla}{\Delta}$

$$= \dfrac{E-1}{1-E^{-1}} - \dfrac{1-E^{-1}}{E-1}$$

$$= \dfrac{(E-1)^2 - (1-E^{-1})^2}{(1-E^{-1})(E-1)}$$

$$= \dfrac{(E^2+1-2E) - (1+E^{-2}-2E^{-1})}{E+E^{-1}-2}$$

$$= \dfrac{E^2 - E^{-2} - 2E + 2E^{-1}}{E+E^{-1}-2}$$

$$= \dfrac{(E+E^{-1})(E-E^{-1}) - 2(E-E^{-1})}{E+E^{-1}-2}$$

$$= \dfrac{(E-E^{-1})(E+E^{-1}-2)}{E+E^{-1}-2}$$

$$= E - E^{-1} = \text{R.H.S.}$$

Example 6 Prove that $E = 1 + \dfrac{1}{2}\delta^2 + \delta\sqrt{1 + \dfrac{1}{4}\delta^2}$

Solution: R.H.S. $= 1 + \dfrac{1}{2}\delta^2 + \delta\sqrt{1 + \dfrac{1}{4}\delta^2}$

$$= 1 + \dfrac{1}{2}\left(E^{\frac{1}{2}} - E^{-\frac{1}{2}}\right)^2 + \left(E^{\frac{1}{2}} - E^{-\frac{1}{2}}\right)\sqrt{1 + \dfrac{1}{4}\left(E^{\frac{1}{2}} - E^{-\frac{1}{2}}\right)^2}$$

$$\because \delta \equiv E^{\frac{1}{2}} - E^{-\frac{1}{2}}$$

$$= 1 + \dfrac{1}{2}(E + E^{-1} - 2) + \left(E^{\frac{1}{2}} - E^{-\frac{1}{2}}\right)\sqrt{1 + \dfrac{1}{4}(E + E^{-1} - 2)}$$

$$= 1 + \dfrac{1}{2}(E + E^{-1} - 2) + \left(E^{\frac{1}{2}} - E^{-\frac{1}{2}}\right)\sqrt{\dfrac{1}{4}(E + E^{-1} + 2)}$$

$$= 1 + \dfrac{1}{2}(E + E^{-1} - 2) + \left(E^{\frac{1}{2}} - E^{-\frac{1}{2}}\right)\sqrt{\dfrac{1}{4}\left(E^{\frac{1}{2}} + E^{-\frac{1}{2}}\right)^2}$$

$$= 1 + \dfrac{1}{2}(E + E^{-1} - 2) + \dfrac{1}{2}\left(E^{\frac{1}{2}} - E^{-\frac{1}{2}}\right)\left(E^{\frac{1}{2}} + E^{-\frac{1}{2}}\right)$$

$$= \dfrac{1}{2}(E + E^{-1}) + \dfrac{1}{2}(E - E^{-1})$$

$$= E = \text{L.H.S.}$$

Example 7 Prove that $\nabla = -\frac{1}{2}\delta^2 + \delta\sqrt{1+\frac{1}{4}\delta^2}$

Solution: R.H.S. $= -\frac{1}{2}\delta^2 + \delta\sqrt{1+\frac{1}{4}\delta^2}$

$$= -\frac{1}{2}\left(E^{\frac{1}{2}} - E^{-\frac{1}{2}}\right)^2 + \left(E^{\frac{1}{2}} - E^{-\frac{1}{2}}\right)\sqrt{1+\frac{1}{4}\left(E^{\frac{1}{2}} - E^{-\frac{1}{2}}\right)^2}$$

$$\because \delta \equiv E^{\frac{1}{2}} - E^{-\frac{1}{2}}$$

$$= -\frac{1}{2}(E + E^{-1} - 2) + \left(E^{\frac{1}{2}} - E^{-\frac{1}{2}}\right)\sqrt{1+\frac{1}{4}(E + E^{-1} - 2)}$$

$$= -\frac{1}{2}(E + E^{-1} - 2) + \left(E^{\frac{1}{2}} - E^{-\frac{1}{2}}\right)\sqrt{\frac{1}{4}(E + E^{-1} + 2)}$$

$$= -\frac{1}{2}(E + E^{-1} - 2) + \left(E^{\frac{1}{2}} - E^{-\frac{1}{2}}\right)\sqrt{\frac{1}{4}\left(E^{\frac{1}{2}} + E^{-\frac{1}{2}}\right)^2}$$

$$= -\frac{1}{2}(E + E^{-1} - 2) + \frac{1}{2}\left(E^{\frac{1}{2}} - E^{-\frac{1}{2}}\right)\left(E^{\frac{1}{2}} + E^{-\frac{1}{2}}\right)$$

$$= -\frac{1}{2}(E + E^{-1} - 2) + \frac{1}{2}(E - E^{-1}) = 1 - E^{-1} = \nabla = \text{L.H.S.}$$

Example 8 Prove that (i) $\Delta - \nabla = \delta^2$ (ii) $\mu = \sqrt{1+\frac{1}{4}\delta^2} = \left(1+\frac{\Delta}{2}\right)(1+\Delta)^{-\frac{1}{2}}$

Solution: (i) $\delta^2 = \left(E^{\frac{1}{2}} - E^{-\frac{1}{2}}\right)^2 = E + E^{-1} - 2$ $\because \delta \equiv E^{\frac{1}{2}} - E^{-\frac{1}{2}}$

$$= (E - 1) - (1 - E^{-1}) = \Delta - \nabla$$

$$\because E - 1 \equiv \Delta \text{ and } 1 - E^{-1} = \nabla$$

(ii) $\sqrt{1+\frac{1}{4}\delta^2} = \sqrt{1+\frac{1}{4}\left(E^{\frac{1}{2}} - E^{-\frac{1}{2}}\right)^2}$ $\because \delta \equiv \left(E^{\frac{1}{2}} - E^{-\frac{1}{2}}\right)$

$$= \sqrt{1+\frac{1}{4}(E + E^{-1} - 2)}$$

$$= \sqrt{\frac{1}{4}(E + E^{-1} + 2)}$$

$$= \sqrt{\frac{1}{4}\left(E^{\frac{1}{2}} + E^{-\frac{1}{2}}\right)^2}$$

$$= \frac{1}{2}\left(E^{\frac{1}{2}} + E^{-\frac{1}{2}}\right) = \mu \quad \because \mu \equiv \frac{1}{2}\left(E^{\frac{1}{2}} + E^{-\frac{1}{2}}\right)$$

Also $\left(1+\frac{\Delta}{2}\right)(1+\Delta)^{-\frac{1}{2}} = \left(1+\frac{E-1}{2}\right)(1+E-1)^{-\frac{1}{2}}$

$$\because \Delta \equiv E - 1$$

$$= \left(\frac{E+1}{2}\right)E^{-\frac{1}{2}} = \frac{1}{2}\left(E^{-\frac{1}{2}} + E^{\frac{1}{2}}\right) = \mu$$

Example 9 Prove that (i) $\Delta \equiv E\nabla \equiv \nabla E = \delta E^{\frac{1}{2}}$ (ii) $E^r = \left(\mu + \frac{\delta}{2}\right)^{2r}$

Solution: (i) $E\nabla = E(1 - E^{-1}) = E - 1 = \Delta$ $\because \nabla \equiv 1 - E^{-1}$

$$\nabla E = (1 - E^{-1})E = E - 1 = \Delta$$

$$\delta E^{\frac{1}{2}} = \left(E^{\frac{1}{2}} - E^{-\frac{1}{2}}\right)E^{\frac{1}{2}} = E - 1 = \Delta \qquad \because \delta \equiv \left(E^{\frac{1}{2}} - E^{-\frac{1}{2}}\right)$$

(ii) R.H.S. $= \left(\mu + \frac{\delta}{2}\right)^{2r} = \left(\frac{1}{2}\left(E^{\frac{1}{2}} + E^{-\frac{1}{2}}\right) + \frac{1}{2}\left(E^{\frac{1}{2}} - E^{-\frac{1}{2}}\right)\right)^{2r}$

$$\because \mu \equiv \frac{1}{2}\left(E^{\frac{1}{2}} + E^{-\frac{1}{2}}\right) \text{ and } \delta \equiv E^{\frac{1}{2}} - E^{-\frac{1}{2}}$$

$$= \left(\frac{1}{2}\left(2E^{\frac{1}{2}}\right)\right)^{2r} = \left(E^{\frac{1}{2}}\right)^{2r} = E^r = \text{L.H.S}$$

Example 10 Prove that (i) $D \equiv \frac{1}{h} \log E$ (ii) $hD \equiv \log(1+\Delta) \equiv -\log(1-\nabla)$

(iii) $\nabla^2 \equiv h^2 D^2 - h^3 D^3 + \frac{7}{12} h^4 D^4 + \cdots$

Solution: (i) We know that $E \equiv e^{hD}$

$\Rightarrow \log E \equiv \log e^{hD}$

$\Rightarrow \log E \equiv hd \log e$

$\Rightarrow D \equiv \frac{1}{h} \log E \qquad \because \log e = 1$

(ii) $hD \equiv \log E \qquad$ From relation (i)

$\equiv \log(1+\Delta) \qquad \because E \equiv 1 + \Delta$

Also $hD \equiv \log E \equiv -\log E^{-1}$

$\equiv -\log(1-\nabla) \quad \because \nabla \equiv 1 - E^{-1}$

(iii) We know that $\nabla \equiv 1 - E^{-1}$

$\Rightarrow \nabla \equiv 1 - \frac{1}{E} \equiv 1 - e^{-hD} \qquad \because E = e^{hD}$

$\Rightarrow \nabla \equiv 1 - \left(1 - hd + \frac{h^2 D^2}{2!} - \frac{h^3 D^3}{3!} + \cdots\right)$

$\Rightarrow \nabla \equiv hd - \frac{h^2 D^2}{2!} + \frac{h^3 D^3}{3!} + \cdots$

$\therefore \nabla^2 \equiv \left(hd - \frac{h^2 D^2}{2!} + \frac{h^3 D^3}{3!} + \cdots\right)^2$

$\Rightarrow \nabla^2 \equiv h^2 D^2 + \left(\frac{h^2 D^2}{2!}\right)^2 + \cdots - 2(hd)\left(\frac{h^2 D^2}{2!}\right) + 2(hd)\left(\frac{h^3 D^3}{3!}\right) - \cdots$

$\Rightarrow \nabla^2 \equiv h^2 D^2 - h^3 D^3 + \left(\frac{h^4 D^4}{4} + \frac{h^4 D^4}{3}\right) - \cdots$

$\Rightarrow \nabla^2 \equiv h^2 D^2 - h^3 D^3 + \frac{7}{12} h^4 D^4 - \cdots$

Remark: In order to prove any relation, we can express the operators (Δ, ∇, δ) in terms of fundamental operator E.

Example 11 Form the forward difference table for the function
$f(x) = x^3 - 2x^2 - 3x - 1$ for $x = 0, 1, 2, 3, 4$.
Hence or otherwise find $\Delta^3 f(x)$, also show that $\Delta^4 f(x) = 0$

Solution: $f(0) = -1, f(1) = -5, f(2) = -7, f(3) = -1, f(4) = 19$

Constructing the forward difference table:

x	$f(x)$	Δ	Δ^2	Δ^3	Δ^4
0	−1				
		−4			
1	−5		2		
		−2		6	
2	−7		8		0
		6		6	
3	−1		14		
		20			
4	19				

From the table, we see that $\Delta^3 f(x) = 6$ and $\Delta^4 f(x) = 0$

Note: Using the formula $\Delta^n f(x) = a_0 n! \, h^n$, $\Delta^3 f(x) = 1.3!.1^n = 6$

Also $\Delta^{n+1} f(x) = 0$ for a polynomial of degree n, $\therefore \Delta^4 f(x) = 0$

Example 12 If for a polynomial, five observations are recorded as:
$y_0 = -8, \; y_1 = -6, \; y_2 = 22, \; y_3 = 148, \; y_4 = 492$, find y_5.

Solution: $y_5 = E^5 y_0 = (1 + \Delta)^5 y_0 \quad \because E \equiv 1 + \Delta$

$= y_0 + {}^5C_1 \Delta y_0 + {}^5C_2 \Delta^2 y_0 + {}^5C_3 \Delta^3 y_0 + {}^5C_4 \Delta^4 y_0 + \Delta^5 y_0 \quad \ldots \text{①}$

Constructing the forward difference table:

x	y	Δ	Δ^2	Δ^3	Δ^4
x_0	−8				
		2			
x_1	−6		26		
		28		72	
x_2	22		98		48
		126		120	
x_3	148		218		
		344			
x_4	492				

From table $\Delta y_0 = 2, \Delta^2 y_0 = 26, \Delta^3 y_0 = 72, \Delta^3 y_0 = 48 \quad \ldots \text{②}$

$\Rightarrow y_5 = -8 + 5(2) + 10(26) + 10(72) + 5(48) = 1222$ using ② in ①

10.4 Missing values of Data

Missing data or missing values occur when an observation is missing for a particular variable in a data sample. Concept of finite differences can help to locate the requisite value using known concepts of curve fitting.

To determine the equation of a line (equation of degree one), we need at least two given points. Similarly to trace a parabola (equation of degree two), at least three points are imperative. Thus we essentially require $(n + 1)$ known observations to determine a polynomial of nth degree.

To find missing values of data using finite differences, we presume the degree of the polynomial by the number of known observations and use the result $\Delta^{n+1} f(x) = 0$ for a polynomial of degree n.

Example 13 Find the missing values in the following table

x	0	5	10	15	20	25
$f(x)$	6	?	13	17	22	?

Solution: Since there are 4 known values of $f(x)$ in the given data, let us assume the polynomial represented by the given data to be of 3rd degree.

Constructing the forward difference table taking missing values as a and b.

x	y	Δ	Δ^2	Δ^3	Δ^4
0	6				
		$a - 6$			
5	a		$19 - 2a$		
		$13 - a$		$3a - 28$	
10	13		$a - 9$		$38 - 4a$
		4		$10 - a$	
15	17		1		$a + b - 38$
		5		$b - 28$	
20	22		$b - 27$		
		$b - 22$			
25	b				

Since the polynomial represented by the given data is considered to be of 3rd degree, 4th and higher order differences are zero i.e. $\Delta^4 y = 0$

$\therefore 38 - 4a = 0$ and $a + b - 38 = 0$

Solving these two equations we get,

$a = 9.5 \quad b = 28.5$

Example 14 Use the concept of missing data to find y_5 given that

$y_0 = -8, \ y_1 = -6, \ y_2 = 22, \ y_3 = 148, \ y_4 = 492$

Solution: Constructing the forward difference table taking y_5 as missing value

x	y	Δ	Δ^2	Δ^3	Δ^4	Δ^5
x_0	-8					
		2				
x_1	-6		26			
		28		72		
x_2	22		98		48	
		126		120		$y_5 - 1222$
x_3	148		218		$y_5 - 1174$	
		344		$y_5 - 1054$		
x_4	492		$y_5 - 836$			
		$y_5 - 492$				
x_5	y_5					

Since 5 observations are known, let us assume that the polynomial represented by given data is of 4th degree. $\therefore \Delta^5 y = 0 \Rightarrow y_5 - 1222 = 0$ or $y_5 = 1222$

10.5 Finding Differences Using Factorial Notation

We can conveniently find the forward differences of a polynomial using factorial notation.

10.5.1 Factorial Notation of a Polynomial

A product of the form $x(x-1)(x-2)\ldots(x-r+1)$ is called a factorial polynomial and is denoted by $[x]^r$

$\therefore \quad [x] = x$

$[x]^2 = x(x-1)$

$[x]^3 = x(x-1)(x-2)$

$\vdots$

$[x]^n = x(x-1)(x-2)\ldots(x-n+1)$

In case, the interval of differencing is h, then

$[x]^n = x(x-h)(x-h)\ldots\left(x - \overline{n-1}\,h\right)$

The results of differencing $[x]^r$ are analogous to that differentiating x^r

$\therefore \quad \Delta[x]^n = n[x]^{n-1}$

$\Delta^2[x]^n = n(n-1)[x]^{n-2}$

$\Delta^3[x]^n = n(n-1)(n-2)[x]^{n-3}$

$\vdots$

$\Delta^n[x]^n = n(n-1)(n-2)\ldots 3.2.1 = n!$

$\Delta^{n+1}[x]^n = 0$

Also $\dfrac{1}{\Delta}[x] = \dfrac{[x]^2}{2}, \dfrac{1}{\Delta}[x]^2 = \dfrac{[x]^3}{3}$ and so on

$\dfrac{1}{\Delta^2}[x] = \dfrac{1}{\Delta}\left[\dfrac{[x]^2}{2}\right] = \dfrac{[x]^3}{6}$

$\vdots$

Remarks:
  i. Every polynomial of degree n can be expressed as a factorial polynomial of the same degree and vice-versa.
  ii. The coefficient of highest power of x and also the constant term remains unchanged while transforming a polynomial to factorial notation.

Example 15 Express the polynomial $2x^2 + 3x + 1$ in factorial notation.
Solution: $2x^2 - 3x + 1 = 2x^2 - 2x + 5x + 1$
$$= 2x(x-1) + 5x + 1 = 2[x]^2 + 5[x] + 1$$

Example 16 Express the polynomial $2x^3 - x^2 + 3x - 4$ in factorial notation.
Solution: $2x^3 - x^2 + 3x - 4 = 2[x]^3 + A[x]^2 + B[x] - 4$
Using remarks i and ii
$$= 2x(x-1)(x-2) + Ax(x-1) + Bx - 4$$
$$= 2x^3 + (A-6)x^2 + (-A+B+4)x - 4$$
Comparing the coefficients on both sides
$$A - 6 = -1, \quad -A + B + 4 = 3$$
$$\Rightarrow A = 5, \quad B = 4$$
$$\therefore 2x^3 - x^2 + 3x - 4 = 2[x]^3 + 5[x]^2 + 4[x] - 4$$

➢ We can also find factorial polynomial using synthetic division as shown:

Coefficients A and B can be found as remainders under x^2 and x columns

	x^3	x^2	x	
1	2	−1	3	−4
	−	2	1	
2	2	1	4 = B	
	−	4		
	2	5 = A		

Example 17 Find $\Delta^3 f(x)$ for the polynomial $f(x) = x^3 - 2x^2 - 3x - 1$
Also show that $\Delta^4 f(x) = 0$

Solution: Finding factorial polynomial of $f(x)$ as shown:
Let $x^3 - 2x^2 - 3x - 1 = [x]^3 + A[x]^2 + B[x] - 1$
Coefficients A and B can be found as remainders under x^2 and x columns

	x^3	x^2	x	
1	1	−2	−3	−1
	−	1	−1	
2	1	−1	−4 = B	
	−	2		
	1	1 = A		

$$\therefore \quad f(x) = x^3 - 2x^2 - 3x - 1 = [x]^3 + [x]^2 - 4[x] - 1$$
$$\Delta^3 f(x) = \Delta^3[[x]^3 + [x]^2 - 4[x] - 1]$$
$$= 3! + 0 = 6 \qquad \because \Delta^n[x]^n = n! \text{ and } \Delta^{n+1}[x]^n = 0$$
Also $\Delta^4 f(x) = \Delta^4[[x]^3 + [x]^2 - 4[x] - 1] = 0$

Note: Results obtained are same as in Example 11, where we have used forward difference table to compute the differences.

Example 18: Obtain the function whose first difference is $8x^3 - 3x^2 + 3x - 1$

Solution: Let $f(x)$ be the function whose first difference is $8x^3 - 3x^2 + 3x - 1$

$\Rightarrow \Delta f(x) = 8x^3 - 3x^2 + 3x - 1$

Let $8x^3 - 3x^2 + 3x - 1 = 8[x]^3 + A[x]^2 + B[x] - 1$

Coefficients A and B can be found as remainders under x^2 and x columns

	x^3	x^2	x	
1	8	-3	3	-1
	-	8	5	
2	8	5	8 = B	
	-	16		
	8	21 = A		

$\therefore \Delta f(x) = 8x^3 - 3x^2 + 3x - 1 = 8[x]^3 + 21[x]^2 + 8[x] - 1$

$f(x) = \frac{1}{\Delta}[8[x]^3 + 21[x]^2 + 8[x] - 1]$

$= \frac{8[x]^4}{4} + \frac{21[x]^3}{3} + \frac{8[x]^2}{2} - [x] \quad \because \frac{1}{\Delta}[x] = \frac{[x]^2}{2}, \frac{1}{\Delta}[x]^2 = \frac{[x]^3}{3}, \ldots$

$= 2[x]^4 + 7[x]^3 + 4[x]^2 - [x]$

$= 2x(x-1)(x-2)(x-3) + 7x(x-1)(x-2) + 4x(x-1) - x$

$= x[2(x-1)(x-2)(x-3) + 7(x-1)(x-2) + 4(x-1) - 1]$

$= x[2x^3 - 5x^2 + 5x - 3] = 2x^4 - 5x^3 + 5x^2 - 3x$

$\Rightarrow f(x) = 2x^4 - 5x^3 + 5x^2 - 3x$

10.6 Series Summation Using Finite Differences

The method of finite differences may be used to find sum of a given series by applying the following algorithm:

1. Let the series be represented by $u_0, u_1, u_2, u_3, \ldots$
2. Use the relation $u_r = E^r u_0$ to introduce the operator E in the series.
3. Replace E by Δ by substituting $E \equiv 1 + \Delta$ and find the sum the series by any of the applicable methods like sum of a G.P. exponential or logarithmic series or by binomial expansion and operate term by term on u_0 to find the required sum.

Example 19 Prove the following using finite differences:

i. $u_0 + u_1 \frac{x}{1!} + u_2 \frac{x^2}{2!} + \cdots = e^x \left[u_0 + x \frac{\Delta u_0}{1!} + x^2 \frac{\Delta^2 u_0}{2!} + \cdots \right]$

ii. $u_0 - u_1 + u_2 - u_3 + \cdots = \frac{1}{2} u_0 - \frac{1}{4} \Delta u_0 + \frac{1}{8} \Delta^2 u_0 - \cdots$

Solution: i. $u_0 + u_1 \frac{x}{1!} + u_2 \frac{x^2}{2!} + \cdots = u_0 + \frac{x}{1!} E u_0 + \frac{x^2}{2!} E^2 u_0 + \cdots$

$$= \left[1 + \frac{xE}{1!} + \frac{x^2 E^2}{2!} u_0 + \cdots \right] u_0$$

$$= e^{xE} u_0 = e^{x(1+\Delta)} u_0$$

$$= e^x e^{x\Delta} u_0$$

$$= e^x \left[1 + \frac{x\Delta}{1!} + \frac{x^2 \Delta^2}{2!} + \cdots \right] u_0$$

$$= e^x \left[u_0 + x \frac{\Delta u_0}{1!} + x^2 \frac{\Delta^2 u_0}{2!} + \cdots \right]$$

ii. $u_0 - u_1 + u_2 - u_3 + \cdots = u_0 - E u_0 + E^2 u_0 - E^3 u_0 + \cdots$

$$= [1 - E + E^2 - E^3 + \cdots] u_0$$

$$= [1 + E]^{-1} u_0$$

$$= [2 + \Delta]^{-1} u_0$$

$$= 2^{-1} \left[1 + \frac{\Delta}{2}\right]^{-1} u_0$$

$$= \frac{1}{2} \left[1 - \frac{\Delta}{2} + \frac{\Delta^2}{4} - \cdots \right] u_0$$

$$= \frac{1}{2} u_0 - \frac{1}{4} \Delta u_0 + \frac{1}{8} \Delta^2 u_0 - \cdots$$

Example 20 Sum the series $1^2, 2^2, 3^2, \ldots, n^2$ using finite differences.

Solution: Let the series $1^2, 2^2, 3^2, \ldots, n^2$ be represented by $u_0, u_1, u_2, \ldots, u_{n-1}$

$\therefore S = u_0 + u_1 + u_2 + \cdots + u_{n-1}$

$\Rightarrow S = u_0 + E u_0 + E^2 u_0 + \cdots + E^{n-1} u_0$

$= (1 + E + E^2 + \cdots + E^{n-1}) u_0$

$= \frac{1 - E^n}{1 - E} u_0 = \frac{E^n - 1}{E - 1} u_0 \qquad \because S_n = a \frac{1 - r^n}{1 - r}$

$\Rightarrow S = \frac{(1+\Delta)^n - 1}{(1+\Delta) - 1} u_0$

$= \frac{1}{\Delta} \left\{ \left[1 + n\Delta + \frac{n(n-1)}{2!} \Delta^2 + \frac{n(n-1)(n-2)}{3!} \Delta^3 + \cdots \right] - 1 \right\} u_0$

$= n u_0 + \frac{n(n-1)}{2!} \Delta u_0 + \frac{n(n-1)(n-2)}{3!} \Delta^2 u_0 + \cdots$

Now $u_0 = 1^2 = 1$

$\Delta u_0 = u_1 - u_0 = 2^2 - 1^2 = 3$

$\Delta^2 u_0 = \Delta u_1 - \Delta u_0 = u_2 - 2u_1 + u_0 = 3^2 - 2(2^2) + 1^2 = 2$

$\Delta^3 u_0, \Delta^4 u_0 \ldots$ are all zero as given series is an expression of degree 2

$\therefore S = n + \frac{n(n-1)}{2!}(3) + \frac{n(n-1)(n-2)}{3!}(2) + 0 = n + \frac{3n(n-1)}{2} + \frac{n(n-1)(n-2)}{3}$

$= \frac{1}{6}\left(6n + 9n(n-1) + 2n(n-1)(n-2)\right)$

$$= \frac{1}{6}n(6 + 9n - 9 + 2n^2 - 6n + 4)$$
$$= \frac{1}{6}n(2n^2 + 3n + 1) = \frac{1}{6}n(n+1)(2n+1)$$

Example 21 Prove that $u_0 + u_1 x + u_2 x^2 + \cdots = \frac{u_0}{1-x} + \frac{x \Delta u_0}{(1-x)^2} + \frac{x^2 \Delta^2 u_0}{(1-x)^3} + \cdots$

and hence evaluate $1.2 + 2.3x + 3.4x^2 + 4.5x^3 + \cdots$

Solution: $u_0 + u_1 x + u_2 x^2 + \cdots = u_0 + xEu_0 + x^2 E^2 u_0 + \cdots$

$$= (1 + xE + x^2 E^2 + \cdots) u_0$$

$$= \frac{1}{1 - xE} u_0 \qquad \because S_\infty = \frac{a}{1-r}$$

$$= \frac{1}{1 - x(1+\Delta)} u_0 = \frac{1}{(1-x) - x\Delta} u_0$$

$$= \frac{1}{1-x} \frac{1}{1 - \frac{x\Delta}{1-x}} u_0$$

$$= \frac{1}{1-x} \left(1 - \frac{x\Delta}{1-x}\right)^{-1} u_0$$

$$= \frac{1}{1-x} \left(1 + \frac{x\Delta}{1-x} + \frac{x^2 \Delta^2}{(1-x)^2} + \cdots\right) u_0$$

$$= \frac{u_0}{1-x} + \frac{x \Delta u_0}{(1-x)^2} + \frac{x^2 \Delta^2 u_0}{(1-x)^3} + \cdots = \text{R.H.S.}$$

Now to evaluate the series $1.2 + 2.3x + 3.4x^2 + 4.5x^3 + \cdots$

Let $u_0 = 1.2 = 2$, $u_1 = 2.3 = 6$, $u_2 = 3.4 = 12$, $u_3 = 4.5 = 20, \ldots$

Forming forward difference table to calculate the differences

u	Δ	Δ^2	Δ^3	Δ^4
$u_0 = 1.2 = 2$				
	4			
$u_1 = 2.3 = 6$		2		
	6		0	
$u_2 = 3.4 = 12$		2		0
	8		0	
$u_3 = 4.5 = 20$		2		
	10			
$u_4 = 5.6 = 30$				

$$\therefore 1.2 + 2.3x + 3.4x^2 + 4.5x^3 + \cdots = \frac{u_0}{1-x} + \frac{x \Delta u_0}{(1-x)^2} + \frac{x^2 \Delta^2 u_0}{(1-x)^3} + \cdots$$

$$= \frac{2}{1-x} + \frac{4x}{(1-x)^2} + \frac{2x^2}{(1-x)^3} + 0 = \frac{2}{(1-x)^3}$$

Exercise 10

1. Express y_4 in terms of successive forward differences.
2. Prove that $\Delta^n e^{3x+5} = (e^3 - 1)^n e^{3x+5}$
3. Evaluate i. $\Delta^2 \left(\frac{5x+12}{x^2+5x+6}\right)$ ii. $\Delta^2 \cos 2x$
4. If $u_0 = 3$, $u_1 = 12$, $u_2 = 81$, $u_3 = 2000$, $u_4 = 100$, calculate $\Delta^4 u_0$.
5. Prove that $\mu = \frac{2+\Delta}{2\sqrt{1+\Delta}} = \sqrt{1 + \frac{1}{4}\delta^2}$
6. Find the missing value in the following table

x	0	5	10	15	20	25
y	6	10	-	17	-	31

7. Assuming that following values of y belong to a polynomial of degree 4, compute the missing values

x	0	1	2	3	4	5	6	7
y	1	-1	1	-1	1	-	-	-

8. Express $f(x) = 2x^3 - 3x^2 + 3x - 10$ in factorial notation and hence show that $\Delta^3 f(x) = 12$, $\Delta^4 f(x) = 0$.
9. Prove that $u_1 x + u_2 x^2 + u_3 x^3 + \cdots = \frac{x}{1-x} u_1 + \left(\frac{x}{1-x}\right)^2 \Delta u_1 + \left(\frac{x}{1-x}\right)^3 \Delta^2 u_1 + \cdots$
10. Sum the series $1^3, 2^3, 3^3, \ldots, n^3$ using finite differences.

Answers

1. $y_4 = y_0 + 4\Delta y_0 + 6\Delta^2 y_0 + 4\Delta^3 y_0 + \Delta^4 y_0$
3. i. $\frac{-3(x^2+9x+15)}{x(x+1)(x+4)(x+5)(x+8)(x+9)}$ ii. $-4\sin^2 h \cos 2(x+h)$
4. -7459
6. $13.25, 22.5$
7. $31, 129, 351$
8. $2[x]^3 + 3[x]^2 + 2[x] - 10$
10. $\left[\frac{n(n+1)}{2}\right]^2$

Chapter 11: Interpolation

Interpolation

11.1 Introduction

Interpolation literally refers to introducing something additional or extraneous between other things or parts. In numerical analysis, interpolation is a method of constructing new data points within a discrete set of known data points, using finite differences. The process of obtaining function values outside (in the vicinity) the given range is called extrapolation.

In this chapter we shall extend the applications of differencing techniques to interpolate and extrapolate data points within a given range, for equal as well as well us unequal interval lengths.

11.2 Interpolation within Equal Intervals

Let $y = f(x)$ be an explicitly unknown function, with given discrete set of points (x_i, y_i), $i = 0,1,2,3,\ldots,n$, where x_i's are equispaced. The process of obtaining the values $f(x_i + ph)$, $-1 < p < 1$, where height of the interval (h) is fixed, is known as interpolation within equal intervals. There are several methods of interpolating data points within a given range, depending upon the location where the value is to be interpolated as given below:

i. Newton's forward interpolation formula
ii. Newton's backward interpolation formula
iii. Gauss's forward difference formula
iv. Gauss's backward difference formula
v. Stirling's central difference formula
vi. Bessels's interpolation formula

We shall discuss these methodologies one by one in the coming sections.

11.2.1 Newton's Forward Interpolation Formula

Newton's forward interpolation formula is used to interpolate the values of the function $y = f(x)$ near the beginning $(x > x_0)$ and to extrapolate the values when $(x < x_0)$, within the range of given data points (x_i, y_i), $i = 0,1,2,3,\ldots,n$.

Let $f(x)$ take the values y_0, y_1, y_2, $\ldots y_n$; for the independent variable x taking values x_0, x_1, x_2, $\ldots x_n$, where height of the interval (h) is fixed, such that $x_1 = x_0 + h$, $x_2 = x_0 + 2h, \ldots, x_n = x_0 + nh$.

Then to evaluate $f(x)$ for $x = x_0 + ph$, $-1 < p < 1$

We have $f(x) = f(x_0 + ph) = E^p f(x_0) \equiv (1 + \Delta)^p y_0 \qquad \because E \equiv 1 + \Delta$ and $f(x_0) = y_0$

$$\Rightarrow f(x) \equiv \left(1 + p\Delta + \frac{p(p-1)}{2!}\Delta^2 + \frac{p(p-1)(p-2)}{3!}\Delta^3 + \cdots\right) y_0, \qquad x = x_0 + ph$$

$$\therefore f(x) \equiv y_0 + p\Delta y_0 + \frac{p(p-1)}{2!}\Delta^2 y_0 + \frac{p(p-1)(p-2)}{3!}\Delta^3 y_0 + \cdots, \qquad p = \frac{x-x_0}{h}$$

This is Newton's forward interpolation formula and is used to interpolate or extrapolate values near the beginning of the table.

➤ In Newton's forward method, p is taken as 0 where $x = x_0$ and all the differences are evaluated taking $p = 0$ as reference point.
➤ Value which is to be interpolated i.e. $f(x)$ may be denoted by y_p, $|p < 1|$

Example1 Use Newton's forward interpolation formula to find the values of
    i. $f(1.4)$ ii. $f(0.9)$ for the given set of values

x	1	2	3	4
f(x)	6	11	18	27

Solution: Function has to be evaluated near the starting of the table, thereby constructing forward difference table for the function $y = f(x)$

x	p	f(x)	Δ	Δ²	Δ³
1	0	6			
			5		
2	1	11		2	
			7		0
3	2	18		2	
			9		
4	3	27			

Newton's forward interpolation formula given by:

$$f(x) \equiv y_0 + p\Delta y_0 + \frac{p(p-1)}{2!}\Delta^2 y_0 + \frac{p(p-1)(p-2)}{3!}\Delta^3 y_0 + \cdots, \quad p = \frac{x - x_0}{h} \quad \ldots ①$$

i. To find $f(1.4)$

$x_0 = 1, y_0 = 6, x = 1.4, h = 1 \therefore p = \frac{1.4 - 1}{1} = 0.4$

Also from table $\Delta y_0 = 5, \Delta^2 y_0 = 2, \Delta^3 y_0 = 0$

Substituting these values in ①, we get

$f(1.4) \equiv 6 + (0.4)(5) + \frac{0.4(0.4-1)}{2}(2) + 0 = 7.76$

ii. To find $f(0.9)$

$x_0 = 1, y_0 = 6, x = 0.9, h = 1 \therefore p = \frac{0.9 - 1}{1} = -0.1$

Also from table $\Delta y_0 = 5, \Delta^2 y_0 = 2, \Delta^3 y_0 = 0$

Substituting these values in ①, we get

$f(0.9) \equiv 6 + (-0.1)(5) + \frac{-0.1(-0.1-1)}{2}(2) + 0 = 5.61$

Example 2 From the following data, estimate the number of students who obtained marks between 40 and 45.

Marks	30–40	40–50	50–60	60–70	70–80
Number of Students	31	42	51	35	31

Solution: Preparing the cumulative frequency table,

Marks less than (x)	40	50	60	70	80
Number of Students (y_x)	31	73	124	159	190

Function has to be evaluated near the starting of the table, thereby constructing forward difference table for the function $y = f(x)$

Marks (x)	p	Number of Students (y)	Δy	$\Delta^2 y$	$\Delta^3 y$	$\Delta^4 y$
less than 40	0	31				
			42			
less than 50	1	73		9		
			51		−25	
less than 60	2	124		−16		37
			35		12	
less than 70	3	159		−4		
			31			
less than 80	4	190				

Newton's forward interpolation formula given by:

$$f(x) \equiv y_0 + p\Delta y_0 + \frac{p(p-1)}{2!}\Delta^2 y_0 + \frac{p(p-1)(p-2)}{3!}\Delta^3 y_0 + \frac{p(p-1)(p-2)(p-3)}{4!}\Delta^4 y_0 + \cdots$$

$$p = \frac{x - x_0}{h} \quad \ldots \text{①}$$

To find the number of students having marks less than 45 i.e. $f(45)$

$x_0 = 40$, $y_0 = 31$, $x = 45$, $h = 10$ $\therefore p = \frac{45-40}{10} = 0.5$

Also from table $\Delta y_0 = 42, \Delta^2 y_0 = 9, \Delta^3 y_0 = -25, \Delta^4 y_0 = 37$

Substituting these values in ①, we get

$$f(45) \equiv 31 + (0.5)(42) + \frac{0.5(0.5-1)}{2}(9) + \frac{0.5(0.5-1)(0.5-2)}{6}(-25) + \frac{0.5(0.5-1)(0.5-2)(0.5-3)}{24}(37)$$

$\Rightarrow f(45) \equiv 31 + 21 - 1.125 - 1.5625 - 1.4453 = 47.87 \approx 48$

$\therefore$ Number of students having marks less than 45 is 48.

Also number of students having marks less than 40 is 31. Hence number of students who obtained marks between 40 and 45 is $(48 - 31)$ i.e. 17

Example 3 Find the cubic polynomial with given set of points

x	0	1	2	3
$f(x)$	5	6	3	14

Hence or otherwise evaluate $f(0.5)$.

Solution: Constructing forward difference table for the function $y = f(x)$

x	p	y	Δ	Δ²	Δ³
0	0	5			
			1		
1	1	6		−4	
			−3		18
2	2	3		14	
			11		
3	3	14			

Newton's forward interpolation formula given by:

$$f(x) \equiv y_0 + p\Delta y_0 + \frac{p(p-1)}{2!}\Delta^2 y_0 + \frac{p(p-1)(p-2)}{3!}\Delta^3 y_0 + \cdots, \quad p = \frac{x - x_0}{h} \quad \ldots \text{①}$$

$x_0 = 0, \ y_0 = 5, \ h = 1 \therefore p = \frac{x-0}{1} = x$

Also from table $\Delta y_0 = 1, \Delta^2 y_0 = -4, \Delta^3 y_0 = 18$

Substituting these values in ①, we get

$$f(x) \equiv 5 + x(1) + \frac{x(x-1)}{2!}(-4) + \frac{x(x-1)(x-2)}{3!}(18)$$

$\Rightarrow f(x) \equiv 5 + x + (-2)x(x-1) + (3)x(x-1)(x-2)$

$\Rightarrow f(x) \equiv 3x^3 - 11x^2 + 9x + 5$

Also $f(0.5) \equiv 3(0.5)^3 - 11(0.5)^2 + 9(0.5) + 5 = 7.125$

11.2.2 Newton's Backward Interpolation Formula

Newton's backward interpolation formula is used to interpolate the values of $y = f(x)$ near the end $(x < x_n)$ and to extrapolate the values when $(x > x_n)$, within the range of given data points (x_i, y_i), $i = 0, 1, 2, 3, \ldots, n$.

Let $f(x)$ take the values $y_0, y_1, y_2, \ldots y_n$; for the independent variable x taking values $x_0, x_1, x_2, \ldots x_n$, where height of the interval (h) is fixed, such that $x_1 = x_0 + h$, $x_2 = x_0 + 2h, \ldots, x_n = x_0 + nh$.

Then to evaluate $f(x)$ for $x = x_n + ph$, $-1 < p < 1$

We have $f(x) = f(x_n + ph) = E^p f(x_n) \equiv (1 - \nabla)^{-p} y_n$

$\because E \equiv (1 - \nabla)^{-1}$ and $f(x_n) = y_n$

$$\Rightarrow f(x) \equiv \left(1 + p\nabla + \frac{p(p+1)}{2!}\nabla^2 + \frac{p(p+1)(p+2)}{3!}\nabla^3 + \cdots \right) y_n, \quad x = x_n + ph$$

$$\therefore f(x) \equiv y_n + p\nabla y_n + \frac{p(p+1)}{2!}\nabla^2 y_n + \frac{p(p+1)(p+2)}{3!}\nabla^3 y_n + \cdots, \quad p = \frac{x - x_n}{h}$$

This is Newton's backward interpolation formula and is used to interpolate or extrapolate values near the end of the table.

> p is the index which is 0, where $x = x_n$

Example4 Following table gives the census population of a state for the years 1971 to 2011. Estimate the population for the years 1974 and 2005 by using appropriate interpolation technique.

Year	1971	1981	1991	2001	2011
Population (Million)	46	66	81	93	101

Solution: Function has to be evaluated near the beginning and also near the end of the table, thereby constructing difference table for the function $y = f(x)$

Year x	Population $f(x)$	1st diff	2nd diff	3rd diff	4th diff
1971	46				
		20			
1981	66		−5		
		15		2	
1991	81		−3		−3
		12		−1	
2001	93		−4		
		8			
2011	101				

To calculate the population for the year 1974, using Newton's forward interpolation formula given by: $f(x) \equiv y_0 + p\Delta y_0 + \frac{p(p-1)}{2!}\Delta^2 y_0 + \frac{p(p-1)(p-2)}{3!}\Delta^3 y_0 + \frac{p(p-1)(p-2)(p-3)}{4!}\Delta^4 y_0 \ldots$

$$p = \frac{x - x_0}{h} \quad \ldots ①$$

$x_0 = 1971, y_0 = 46, x = 1974, h = 10 \therefore p = \frac{1974 - 1971}{10} = 0.3$

Also from table $\Delta y_0 = 20, \Delta^2 y_0 = -5, \Delta^3 y_0 = 2, \Delta^4 y_0 = -3$

Substituting these values in ①, we get

$f(1974) \equiv 46 + (0.3)(20) + \frac{0.3(0.3-1)}{2}(-5) + \frac{0.3(0.3-1)(0.3-2)}{6}(2) + \frac{0.3(0.3-1)(0.3-2)(0.3-3)}{24}(-3)$

$\equiv 46 + 6 + 0.525 + 0.119 + 0.1205 = 52.7645$ Million

To calculate the population for the year 2005, using Newton's backward interpolation formula given by: $f(x) \equiv y_n + p\nabla y_n + \frac{p(p+1)}{2!}\nabla^2 y_n + \frac{p(p+1)(p+2)}{3!}\nabla^3 y_n + \frac{p(p+1)(p+2)(p+3)}{4!}\nabla^4 y_n \ldots$

$$p = \frac{x - x_n}{h} \quad \ldots ②$$

$x_n = 2011, y_n = 101, x = 2005, h = 10 \therefore p = \frac{2005 - 2011}{10} = -0.6$

Also from table $\Delta y_n = 8, \Delta^2 y_n = -4, \Delta^3 y_n = -1, \Delta^4 y_n = -3$

Substituting these values in ②, we get

$f(2005) \equiv 101 + (-0.6)(8) + \frac{-0.6(-0.6+1)}{2}(-4) + \frac{-0.6(-0.6+1)(-0.6+2)}{6}(-1) +$

$\frac{-0.6(-0.6+1)(-0.6+2)(-0.6+3)}{24}(-3)$

$$\equiv 101 - 4.8 + 0.48 + 0.056 + 0.1008 = 96.837 \text{ Million}$$

Example5 Given a set of points for the function $f(x)$, evaluate $f(2.8)$ and $f(3.5)$.

x	0	1	2	3
$f(x)$	1	2	1	10

Solution: Function has to be evaluated near the end of the table, thereby constructing backward difference table for the function $y = f(x)$

x	$f(x)$	∇	∇^2	∇^3
0	1			
		1		
1	2		−2	
		−1		12
2	1		10	
		9		
3	10			

Newton's backward interpolation formula given by:

$$f(x) \equiv y_n + p\nabla y_n + \frac{p(p+1)}{2!}\nabla^2 y_n + \frac{p(p+1)(p+2)}{3!}\nabla^3 y_n + \cdots, p = \frac{x-x_n}{h} \quad \text{①}$$

i. To find $f(2.8)$

$x_n = 3, y_n = 10, x = 2.8, h = 1 \therefore p = \frac{2.8-3}{1} = -0.2$

Also from table $\Delta y_n = 9, \Delta^2 y_n = 10, \Delta^3 y_n = 12$

Substituting these values in ①, we get

$$f(2.8) \equiv 10 + (-.2)(9) + \frac{-.2(-.2+1)}{2}(10) + \frac{-.2(-.2+1)(-.2+2)}{6}(12) = 6.824$$

ii. To find $f(3.5)$

$x_n = 3, y_n = 10, x = 3.5, h = 1 \therefore p = \frac{3.5-3}{1} = 0.5$

Also from table $\Delta y_n = 9, \Delta^2 y_n = 10, \Delta^3 y_n = 12$

Substituting these values in ①, we get

$$f(3.5) \equiv 10 + (0.5)(9) + \frac{.5(.5+1)}{2}(10) + \frac{.5(.5+1)(.5+2)}{6}(12) = 22$$

Example6 For the given set of values, evaluate $\cos 22°$ and $\cos 73°$, using suitable interpolation techniques.

x	10°	20°	30°	40°	50°	60°	70°	80°
$\cos x$	0.9848	0.9397	0.8660	0.7660	0.6428	0.5000	0.3420	0.1737

Solution: Function has to be evaluated near the beginning and also near the end of the table, thereby constructing difference table for the function $y = f(x)$.

x	$y = \cos x$	1st diff	2nd diff	3rd diff	4th diff	5th diff
10°	0.9848					
		-0.0451				
20°	0.9397		-0.0286			
		-0.0737		0.0023		
30°	0.8660		-0.0263		0.0008	
		-0.1000		0.0031		-0.0003
40°	0.7660		-0.0232		0.0005	
		-0.1232		0.0036		0.0003
50°	0.6428		-0.0196		0.0008	
		-0.1428		0.0044		-0.0003
60°	0.5000		-0.0152		0.0005	
		-0.1580		0.0049		
70°	0.3420		-0.0103			
		-0.1683				
80°	0.1737					

To calculate the value of $\cos 22°$, taking $x_0 = 20°$ as $22°$ is nearest to this point and applying Newton's forward interpolation formula given by:

$$f(x) \equiv y_0 + p\Delta y_0 + \frac{p(p-1)}{2!}\Delta^2 y_0 + \frac{p(p-1)(p-2)}{3!}\Delta^3 y_0 + \frac{p(p-1)(p-2)(p-3)}{4!}\Delta^4 y_0 + \cdots$$

$$p = \frac{x - x_0}{h} \quad \ldots \text{①}$$

$x_0 = 20°$, $y_0 = 0.9397$, $x = 22°$, $h = 10°$ $\therefore p = \frac{22° - 20°}{10°} = 0.2$

Also from table $\Delta y_0 = -0.0737$, $\Delta^2 y_0 = -0.0263$, $\Delta^3 y_0 = 0.0031$,
$\Delta^4 y_0 = 0.0005$, $\Delta^5 y_0 = 0.0003$

Substituting these values in ①, we get

$$f(22°) \equiv 0.9397 + (0.2)(-0.0737) + \frac{0.2(0.2-1)}{2}(-0.0263) + \frac{0.2(0.2-1)(0.2-2)}{6}(0.0031) +$$
$$\frac{0.2(0.2-1)(0.2-2)(0.2-3)}{24}(0.0005) + \frac{0.2(0.2-1)(0.2-2)(0.2-3)(0.2-4)}{120}(0.00003)$$

$$\equiv 0.9397 - 0.0147 + 0.0021 + 0.0002 - 0.00002 + 0.000001 = 0.9272$$

To calculate the value of $\cos 73°$, taking $x_n = 70°$ as $73°$ is nearest to this point and applying Newton's backward interpolation formula given by:

$$f(x) \equiv y_n + p\nabla y_n + \frac{p(p+1)}{2!}\nabla^2 y_n + \frac{p(p+1)(p+2)}{3!}\nabla^3 y_n + \frac{p(p+1)(p+2)(p+3)}{4!}\nabla^4 y_n \ldots$$

$$p = \frac{x - x_n}{h} \quad \ldots \text{②}$$

$x_n = 70°$, $y_n = 0.3420$, $x = 73°$, $h = 10°$ $\therefore p = \frac{73° - 70°}{10°} = 0.3$

Also from table $\Delta y_n = -0.158$, $\Delta^2 y_n = -.0152$, $\Delta^3 y_n = 0.0044$,
$\Delta^4 y_n = 0.0008$, $\Delta^5 y_n = 0.0003$

Substituting these values in ②, we get

$$f(73°) \equiv 0.3420 + (0.3)(-0.158) + \frac{0.3(0.3+1)}{2}(-.0152) + \frac{0.3(0.3+1)(0.3+2)}{6}(0.0044) +$$

$$\frac{0.3(0.3+1)(0.3+2)(0.3+3)}{24}(0.0008) + \frac{0.3(0.3+1)(0.3+2)(0.3+3)(0.3+4)}{120}(0.0003)$$

$$\equiv 0.3420 - 0.0474 - 0.003 + 0.0007 + 0.0001 + 0.00003 = 0.2924$$

11.2.3 Gauss's Forward and Backward Difference Formulae

Gauss central difference formula is used to interpolate the values of y near the middle of the table.

Newton's forward difference formula is given by:

$$f(x) \equiv y_0 + p\Delta y_0 + \frac{p(p-1)}{2!}\Delta^2 y_0 + \frac{p(p-1)(p-2)}{3!}\Delta^3 y_0 + \frac{p(p-1)(p-2)(p-3)}{4!}\Delta^4 y_0 + \cdots$$

$$p = \frac{x - x_0}{h} \quad \ldots ①$$

Now $\Delta^3 y_{-1} = \Delta^2 y_0 - \Delta^2 y_{-1}$

$\Rightarrow \Delta^2 y_0 = \Delta^2 y_{-1} + \Delta^3 y_{-1}$ ... ②

Similarly $\Delta^3 y_0 = \Delta^3 y_{-1} + \Delta^4 y_{-1}$ ... ③

$\Delta^4 y_0 = \Delta^4 y_{-1} + \Delta^5 y_{-1}$ ... ④

Substituting $\Delta^2 y_0, \Delta^3 y_0, \Delta^4 y_0$ from ② ③ ④ in ①, we get

$$f(x) \equiv y_0 + p\Delta y_0 + \frac{p(p-1)}{2!}(\Delta^2 y_{-1} + \Delta^3 y_{-1}) + \frac{p(p-1)(p-2)}{3!}(\Delta^3 y_{-1} + \Delta^4 y_{-1}) +$$

$$\frac{p(p-1)(p-2)(p-3)}{4!}(\Delta^4 y_{-1} + \Delta^5 y_{-1}) + \cdots$$

Rewriting by collecting the coefficients of $\Delta^2 y_{-1}, \Delta^3 y_{-1}, \Delta^4 y_{-1}$ ..., we get

$$f(x) \equiv y_0 + p\Delta y_0 + \frac{p(p-1)}{2!}\Delta^2 y_{-1} + \frac{(p+1)p(p-1)}{3!}\Delta^3 y_{-1} +$$

$$\frac{(p+1)p(p-1)(p-2)}{4!}\Delta^4 y_{-1} + \cdots \quad \ldots ⑤$$

Again $\Delta^4 y_{-1} = \Delta^4 y_{-2} + \Delta^5 y_{-2}$ ... ⑥

Using ⑥ in ⑤, we get

$$f(x) \equiv y_0 + p\Delta y_0 + \frac{p(p-1)}{2!}\Delta^2 y_{-1} + \frac{(p+1)p(p-1)}{3!}\Delta^3 y_{-1} +$$

$$\frac{(p+1)p(p-1)(p-2)}{4!}\Delta^4 y_{-2} + \cdots \quad \ldots ⑦$$

Expression given by ⑦ is known as **Gauss forward interpolation formula**

Again $\Delta^2 y_{-1} = \Delta y_0 - \Delta y_{-1}$

$\Rightarrow \Delta y_0 = \Delta y_{-1} + \Delta^2 y_{-1}$ ... ②'

Similarly $\Delta^2 y_0 = \Delta^2 y_{-1} + \Delta^3 y_{-1}$ ... ③'

$\Delta^3 y_0 = \Delta^3 y_{-1} + \Delta^4 y_{-1}$ ... ④'

Substituting $\Delta y_0, \Delta^2 y_0, \Delta^3 y_0$ from ②′, ③′, ④′ in ①, we get

$$f(x) \equiv y_0 + p(\Delta y_{-1} + \Delta^2 y_{-1}) + \frac{p(p-1)}{2!}(\Delta^2 y_{-1} + \Delta^3 y_{-1}) +$$

$$\frac{p(p-1)(p-2)}{3!}(\Delta^3 y_{-1} + \Delta^4 y_{-1}) + \frac{p(p-1)(p-2)(p-3)}{4!}(\Delta^4 y_{-1} + \Delta^5 y_{-1}) + \cdots$$

Rewriting by collecting the coefficients of $\Delta y_{-1}, \Delta^2 y_{-1}, \Delta^3 y_{-1}, \Delta^4 y_{-1}$, we get

$$f(x) \equiv y_0 + p\Delta y_{-1} + \frac{(p+1)p}{2!}\Delta^2 y_{-1} + \frac{(p+1)p(p-1)}{3!}\Delta^3 y_{-1} +$$

$$\frac{(p+1)p(p-1)(p-2)}{4!}\Delta^4 y_{-1} + \frac{p(p-1)(p-2)(p-3)}{4!}\Delta^5 y_{-1} + \cdots \quad \ldots ⑤'$$

Again $\Delta^3 y_{-1} = \Delta^3 y_{-2} + \Delta^4 y_{-2}$ and $\Delta^4 y_{-1} = \Delta^4 y_{-2} + \Delta^5 y_{-2}$ ... ⑥′

Using ⑥′ in ⑤′, we get

$$f(x) \equiv y_0 + p\Delta y_{-1} + \frac{(p+1)p}{2!}\Delta^2 y_{-1} + \frac{(p+1)p(p-1)}{3!}(\Delta^3 y_{-2} + \Delta^4 y_{-2}) +$$

$$\frac{(p+1)p(p-1)(p-2)}{4!}(\Delta^4 y_{-2} + \Delta^5 y_{-2}) + \cdots$$

$$\Rightarrow f(x) \equiv y_0 + p\Delta y_{-1} + \frac{(p+1)p}{2!}\Delta^2 y_{-1} + \frac{(p+1)p(p-1)}{3!}\Delta^3 y_{-2} +$$

$$\frac{(p+2)(p+1)p(p-1)}{4!}\Delta^4 y_{-2} + \cdots \quad \ldots ⑦'$$

Expression given by ⑦′ is known as **Gauss backward interpolation formula**

Example 7 Given a set of points for the function $y = f(x)$, evaluate $f(33)$ using
i. Gauss's forward ii. Gauss's backward interpolation formulae

x	25	30	35	40
$f(x)$	0.25	0.3	0.33	0.37

Solution: Function has to be evaluated near centre of the table, thereby constructing difference table for the function $y = f(x)$, taking $x_0 = 30$

x	p	$f(x)$	Δ	Δ^2	Δ^3
25	-1	0.25			
			0.05		
30	0	0.3		-0.02	
			0.03		0.03
35	1	0.33		0.01	
			0.04		
40	2	0.37			

i. Gauss forward interpolation formula is given by

$$f(x) \equiv y_0 + p\Delta y_0 + \frac{p(p-1)}{2!}\Delta^2 y_{-1} + \frac{(p+1)p(p-1)}{3!}\Delta^3 y_{-1} + \cdots , \quad p = \frac{x - x_0}{h} \quad \ldots ①$$

To find $f(33)$, $x_0 = 30$, $y_0 = 0.3$, $x = 33$, $h = 5$ $\therefore p = \frac{33-30}{5} = 0.6$

∴ From table $\Delta y_0 = 0.03, \Delta^2 y_{-1} = -0.02, \Delta^3 y_{-1} = 0.03$

Substituting these values in ①, we get

$$f(33) \equiv 0.3 + (0.6)(0.03) + \frac{0.6(0.6-1)}{2}(-0.02) + \frac{(0.6+1)0.6(0.6-1)}{6}(0.03)$$

$\Rightarrow f(33) \equiv 0.3185$ using Gauss forward interpolation formula

ii. Gauss backward interpolation formula is given by

$$f(x) \equiv y_0 + p\Delta y_{-1} + \frac{(p+1)p}{2!}\Delta^2 y_{-1} + \frac{(p+1)p(p-1)}{3!}\Delta^3 y_{-2}, \quad p = \frac{x-x_0}{h} \quad \ldots ②$$

To find $f(33)$, $x_0 = 30$, $y_0 = 0.3$, $x = 33$, $h = 5$ ∴ $p = \frac{33-30}{5} = 0.6$

∴ From table $\Delta y_{-1} = 0.05, \Delta^2 y_{-1} = -0.02, \Delta^3 y_{-2} = 0$

Substituting these values in ②, we get

$$f(33) \equiv 0.3 + (0.6)(0.05) + \frac{0.6(0.6-1)}{2}(-0.02) + \frac{(0.6+1)0.6(0.6-1)}{6}(0) \equiv 0.3324$$

$\Rightarrow f(33) \equiv 0.3324$ using Gauss backward interpolation formula

11.2.4 Stirling's Central Difference Formula

Stirling gave the most general formula for interpolating values near the centre of the table by taking mean of Gauss forward and Gauss backward interpolation formulae.

Taking mean of expressions given by ⑦ and ⑦' respectively, we get

$$f(x) \equiv y_0 + p\left(\frac{\Delta y_0 + \Delta y_{-1}}{2}\right) + \left(\frac{p(p-1)}{2!} + \frac{(p+1)p}{2!}\right)\frac{\Delta^2 y_{-1}}{2} + \frac{(p+1)p(p-1)}{3!}\left(\frac{\Delta^3 y_{-1} + \Delta^3 y_{-2}}{2}\right) +$$

$$\left(\frac{(p+1)p(p-1)(p-2)}{4!} + \frac{(p+2)(p+1)p(p-1)}{4!}\right)\frac{\Delta^2 y_{-2}}{2} + \cdots$$

$$\Rightarrow f(x) \equiv y_0 + p\left(\frac{\Delta y_0 + \Delta y_{-1}}{2}\right) + \frac{p^2}{2!}\Delta^2 y_{-1} + \frac{p(p^2-1)}{3!}\left(\frac{\Delta^3 y_{-1} + \Delta^3 y_{-2}}{2}\right)$$

$$+ \frac{p^2(p^2-1)}{4!}\Delta^4 y_{-2} + \cdots \quad \ldots ⑧$$

Expression given in ⑧ is known as **Stirling's central difference formula**

Putting $\frac{1}{2}(\Delta y_0 + \Delta y_{-1}) = \frac{1}{2}\left(\delta y_{\frac{1}{2}} + \delta y_{-\frac{1}{2}}\right) = \mu\delta y_0$

$\frac{1}{2}(\Delta^3 y_{-1} + \Delta^3 y_{-2}) = \frac{1}{2}\left(\delta^3 y_{\frac{1}{2}} + \delta^3 y_{-\frac{1}{2}}\right) = \mu\delta^3 y_0$

⋮

In terms of central differences, ⑧ takes the form

$$f(x) \equiv y_0 + p\mu\delta y_0 + \frac{p^2}{2!}\delta^2 y_0 + \frac{p(p^2-1^2)}{3!}\mu\delta^3 y_0 + \frac{p^2(p^2-1^2)}{4!}\delta^4 y_0 + \cdots$$

This is another form of **Stirling's central difference formula.**

The difference table used to evaluate $f(x)$ as per Stirling's formula, is shown below. Column wise averages of differences (shown in boxes) are taken while evaluation phase.

x	p	$f(x)$	1st diff	2nd diff	3rd diff	4th diff
x_{-2}	-2	y_{-2}				
			Δy_{-2}			
x_{-1}	-1	y_{-1}				
			$\boxed{\Delta y_{-1}}$		$\boxed{\Delta^3 y_{-2}}$	
$\underline{x_0}$	0	$\underline{y_0}$		$\boxed{\Delta^2 y_{-1}}$		$\boxed{\Delta^4 y_{-2}}$
			$\boxed{\Delta y_0}$		$\boxed{\Delta^3 y_{-1}}$	
x_1	1	y_1		$\Delta^2 y_0$		
			Δy_1			
x_2	2	y_2				

Example 8 Given a set of points for the function $y = f(x)$

x	25	30	35	40	45
$f(x)$	0.25	0.3	0.33	0.37	0.43

Evaluate $f(33)$ using Stirling's central difference formula.

Solution: Function has to be evaluated near centre of the table, thereby constructing difference table for the function $y = f(x)$, taking $x_0 = 35$.

Also $\Delta^n y_0$, $n = 1, 2$ lie along the dotted line as shown.

x	p	$f(x)$	Δ	Δ^2	Δ^3	Δ^4
25	-2	0.25				
			0.05			
30	-1	0.3		-0.02		
			$\boxed{0.03}$		$\boxed{0.03}$	
35	0	<u>0.33</u>		$\boxed{0.01}$		$\boxed{-0.02}$
			$\boxed{0.04}$		$\boxed{0.01}$	
40	1	0.37		0.02		
			0.06			
45	2	0.43				

Stirling's central differences formula is given by

$$f(x) \equiv y_0 + p\left(\frac{\Delta y_0 + \Delta y_{-1}}{2}\right) + \frac{p^2}{2!}\Delta^2 y_{-1} + \frac{p(p^2-1)}{3!}\left(\frac{\Delta^3 y_{-1} + \Delta^3 y_{-2}}{2}\right) +$$

$$\frac{p^2(p^2-1)}{4!}\Delta^4 y_{-2} + \cdots, \quad p = \frac{x - x_0}{h} \quad \ldots ①$$

To find $f(33)$, $x_0 = 35$, $y_0 = 0.33$, $x = 33$, $h = 5$ $\therefore p = \frac{33-35}{5} = -0.4$

Also from the table $\Delta y_0 = 0.04$, $\Delta y_{-1} = 0.03$ $\Delta^2 y_{-1} = 0.01$, $\Delta^3 y_{-1} = 0.01$, $\Delta^3 y_{-2} = 0.03$, $\Delta^4 y_{-2} = -0.02$. All the positions have been shown, enclosed in boxes.

Substituting these values in ①, we get

$$f(33) \equiv .33 + (-.4)\left(\frac{.04+.03}{2}\right) + \frac{(-.4)^2}{2}(.01) + \frac{-.4((-.4)^2-1)}{6}\left(\frac{.01+.03}{2}\right)$$
$$+ \frac{(-.4)^2((-.4)^2-1)}{24}(-0.02)$$

$\Rightarrow f(33) \equiv 0.33 - 0.014 + 0.0008 + 0.0011 + 0.0001 = 0.318$

Example 9 Use Stirling's formula to evaluate $f(1.22)$ given that

x	1.0	1.1	1.2	1.3	1.4
$f(x)$	0.841	0.891	0.932	0.963	0.985

Solution: Function has to be evaluated near centre of the table, thereby constructing difference table for the function $y = f(x)$, taking $x_0 = 1.2$.

Also $\Delta^n y_0$, $n = 1, 2$ lie along the dotted line as shown.

x	p	$f(x)$	Δ	Δ^2	Δ^3	Δ^4
1.0	−2	0.841				
			0.050			
1.1	−1	0.891		−0.009		
			$\boxed{0.041}$		$\boxed{-0.001}$	
1.2	0	0.932		$\boxed{-0.01}$		$\boxed{0.002}$
			$\boxed{0.031}$		$\boxed{0.001}$	
1.3	1	0.963		−0.009		
			0.022			
1.4	2	0.985				

Stirling's central differences formula is given by

$$f(x) \equiv y_0 + p\left(\frac{\Delta y_0 + \Delta y_{-1}}{2}\right) + \frac{p^2}{2!}\Delta^2 y_{-1} + \frac{p(p^2-1)}{3!}\left(\frac{\Delta^3 y_{-1} + \Delta^3 y_{-2}}{2}\right)$$
$$+ \frac{p^2(p^2-1)}{4!}\Delta^4 y_{-2} + \cdots \quad , p = \frac{x-x_0}{h} \quad \ldots ①$$

To find $f(1.22)$, $x_0 = 1.2$, $y_0 = 0.932$, $x = 1.22$, $h = 0.1$ $\therefore p = \frac{1.22-1.2}{0.1} = 0.2$

Also from the table $\Delta y_0 = 0.031$, $\Delta y_{-1} = 0.041$, $\Delta^2 y_{-1} = -0.01$, $\Delta^3 y_{-1} = 0.001$, $\Delta^3 y_{-2} = -0.001$, $\Delta^4 y_{-2} = 0.002$. All the positions have been shown, enclosed in boxes.

Substituting these values in ①, we get

$$f(1.22) \equiv .932 + (0.2)\left(\frac{.031+.041}{2}\right) + \frac{(.2)^2}{2}(-.01) + \frac{.2((.2)^2-1)}{6}\left(\frac{.001-.001}{2}\right)$$
$$+ \frac{(0.2)^2((0.2)^2-1)}{24}(0.002)$$

$$\Rightarrow f(1.22) \equiv 0.932 + 0.0072 - 0.0002 + 0 - 0.0000032 = 0.9390$$

Example 10 Given following data $f(x) = 10^5 u_x$, for values of x in degrees, where $u_x = 1 + \log \sin x$. Use Stirling's formula to compute $u_{11.8°}$.

$x°$	10	11	12	13	14
$f(x) = 10^5 u_x$	23,967	28,060	31,788	35,209	38,368

Solution: Function has to be evaluated near centre of the table, thereby constructing difference table for the function $y = f(x)$, taking $x_0 = 12$. Also $\Delta^n y_0$, $n = 1, 2$ lie along the dotted line as shown.

$x°$	p	$f(x) = 10^5 u_x$	Δ	Δ^2	Δ^3	Δ^4
10	-2	23967				
			4093			
11	-1	28060		-365		
			3728		58	
12	0	31788		-307		-13
			3421		45	
13	1	35209		-262		
			3159			
14	2	38368				

Stirling's central differences formula is given by

$$f(x) \equiv y_0 + p\left(\frac{\Delta y_0 + \Delta y_{-1}}{2}\right) + \frac{p^2}{2!}\Delta^2 y_{-1} + \frac{p(p^2-1)}{3!}\left(\frac{\Delta^3 y_{-1} + \Delta^3 y_{-2}}{2}\right)$$
$$+ \frac{p^2(p^2-1)}{4!}\Delta^4 y_{-2} + \cdots \quad , \quad p = \frac{x-x_0}{h} \quad \ldots ①$$

To find $f(11.8°)$

$x_0 = 12$, $y_0 = 31788$, $x = 11.8$, $h = 1$, $\therefore p = \frac{11.8-12}{1} = -0.2$

Also from the table $\Delta y_0 = 3421$, $\Delta y_{-1} = 3728$, $\Delta^2 y_{-1} = -307$, $\Delta^3 y_{-1} = 45$, $\Delta^3 y_{-2} = 58$, $\Delta^4 y_{-2} = -13$. All the positions have been shown, enclosed in boxes.

Substituting these values in ①, we get

$$f(11.8°) \equiv 31788 + (-0.2)\left(\frac{3421+3728}{2}\right) + \frac{(-0.2)^2}{2}(-307) +$$

$$\frac{-0.2((-0.2)^2-1)}{6}\left(\frac{45+58}{2}\right) + \frac{(-0.2)^2((-0.2)^2-1)}{24}(-13)$$

$\Rightarrow f(11.8°) \equiv 31788 + -714.9 - 6.14 + 1.648 + 0.0208 = 31068.6288$

$\therefore f(11.8°) = 10^5 u_{11.8°} \equiv 31068.6288$

$\Rightarrow u_{11.8°} \equiv 31068.6288(10^{-5}) = 0.31069$

Example 11 Use Stirling's formula to evaluate $\sin 57°$, given that $\sin 45° = .7071$, $\sin 50° = .7660$, $\sin 55° = .8192$, $\sin 60° = .8660$, $\sin 65° = .9063$

Also compare the results by evaluating $\sin 57°$ using Newton's forward interpolation formula.

Solution: Function has to be evaluated near centre of the table, thereby constructing difference table for the function $y = \sin x$, taking $x_0 = 55$. Also $\Delta^n y_0$, $n = 1, 2$ lie along the dotted line as shown.

x	p	$f(x)$	Δ	Δ^2	Δ^3	Δ^4
45°	−2	0.7071				
			0.0589			
50°	−1	0.7660		−0.0057		
			0.0532		−0.0007	
55°	0	0.8192		−0.0064		0.0006
			0.0468		−0.0001	
60°	1	0.8660		−0.0065		
			0.0403			
65°	2	0.9063				

Stirling's central differences formula is given by

$$f(x) \equiv y_0 + p\left(\frac{\Delta y_0 + \Delta y_{-1}}{2}\right) + \frac{p^2}{2!}\Delta^2 y_{-1} + \frac{p(p^2-1)}{3!}\left(\frac{\Delta^3 y_{-1} + \Delta^3 y_{-2}}{2}\right)$$

$$+ \frac{p^2(p^2-1)}{4!}\Delta^4 y_{-2} + \cdots \quad , p = \frac{x - x_0}{h} \quad \ldots \text{①}$$

To find $f(57°)$, $x_0 = 55°$, $y_0 = 0.8192$, $x = 57°$, $h = 5°$ $\therefore p = \frac{57°-55°}{5°} = 0.4$

Also from the table $\Delta y_0 = 0.0468$, $\Delta y_{-1} = 0.0532$, $\Delta^2 y_{-1} = -0.0064$,

$\Delta^3 y_{-1} = -0.0001$, $\Delta^3 y_{-2} = -0.0007$, $\Delta^4 y_{-2} = 0.0006$. All the positions have been shown, enclosed in boxes. Substituting these values in ①, we get

$$f(57°) \equiv 0.8192 + (0.4)\left(\frac{0.0468+0.0532}{2}\right) + \frac{(.4)^2}{2}(-0.0064) +$$

$$\frac{0.4((0.4)^2-1)}{6}\left(\frac{-0.0001-0.0007}{2}\right) + \frac{(0.4)^2((0.4)^2-1)}{24}(0.0006)$$

$\Rightarrow f(57°) \equiv 0.8192 + 0.02 - 0.000512 + 0.0000224 - 0.00000336$

$\therefore f(57°) \equiv 0.83870704$ using Stirling's central differences formula.

To evaluate the value of sin 57°, taking $x_0 = 55°$ as 57° is nearest to this point and applying Newton's forward interpolation formula given by:

$$f(x) \equiv y_0 + p\Delta y_0 + \frac{p(p-1)}{2!}\Delta^2 y_0 + \frac{p(p-1)(p-2)}{3!}\Delta^3 y_0 + \cdots, \quad p = \frac{x - x_0}{h} \quad \ldots ②$$

$x_0 = 55°$, $y_0 = 0.8192$, $x = 57°$, $h = 10°$ $\therefore p = \frac{57° - 55°}{5°} = 0.4$

Also from table $\Delta y_0 = 0.0468$, $\Delta^2 y_0 = -0.0065$, $\Delta^3 y_0 = 0, \ldots$

Substituting these values in ②, we get

$$f(57°) \equiv 0.8192 + (0.4)(0.0468) + \frac{0.4(0.4-1)}{2}(-.0065) + 0 = 0.8387$$

11.3 Interpolation with Unequal Intervals

For the function $y = f(x)$, with given discrete set of points (x_i, y_i), $i = 0, 1, 2, 3, \ldots, n$, where x_i's are not equispaced, common methods of interpolating data points are listed below:
  i. Newton's divided difference formula
  ii. Lagrange's interpolation formula

11.3.1 Newton's Divided Difference Method

Let the function $y = f(x)$ take the values $f(x_0) = y_0, f(x_1) = y_1, f(x_2) = y_2, \ldots, f(x_n) = y_n$; for the argument x taking values $x_0, x_1, x_2, \ldots, x_n$, which are not equally spaced. Divided difference may be defined as the difference between two successive values of the ordinates divided by the difference between the corresponding values of the abscissa.

So the first divided difference denoted by $[x_0, x_1]$ or $f(x_0, x_1)$ is defined as:

$$f(x_0, x_1) \equiv [x_0, x_1] = \frac{y_1 - y_0}{x_1 - x_0} = \frac{f(x_1) - f(x_0)}{x_1 - x_0}$$

Similarly $f(x_1, x_2) \equiv [x_1, x_2] = \frac{y_2 - y_1}{x_2 - x_1} = \frac{f(x_2) - f(x_1)}{x_2 - x_1}$

$\vdots$

Second divided differences, denoted by $[x_0, x_1, x_2]$ or $f(x_0, x_1, x_2)$ are defined as:

$$f(x_0, x_1, x_2) \equiv [x_0, x_1, x_2] = \frac{[x_1, x_2] - [x_0, x_1]}{x_2 - x_0}$$

$$f(x_1, x_2, x_3) \equiv [x_1, x_2, x_3] = \frac{[x_2, x_3] - [x_1, x_2]}{x_3 - x_1}$$

$\vdots$

Third divided differences, denoted by $[x_0, x_1, x_2]$ or $f(x_0, x_1, x_2)$ are defined as:

$$f(x_0, x_1, x_2, x_3) \equiv [x_0, x_1, x_2, x_3] = \frac{[x_1, x_2, x_3] - [x_0, x_1, x_2]}{x_3 - x_0} \quad \text{and so on.}$$

Remark: Divided differences are symmetrical in arguments,

i.e. $[x_r, x_s] = [x_s, x_r]$ or $f(x_r, x_s) = f(x_s, x_r)$

11.3.1.1 Newton's Divided Difference Formula

By definition of divided differences $f(x, x_0) = \frac{f(x_0) - f(x)}{x_0 - x} = \frac{f(x) - f(x_0)}{x - x_0}$

$$\Rightarrow f(x) = f(x_0) + (x - x_o)f(x, x_0) \quad \ldots ①$$

Also $f(x, x_0, x_1) = \frac{f(x_0, x_1) - f(x, x_0)}{x_1 - x} = \frac{f(x, x_0) - f(x_0, x_1)}{x - x_1}$

$$\Rightarrow f(x, x_0) = f(x_0, x_1) + (x - x_1)f(x, x_0, x_1) \quad \ldots ②$$

Similarly $f(x, x_0, x_1) = f(x_0, x_1, x_2) + (x - x_2)f(x, x_0, x_1, x_2) \quad \ldots ③$

$\vdots$

$f(x, x_0, x_1, \ldots x_n) = f(x_0, x_1, x_2, \ldots x_n) + (x - x_n)f(x, x_0, x_1, x_2, \ldots x_n) \quad \ldots ④$

Multiplying ② by $(x - x_0)$, ③ by $(x - x_0)(x - x_1)$, ...

④ by $(x - x_0)(x - x_1) \ldots (x - x_{n-1})$ and adding to equation ①, we get

$f(x) \equiv f(x_0) + (x - x_0)f(x_0, x_1) + (x - x_0)(x - x_1)f(x_0, x_1, x_2) + \cdots$
$\quad + (x - x_0)(x - x_1) \ldots (x - x_{n-1})(x - x_n)f(x_0, x_1, x_2, \ldots x_n) + R^n$

Where R^n denotes remainder terms which vanish being $(n + 1)^{th}$ order divided differences.

∴ Newton's divided difference formula is given by:

$f(x) \equiv f(x_0) + (x - x_0)f(x_0, x_1) + (x - x_o)(x - x_1)f(x_0, x_1, x_2) +$
$(x - x_0)(x - x_1)(x - x_2)f(x_0, x_1, x_2, x_3) + \cdots$

Example 12 Estimate $f(2)$ from the following data, using Newton's divided differences method.

x	0	1	3	6
y	1	3	55	343

Solution: The divided difference table is given as follows:

x	y	1st diff	2nd diff	3rd diff
0	1			
		$\frac{3-1}{1-0} = 2$		
1	3		$\frac{26-2}{3-0} = 8$	
		$\frac{55-3}{3-1} = 26$		$\frac{14-8}{6-0} = 1$
3	55		$\frac{96-26}{6-1} = 14$	
		$\frac{343-55}{6-3} = 96$		
6	343			

Newton's divided difference formula is given by:

$f(x) \equiv f(x_0) + (x - x_0)f(x_0, x_1) + (x - x_o)(x - x_1)f(x_0, x_1, x_2) +$

$$(x-x_0)(x-x_1)(x-x_2)f(x_0, x_1, x_2, x_3) + \cdots \quad \ldots \text{①}$$

Here $x=2$, $x_0=0$, $x_1=1$, $x_2=3$, $f(x_0)=1$, $f(x_0, x_1)=2$, $f(x_0, x_1, x_2)=8$, $f(x_0, x_1, x_2, x_3)=1$. Substituting these values in ①, we get

$$f(2) \equiv 1 + (2-0)(2) + (2-0)(2-1)(8) + (2-0)(2-1)(2-3)(1) = 19$$

Example 13 Find $\log_{10} 656$, given that $\log_{10} 654 = 2.8156$, $\log_{10} 658 = 2.8182$, $\log_{10} 659 = 2.8189$, $\log_{10} 661 = 2.8202$.

Solution: The divided difference table is given as follows:

x	$f(x) = \log_{10} x$	1st diff	2nd diff	3rd diff
654	2.8156			
		$\dfrac{2.8182 - 2.8156}{658 - 654} = .00065$		
658	2.8182		$\dfrac{.0007 - .00065}{659 - 654} = .00001$	
		$\dfrac{2.8189 - 2.8182}{659 - 658} = .0007$		$\dfrac{-.00002 - .00001}{661 - 654} = -.000004$
659	2.8189		$\dfrac{.00065 - .0007}{661 - 658} = -.00002$	
		$\dfrac{2.8202 - 2.8189}{661 - 659} = .00065$		
661	2.8202			

Newton's divided difference formula is given by:

$$f(x) \equiv f(x_0) + (x-x_0)f(x_0, x_1) + (x-x_0)(x-x_1)f(x_0, x_1, x_2) +$$
$$(x-x_0)(x-x_1)(x-x_2)f(x_0, x_1, x_2, x_3) + \cdots \quad \ldots \text{①}$$

Here $x = 656$, $x_o = 654$, $x_1 = 658$, $x_2 = 659$, $f(x_0) = 2.8156$, $f(x_0, x_1) = .00065$, $f(x_0, x_1, x_2) = .00001$, $f(x_0, x_1, x_2, x_3) = -.000004$

Substituting these values in ①, we get

$$f(656) \equiv 2.8156 + (656 - 654)(.00065) + (656 - 654)(656 - 658)(0.00001)$$
$$+ (656 - 654)(656 - 658)(656 - 659)(-.00004)$$
$$\therefore f(656) = \log_{10} 656 = 2.8168$$

Example 13 Employing Newton's divided difference interpolation formula, estimate $f(x)$ from the following data and hence or otherwise find $f(2.5)$.

x	0	1	2	4	5	6
$f(x)$	1	14	15	5	6	9

Solution: Constructing divided difference table for the function $y = f(x)$

x	$f(x)$	1st diff	2nd diff	3rd diff	4th diff
0	1				
		$\dfrac{14-1}{1-0} = 13$			
1	14		$\dfrac{1-13}{2-0} = -6$		
		$\dfrac{15-14}{2-1} = 1$		$\dfrac{-2+6}{4-0} = 1$	
2	15		$\dfrac{-5-1}{4-1} = -2$		$\dfrac{1-1}{5-0} = 0$
		$\dfrac{5-15}{4-2} = -5$		$\dfrac{2+2}{5-1} = 1$	
4	5		$\dfrac{1+5}{5-2} = 2$		$\dfrac{1-1}{6-1} = 0$
		$\dfrac{6-5}{5-4} = 1$		$\dfrac{6-2}{6-2} = 1$	
5	6		$\dfrac{13-1}{6-4} = 6$		
		$\dfrac{19-6}{6-5} = 13$			
6	19				

Newton's divided difference formula is given by

$$f(x) \equiv f(x_0) + (x - x_0)f(x_0, x_1) + (x - x_0)(x - x_1)f(x_0, x_1, x_2) +$$
$$(x - x_0)(x - x_1)(x - x_2)f(x_0, x_1, x_2, x_3) + \cdots \quad \ldots \text{①}$$

Here $x_0 = 0$, $x_1 = 1$, $x_2 = 2$, $f(x_0) = 1$, $f(x_0, x_1) = 13$, $f(x_0, x_1, x_2) = -6$,

$f(x_0, x_1, x_2, x_3) = 1$, $f(x_0, x_1, x_2, x_3, x_4) = 0$

Substituting these values in ①, we get

$f(x) \equiv 1 + (x - 0)(13) + (x - 0)(x - 1)(-6) + (x - 0)(x - 1)(x - 2)(1) + 0$

$\therefore f(x) \equiv 1 + 13x - 6(x^2 - x) + (x^3 - 3x^2 + 2x)$

$\Rightarrow f(x) \equiv x^3 - 9x^2 + 21x + 1$

Also $f(2.5) = (5.5)^3 - 9(5.5)^2 + 21(5.5) + 1 = 12.875$

11.3.2 Lagrange's Interpolation Formula

Let $y = f(x)$ take the values $y_0, y_1, y_2,..., y_n$; for the argument x taking values $x_0, x_1, x_2, ..., x_n$, then the polynomial by Lagrange's interpolation formula is given by

$$f(x) = \sum_{i=0}^{n} L_i y_i = L_0 y_0 + L_1 y_1 + L_2 y_2 + \cdots + L_n y_n$$

where $L_0 = \dfrac{(x-x_1)(x-x_2)\ldots(x-x_n)}{(x_0-x_1)(x_0-x_2)\ldots(x_0-x_n)}$

$L_1 = \dfrac{(x-x_0)(x-x_2)\ldots(x-x_n)}{(x_1-x_0)(x_1-x_2)\ldots(x_1-x_n)}$

$\vdots$

$L_n = \dfrac{(x-x_0)(x-x_1)\ldots(x-x_{n-1})}{(x_n-x_0)(x_n-x_2)\ldots(x_n-x_{n-1})}$

$\therefore f(x) = \left(\dfrac{(x-x_1)(x-x_2)\ldots(x-x_n)}{(x_0-x_1)(x_0-x_2)\ldots(x_0-x_n)}\right) y_0 + \left(\dfrac{(x-x_0)(x-x_2)\ldots(x-x_n)}{(x_1-x_0)(x_1-x_2)\ldots(x_1-x_n)}\right) y_1 +$

$\cdots + \left(\dfrac{(x-x_0)(x-x_1)\ldots(x-x_{n-1})}{(x_n-x_0)(x_n-x_2)\ldots(x_n-x_{n-1})}\right) y_n$

Remarks:

- This formula can be used irrespective of whether the values $x_0, x_1, x_2, ..., x_n$ are equispaced or not.
- It is easy to remember but cumbersome to apply.

Example15 Find the polynomial of the lowest possible degree which assumes the values -21, 15, 12, 3 for x taking the values $-1, 1, 2, 3$ respectively, using Newton's divided difference formula and hence find $f(1.5)$. Compare the results by finding $f(1.5)$ using Lagrange's interpolation formula.

Solution: *i.* The divided difference table is given as follows:

x	y	1st diff	2nd diff	3rd diff
-1	-21			
		$\dfrac{15+21}{1+1} = 18$		
1	15		$\dfrac{-3-18}{2+1} = -7$	
		$\dfrac{12-15}{2-1} = -3$		$\dfrac{-3+7}{3+1} = 1$
2	12		$\dfrac{-9+3}{3-1} = -3$	
		$\dfrac{3-12}{3-2} = -9$		
3	3			

Newton's divided difference formula is given by:

$f(x) \equiv f(x_0) + (x-x_0)f(x_0, x_1) + (x-x_0)(x-x_1)f(x_0, x_1, x_2) +$
$(x-x_0)(x-x_1)(x-x_2)f(x_0, x_1, x_2, x_3) + \cdots \qquad \ldots ①$

Here $x_0 = -1$, $x_1 = 1$, $x_2 = 2$, $f(x_0) = -21$, $f(x_0, x_1) = 18$, $f(x_0, x_1, x_2) = -7$, $f(x_0, x_1, x_2, x_3) = 1$

Substituting these values in ①, we get

$f(x) \equiv -21 + (x+1)(18) + (x+1)(x-1)(-7)$
$\quad + (x+1)(x-1)(x-2)(1) + 0$

$\therefore f(x) \equiv 1 + 13x - 6(x^2 - x) + (x^3 - 3x^2 + 2x)$

$\Rightarrow f(x) \equiv x^3 - 9x^2 + 17x + 6$

Also $f(1.5) \equiv (1.5)^3 - 9(1.5)^2 + 17(1.5) + 6 = 14.625$

ii. To find $f(1.5)$ using Lagrange's interpolation formula:

$f(x) = \sum_{i=0}^{3} L_i y_i = L_0 y_0 + L_1 y_1 + L_2 y_2 + L_3 y_3$

$\Rightarrow f(x) = \left(\frac{(x-x_1)(x-x_2)(x-x_3)}{(x_0-x_1)(x_0-x_2)(x_0-x_3)}\right) y_0 + \left(\frac{(x-x_0)(x-x_2)(x-x_3)}{(x_1-x_0)(x_1-x_2)(x_1-x_3)}\right) y_1 +$

$\left(\frac{(x-x_0)(x-x_1)(x-x_3)}{(x_2-x_0)(x_2-x_1)(x_2-x_3)}\right) y_2 + \left(\frac{(x-x_0)(x-x_1)(x-x_2)}{(x_3-x_0)(x_3-x_1)(x_3-x_2)}\right) y_3$

$\Rightarrow f(1.5) = \left(\frac{(1.5-1)(1.5-2)(1.5-3)}{(-1-1)(-1-2)(-1-3)}\right)(-21) + \left(\frac{(1.5+1)(1.5-2)(1.5-3)}{(1+1)(1-2)(1-3)}\right)(15) +$

$\left(\frac{(1.5+1)(1.5-1)(1.5-3)}{(2+1)(2-1)(2-3)}\right)(12) + \left(\frac{(1.5+1)(1.5-1)(1.5-2)}{(3+1)(3-1)(3-2)}\right)(3)$

$\therefore f(1.5) = 0.328125 + 7.03125 + 7.5 - 0.234375 = 14.625$

Example16 Estimate $f(10)$ from the following data, using Lagrange's interpolation formula.

x	5	6	9	11
y	12	13	14	16

Solution: By Lagrange's interpolation formula

$f(x) = \sum_{i=0}^{3} L_i y_i = L_0 y_0 + L_1 y_1 + L_2 y_2 + L_3 y_3$

$\Rightarrow f(x) = \left(\frac{(x-x_1)(x-x_2)(x-x_3)}{(x_0-x_1)(x_0-x_2)(x_0-x_3)}\right) y_0 + \left(\frac{(x-x_0)(x-x_2)(x-x_3)}{(x_1-x_0)(x_1-x_2)(x_1-x_3)}\right) y_1 +$

$\left(\frac{(x-x_0)(x-x_1)(x-x_3)}{(x_2-x_0)(x_2-x_1)(x_2-x_3)}\right) y_2 + \left(\frac{(x-x_0)(x-x_1)(x-x_2)}{(x_3-x_0)(x_3-x_1)(x_3-x_2)}\right) y_3$

Putting $x = 10$ and remaining values from given data

$f(10) = \frac{(10-6)(10-9)(10-11)}{(5-6)(5-9)(5-11)}(12) + \frac{(10-5)(10-9)(10-11)}{(6-5)(6-9)(6-11)}(13)$

$+ \frac{(10-5)(10-6)(10-11)}{(9-5)(9-6)(9-11)}(14) + \frac{(10-5)(10-6)(10-9)}{(11-5)(11-6)(11-9)}(16)$

$\Rightarrow f(10) = 2 - 4.3333 + 11.6667 + 5.3333 = 14.6667$

Example17 Find the polynomial of the lowest degree which assumes the values 1, 27, 64 for x taking the values 1, 3, 4 respectively, using Lagrange's interpolation formula and hence find $f(2)$.

Solution: To find $f(x)$ using Lagrange's interpolation formula:

$$f(x) = \sum_{i=0}^{3} L_i y_i = L_0 y_0 + L_1 y_1 + L_2 y_2$$

$$\Rightarrow f(x) = \left(\frac{(x-x_1)(x-x_2)}{(x_0-x_1)(x_0-x_2)}\right) y_0 + \left(\frac{(x-x_0)(x-x_2)}{(x_1-x_0)(x_1-x_2)}\right) y_1 + \left(\frac{(x-x_0)(x-x_1)}{(x_2-x_0)(x_2-x_1)}\right) y_2 \quad \ldots \text{①}$$

Here $x_0 = 1$, $x_1 = 3$, $x_2 = 4$, $y_0 = 1$, $y_1 = 27$, $y_2 = 64$

Substituting these values in ①, we get

$$f(x) = \left(\frac{(x-3)(x-4)}{(1-3)(1-4)}\right)(1) + \left(\frac{(x-1)(x-4)}{(3-1)(3-4)}\right)(27) + \left(\frac{(x-1)(x-3)}{(4-1)(4-3)}\right)(64)$$

$$\Rightarrow f(x) = \left(\frac{x^2-7x+12}{(-2)(-3)}\right)(1) + \left(\frac{x^2-5x+4}{(2)(-1)}\right)(27) + \left(\frac{x^2-4x+3}{(3)(1)}\right)(64)$$

$$\Rightarrow f(x) = \frac{1}{6}(48x^2 - 114x + 72) = 8x^2 - 19x + 12$$

$$\therefore f(2) = 8(2)^2 - 19(2) + 12 = 6$$

Exercise 11

1. Find the mean number of men getting wages between Rs. 10 and Rs. 15 from the following data

Wages in Rs.	0-10	10-20	20-30	30-40
Frequency	9	30	35	42

2. Find the cubic polynomial with given set of data points

x	0	1	2	3
$f(x)$	1	2	1	10

   Hence or otherwise evaluate $f(4)$.

3. Find approximate value of $\cos 23^0$ using interpolation from the given data

x	10^0	20^0	30^0	40^0	50^0	60^0	70^0	80^0
$\cos x$	0.9848	0.9397	0.8660	0.7660	0.6428	0.5	0.3420	0.1737

4. The values of y in the given table are consecutive terms of a series of which 23.6 is the 6th term. Find the 1st and 10th terms of the series.

x	3	4	5	6	7	8	9
Y	4.8	8.4	14.5	23.6	36.2	52.8	73.9

5. Find the 4th order divided differences from the given data

x	0.5	1.5	3.0	5.0	6.5	8.0
Y	1.625	5.875	31	131	282.125	521

6. Use Stirling's formula to evaluate $f(32)$ given that

x	22	25	30	35	40
$f(x)$	14.035	13.674	13.257	12.089	11.309

7. Find $\log_{10} 102$, given that $\log_{10} 100 = 2.0$, $\log_{10} 101 = 2.0043$, $\log_{10} 103 = 2.0128$, $\log_{10} 104 = 2.017$.
8. Find the polynomial of the lowest possible degree which assumes the values 1245, 33, 5, 9, 1335 for x taking the values $-4, -1, 0, 2, 5$ respectively, using Newton's divided difference formula and hence find $f(1)$.
9. The pressure p of wind corresponding to velocity v is given by the following data.

v	10	20	30	40
p	1.1	2.0	4.4	7.9

Estimate p, when $v = 25$.

10. Estimate $f(7)$ from the following data, using Lagrange's interpolation formula.

x	5	6	9	11
y	12	13	14	16

Answers

1. 15
2. $2x^3 - 7x^2 + 6x + 1$, $f(4) = 82$
3. 0.9205
4. 3.1, 100
5. 0
6. 13.0622
7. 2.0086
8. $3x^4 - 5x^3 + 6x^2 - 14x + 5$, $f(1) = -5$
9. 3.0375
10. 13.46

Chapter 12: Numerical Differentiation & Integration

Numerical Differentiation
And
Integration

12.1 Introduction

Differentiation and integration are basic mathematical operations with a wide range of applications in various fields of science and engineering. Simple continuous algebraic or transcendental functions can be easily differentiated or integrated directly. However at times there are complicated continuous functions which are tedious to differentiate or integrate directly or in the case of experimental data, where tabulated values of variables are given in discrete form, direct methods of calculus are not applicable.

In this chapter, we develop ways to approximate the derivatives of function $y = f(x)$, when only data points are given and also to integrate definite integrals by splitting the area under the curve in specified ways.

12.2 Numerical Differentiation

Numerical differentiation is the process of computing the value of the derivative of an explicitly unknown function, with given discrete set of points (x_i, y_i), $i = 0, 1, 2, 3, \ldots, n$. To differentiate a function numerically, we first determine an interpolating polynomial and then compute the approximate derivative at the given point.

If x_i's are equispaced
- Newton's forward interpolation formula is used to find the derivative near the beginning of the table.
- Newton's backward interpolation formula is used to compute the derivation near the end of the table.
- Stirling's formula is used to estimate the derivative near the centre of the table.

If x_i's are not equispaced, we may find $f(x)$ using Newton's divided difference method or Lagrange's interpolation formula and then differentiate it as many times as required.

12.2.1 Derivatives Using Newton's Forward Interpolation Formula

Newton's forward interpolation formula for the function $y = f(x)$ is given by

$$y \equiv y_0 + p\Delta y_0 + \frac{p(p-1)}{2!}\Delta^2 y_0 + \frac{p(p-1)(p-2)}{3!}\Delta^3 y_0 + \frac{p(p-1)(p-2)(p-3)}{4!}\Delta^4 y_0 + \cdots,$$

$$p = \frac{x-x_0}{h} \quad \ldots \text{①}$$

Differentiating ① with respect to p

$$\frac{dy}{dp} = \Delta y_0 + \frac{2p-1}{2!}\Delta^2 y_0 + \frac{3p^2-6p+2}{3!}\Delta^3 y_0 + \frac{4p^3-18p^2+22p-6}{4!}\Delta^4 y_0 + \cdots \quad \ldots \text{②}$$

Also $\frac{dp}{dx} = \frac{1}{h}$ ...③

Again $\frac{dy}{dx} = \frac{dy}{dp}\frac{dp}{dx}$ ...④

Using ② and ③ in ④, we get

$$\frac{dy}{dx} = \frac{1}{h}\left[\Delta y_0 + \frac{2p-1}{2!}\Delta^2 y_0 + \frac{3p^2-6p+2}{3!}\Delta^3 y_0 + \frac{4p^3-18p^2+22p-6}{4!}\Delta^4 y_0 + \cdots\right]$$

Now at $x = x_0$, $p = 0$ $\therefore \left.\frac{dy}{dx}\right]_{x=x_0} = \frac{1}{h}\left[\Delta y_0 - \frac{\Delta^2 y_0}{2} + \frac{\Delta^3 y_0}{3} - \frac{\Delta^4 y_0}{4} + \cdots\right]$

Also $\frac{d^2 y}{dx^2} = \frac{d}{dx}\left(\frac{dy}{dx}\right) = \frac{d}{dp}\left(\frac{dy}{dx}\right)\cdot\frac{dp}{dx}$

or $\frac{d^2 y}{dx^2} = \frac{1}{h^2}\frac{d}{dp}\left[\Delta y_0 + \frac{2p-1}{2!}\Delta^2 y_0 + \frac{3p^2-6p+2}{3!}\Delta^3 y_0 + \frac{4p^3-18p^2+22p-6}{4!}\Delta^4 y_0 + \cdots\right]$

$\therefore \frac{d^2 y}{dx^2} = \frac{1}{h^2}\left[\Delta^2 y_0 + \frac{6p-6}{6}\Delta^3 y_0 + \frac{12p^2-36p+22}{24}\Delta^4 y_0 + \cdots\right]$

$\left.\frac{d^2 y}{dx^2}\right]_{x=x_0} = \frac{1}{h^2}\left[\Delta^2 y_0 - \Delta^3 y_0 + \frac{11}{12}\Delta^4 y_0 - \frac{5}{6}\Delta^5 y_0 + \cdots\right]$

12.2.2 Derivatives Using Newton's Backward Interpolation Formula

Newton's backward interpolation formula for the function $y = f(x)$ is given by

$$y \equiv y_n + p\nabla y_n + \frac{p(p+1)}{2!}\nabla^2 y_n + \frac{p(p+1)(p+2)}{3!}\nabla^3 y_n + \frac{p(p+1)(p+2)(p+3)}{4!}\nabla^4 y_n + \cdots,$$
$$p = \frac{x-x_n}{h} \qquad \text{...①}$$

Differentiating ① with respect to p

$\frac{dy}{dp} = \nabla y_n + \frac{2p+1}{2!}\nabla^2 y_n + \frac{3p^2+6p+2}{3!}\nabla^3 y_n + \frac{4p^3+18p^2+22p+6}{4!}\nabla^4 y_n + \cdots$ ...②

Also $\frac{dp}{dx} = \frac{1}{h}$ ...③

Again $\frac{dy}{dx} = \frac{dy}{dp}\frac{dp}{dx}$ ...④

Using ② and ③ in ④, we get

$$\frac{dy}{dx} = \frac{1}{h}\left[\nabla y_n + \frac{2p+1}{2!}\nabla^2 y_n + \frac{3p^2+6p+2}{3!}\nabla^3 y_n + \frac{4p^3+18p^2+22p+6}{4!}\nabla^4 y_n + \cdots\right]$$

Now at $x = x_n$, $p = 0$

$\therefore \left.\frac{dy}{dx}\right]_{x=x_n} = \frac{1}{h}\left[\nabla y_n + \frac{\nabla^2 y_n}{2} + \frac{\nabla^3 y_n}{3} + \frac{\nabla^4 y_n}{4} + \cdots\right]$

Similarly $\left.\frac{d^2 y}{dx^2}\right]_{x=x_n} = \frac{1}{h^2}\left[\nabla^2 y_n + \nabla^3 y_n + \frac{11}{12}\nabla^4 y_n + \frac{5}{6}\nabla^5 y_n + \cdots\right]$

12.2.3 Derivatives Using Stirling's Interpolation Formula

Stirling's central difference interpolation formula (taking x_0 as the middle value of the table) is given by:

$$y \equiv y_0 + p\left(\frac{\Delta y_0 + \Delta y_{-1}}{2}\right) + \frac{p^2}{2!}\Delta^2 y_{-1} + \frac{p(p^2-1)}{3!}\left(\frac{\Delta^3 y_{-1} + \Delta^3 y_{-2}}{2}\right)$$

$$+ \frac{p^2(p^2-1)}{4!}\Delta^4 y_{-2} + \cdots \qquad p = \frac{x-x_0}{h} \qquad \ldots ①$$

Differentiating ① with respect to p

$$\frac{dy}{dp} = \left(\frac{\Delta y_0 + \Delta y_{-1}}{2}\right) + p\Delta^2 y_{-1} + \frac{3p^2-1}{3!}\left(\frac{\Delta^3 y_{-1}+\Delta^3 y_{-2}}{2}\right) + \frac{4p^3-2p}{4!}\Delta^4 y_{-2} + \cdots \quad \ldots ②$$

Also $\frac{dp}{dx} = \frac{1}{h}$ ... ③

Again $\frac{dy}{dx} = \frac{dy}{dp}\frac{dp}{dx}$ ... ④

Using ② and ③ in ④, we get

$$\frac{dy}{dx} = \frac{1}{h}\left[\left(\frac{\Delta y_0 + \Delta y_{-1}}{2}\right) + p\Delta^2 y_{-1} + \frac{3p^2-1}{3!}\left(\frac{\Delta^3 y_{-1}+\Delta^3 y_{-2}}{2}\right) + \frac{4p^3-2p}{4!}\Delta^4 y_{-2} + \cdots\right]$$

Now at $x = x_0, p = 0$

$$\therefore \left.\frac{dy}{dx}\right]_{x=x_0} = \frac{1}{h}\left[\left(\frac{\Delta y_0 + \Delta y_{-1}}{2}\right) - \frac{1}{6}\left(\frac{\Delta^3 y_{-1}+\Delta^3 y_{-2}}{2}\right) + \frac{1}{30}\left(\frac{\Delta^5 y_{-2}+\Delta^5 y_{-3}}{2}\right) - \cdots\right]$$

Similarly $\left.\frac{d^2y}{dx^2}\right]_{x=x_0} = \frac{1}{h^2}\left[\Delta^2 y_{-1} - \frac{1}{12}\Delta^4 y_{-2} + \frac{1}{90}\Delta^6 y_{-3} - \cdots\right]$

12.2.4 Derivatives of Polynomial with Unequispaced x_i's

In case where x_i's are not equispaced in a given data, the polynomial $y = f(x)$ may be found using Newton's divided difference method or Lagrange's interpolation formula and then direct differentiation may be applied.

Remark:
- In case derivative has to be found at a point, whose value needs to be interpolated, then first apply applicable interpolation technique and then differentiate the function.
- In all the cases irrespective of data points being equispaced or not, the polynomial $y = f(x)$ may be found using the applicable interpolation formulae and then direct differentiation can be done using usual calculus techniques.

Example 1 Given a cubic polynomial with following data points

x	0	1	2	3
$f(x)$	5	6	3	8

Find $\frac{dy}{dx}$ and $\frac{d^2y}{dx^2}$ at $x = 0$

Solution: Derivative has to be evaluated near the starting of the table, thereby constructing forward difference table for the function $y = f(x)$

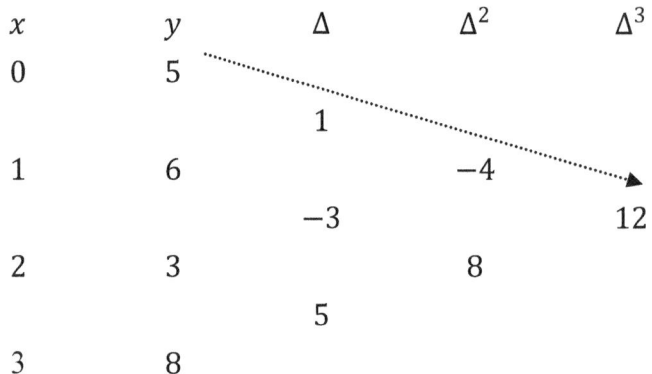

x	y	Δ	Δ^2	Δ^3
0	5			
		1		
1	6		-4	
		-3		12
2	3		8	
		5		
3	8			

To find the derivative at $x = 0$, taking $x_0 = 0$ and applying the relation:

$$\left.\frac{dy}{dx}\right]_{x=x_0} = \frac{1}{h}\left[\Delta y_0 - \frac{\Delta^2 y_0}{2} + \frac{\Delta^3 y_0}{3} - \frac{\Delta^4 y_0}{4} + \cdots\right] \quad \ldots \text{①}$$

From table $h = 1$, $\Delta y_0 = 1$, $\Delta^2 y_0 = -4$, $\Delta^3 y_0 = 12$, $\Delta^4 y_0 = 0$

Substituting these values in ①, we get

$$\left.\frac{dy}{dx}\right]_{x=0} = \frac{1}{1}\left[1 - \frac{(-4)}{2} + \frac{12}{3} + 0\right] = 7$$

Also $\left.\dfrac{d^2y}{dx^2}\right]_{x=x_0} = \dfrac{1}{h^2}\left[\Delta^2 y_0 - \Delta^3 y_0 + \dfrac{11}{12}\Delta^4 y_0 - \cdots\right]$

$$\therefore \left.\frac{d^2y}{dx^2}\right]_{x=0} = \frac{1}{1^2}[-4 - 12 + 0] = -16$$

Aliter

Newton's forward interpolation formula given by:

$$y = y_0 + p\Delta y_0 + \frac{p(p-1)}{2!}\Delta^2 y_0 + \frac{p(p-1)(p-2)}{3!}\Delta^3 y_0 + \cdots, \quad p = \frac{x-x_0}{h} \quad \ldots \text{①}$$

$x_0 = 0$, $y_0 = 5$, $h = 1$ $\therefore p = \dfrac{x-0}{1} = x$

Also from table $\Delta y_0 = 1, \Delta^2 y_0 = -4, \Delta^3 y_0 = 12$

Substituting these values in ①, we get

$$y = 5 + x(1) + \frac{x(x-1)}{2}(-4) + \frac{x(x-1)(x-2)}{6}(12)$$

$$\Rightarrow y = 2x^3 - 8x^2 + 7x + 5$$

$$\therefore \frac{dy}{dx} = 6x^2 - 16x + 7$$

$$\Rightarrow \left.\frac{dy}{dx}\right]_{x=0} = 7$$

Also $\dfrac{d^2y}{dx^2} = 12x - 16$

$$\therefore \left.\frac{d^2y}{dx^2}\right]_{x=0} = -16$$

Example 2 Given a polynomial with following data points:

x	1.0	1.1	1.2	1.3	1.4	1.5	1.6
f(x)	7.989	8.403	8.781	9.129	9.451	9.750	10.031

Find $\frac{dy}{dx}$ and $\frac{d^2y}{dx^2}$ at $x = 1.1$ and $x = 1.5$

Solution: Derivatives has to be evaluated near the starting as well as towards the end of the table, thereby constructing difference table for the function $y = f(x)$

x	y = f(x)	1st diff	2nd diff	3rd diff	4th diff	5th diff	6th diff
1.0	7.989						
		0.414					
1.1	8.403		−0.036				
		0.378		0.006			
1.2	8.781		−0.030		−0.002		
		0.348		0.004		0.001	
1.3	9.129		−0.026		−0.001		0.002
		0.322		0.003		0.003	
1.4	9.451		−0.023		0.002		
		0.299		0.005			
1.5	9.750		−0.018				
		0.281					
1.6	10.031						

To find the derivative at $x = 1.1$, taking $x_0 = 1.1$ and applying the relation:

$$\left.\frac{dy}{dx}\right]_{x = x_0} = \frac{1}{h}\left[\Delta y_0 - \frac{\Delta^2 y_0}{2} + \frac{\Delta^3 y_0}{3} - \frac{\Delta^4 y_0}{4} + \frac{\Delta^5 y_0}{5} - \cdots\right] \quad \ldots \text{①}$$

From table $h = 0.1$, $\Delta y_0 = 0.378$, $\Delta^2 y_0 = -0.03$, $\Delta^3 y_0 = 0.004$, $\Delta^4 y_0 = -0.001$, $\Delta^5 y_0 = 0.003$

Substituting these values in ①, we get

$$\left.\frac{dy}{dx}\right]_{x=1.1} = \frac{1}{0.1}\left[0.378 - \frac{(-0.03)}{2} + \frac{0.004}{3} - \frac{(-0.001)}{4} + \frac{0.003}{5}\right] = 3.952$$

Also $\left.\frac{d^2y}{dx^2}\right]_{x = x_0} = \frac{1}{h^2}\left[\Delta^2 y_0 - \Delta^3 y_0 + \frac{11}{12}\Delta^4 y_0 - \frac{5}{6}\Delta^5 y_0 + \cdots\right]$

$$\therefore \left.\frac{d^2y}{dx^2}\right]_{x=0} = \frac{1}{(0.1)^2}\left[-0.03 - 0.004 + \frac{11}{12}(-0.001) - \frac{5}{6}(0.003)\right] = -3.74$$

To find the derivative at $x = 1.5$, taking $x_n = 1.5$ and applying the relation:

$$\left.\frac{dy}{dx}\right]_{x=x_n} = \frac{1}{h}\left[\nabla y_n + \frac{\nabla^2 y_n}{2} + \frac{\nabla^3 y_n}{3} + \frac{\nabla^4 y_n}{4} + \frac{\nabla^4 y_n}{4} + \cdots\right] \quad \ldots ②$$

From table $h = 0.1$, $\nabla y_n = 0.299$, $\nabla^2 y_n = -0.023$, $\nabla^3 y_n = 0.003$,

$\nabla^4 y_n = -0.001$, $\nabla^5 y_n = 0.001$

Substituting these values in ②, we get

$$\left.\frac{dy}{dx}\right]_{x=1.5} = \frac{1}{0.1}\left[0.299 + \frac{(-0.023)}{2} + \frac{0.003}{3} + \frac{(-0.001)}{4} + \frac{0.001}{5}\right] = 2.8845$$

Also $\left.\frac{d^2 y}{dx^2}\right]_{x=x_n} = \frac{1}{h^2}\left[\nabla^2 y_n + \nabla^3 y_n + \frac{11}{12}\nabla^4 y_n + \frac{5}{6}\nabla^5 y_n + \cdots\right]$

$$\therefore \left.\frac{d^2 y}{dx^2}\right]_{x=n} = \frac{1}{(0.1)^2}\left[-0.023 + 0.003 + \frac{11}{12}(-0.001) + \frac{5}{6}(0.001)\right] = -2.0083$$

Example 3 Following table gives the census population of a state for the years 1961 to 2001.

Year	1961	1971	1981	1991	2001
Population (Million)	19.96	36.65	58.81	77.21	94.61

Find the rate of growth of the population in the year 2001

Solution: Derivative has to be evaluated near the end of the table, thereby constructing backward difference table for the function $y = f(x)$

Year x	Population $f(x)$	∇	∇^2	∇^3	∇^4
1961	19.96				
		16.69			
1971	36.65		5.47		
		22.16		−9.23	
1981	58.81		−3.76		11.99
		18.40		2.76	
1991	77.21		−1.00		
		17.40			
2001	94.61				

To find rate of growth (derivative) in the year 2001, taking $x_n = 2001$ and applying the relation:

$$\left.\frac{dy}{dx}\right]_{x=x_n} = \frac{1}{h}\left[\nabla y_n + \frac{\nabla^2 y_n}{2} + \frac{\nabla^3 y_n}{3} + \frac{\nabla^4 y_n}{4} + \cdots\right] \quad \ldots ①$$

From table $h = 10$, $\nabla y_n = 17.40$, $\nabla^2 y_n = -1$, $\nabla^3 y_n = 2.76$, $\nabla^4 y_n = 11.99$

Substituting these values in ①, we get

$$\left.\frac{dy}{dx}\right]_{x=2011} = \frac{1}{10}\left[17.40 + \frac{(-1)}{2} + \frac{2.76}{3} + \frac{(11.99)}{4}\right] = 2.08175$$

Example 4 A slider in a machine moves along a fixed straight rod. Its distance x cm along the rod is given below for various values of the time. Find the velocity and acceleration of the slider when $t = 0.3$ seconds.

t(seconds)	0	0.1	0.2	0.3	0.4	0.5	0.6
x(cm)	30.13	31.62	32.87	33.64	33.95	33.81	33.24

Solution: Derivatives has to be evaluated towards the centre of the table, thereby constructing central difference table for the function $x = f(t)$. Also $\Delta^n x_0$, $n = 1, 2, 3, 3, 4, 5, 6$ lie along the dotted line as shown.

t	x = f(t)	1st diff	2nd diff	3rd diff	4th diff	5th diff	6th diff
0	30.13						
		1.49					
0.1	31.62		−0.24				
		1.25		−0.24			
0.2	32.87		−0.48		0.26		
		0.77		0.02		−0.27	
0.3	33.64		−0.46		−0.01		0.29
		0.31		0.01		0.02	
0.4	33.95		−0.45		0.01		
		−0.14		0.02			
0.5	33.81		−0.43				
		−0.57					
0.6	33.24						

To find the velocity (derivative) at $t = 0.3$, taking $t_0 = 0.3$ and applying the relation:

$$\left.\frac{dx}{dt}\right]_{t=t_0} = \frac{1}{h}\left[\left(\frac{\Delta x_0 + \Delta x_{-1}}{2}\right) - \frac{1}{6}\left(\frac{\Delta^3 x_{-1} + \Delta^3 x_{-2}}{2}\right) + \frac{1}{30}\left(\frac{\Delta^5 x_{-2} + \Delta^5 x_{-3}}{2}\right) - \cdots\right] \quad \ldots ①$$

From table $h = 0.1$, $\Delta x_0 = 0.31$, $\Delta x_{-1} = 0.77$, $\Delta^3 x_{-1} = 0.01$, $\Delta^3 x_{-2} = 0.02$, $\Delta^5 x_{-2} = 0.02$, $\Delta^5 x_{-3} = -0.27$. All the positions have been shown, enclosed in boxes.

Substituting these values in ①, we get

$$\left.\frac{dx}{dt}\right]_{t=0.3} = \frac{1}{0.1}\left[\left(\frac{0.31+0.77}{2}\right) - \frac{1}{6}\left(\frac{0.01+0.02}{2}\right) + \frac{1}{30}\left(\frac{0.02-0.27}{2}\right)\right] = 5.33 \text{ cm/second}$$

To find the acceleration (second derivative) at $t = 0.3$ and applying the relation

$$\left.\frac{d^2x}{dt^2}\right]_{t=t_0} = \frac{1}{h^2}\left[\Delta^2 x_{-1} - \frac{1}{12}\Delta^4 x_{-2} + \frac{1}{90}\Delta^6 x_{-3} - \cdots\right] \quad \ldots ②$$

Also from table, $\Delta^2 x_{-1} = -0.46$, , $\Delta^4 x_{-2} = -0.01$, $\Delta^6 x_{-3} = 0.29$

Substituting these values in ②, we get

$$\left.\frac{d^2x}{dt^2}\right]_{t=0.3} = \frac{1}{(0.1)^2}\left[(-0.46) - \frac{1}{12}(-0.01) + \frac{1}{90}(0.29)\right] = -45.60 \text{ cm/second}^2$$

Example 5 The following data gives corresponding values of pressure 'P' and specific volume 'V' of a super heated expandable system

V	2	4	6	8	10
P	105	42.7	25.3	16.7	13

Find the rate of change of volume with respect to pressure, when $P = 105$ units.

Solution: Values of P are not equispaced, thereby constructing divided difference table for the function $V = f(P)$

P	$V = f(P)$	1st diff	2nd diff	3rd diff	4th diff
105	2				
		-0.032			
42.7	4		0.001		
		-0.115		-0.00005	
25.3	6		0.005		0.000007
		-0.233		-0.0007	
16.7	8		0.025		
		-0.541			
13	10				

Newton's divided difference formula is given by

$$V = f(P) \equiv f(P_0) + (P - P_0)f(P_0, P_1) + (P - P_0)(P - P_1)f(P_0, P_1, P_2) +$$
$$(P - P_0)(P - P_1)(P - P_2)f(P_0, P_1, P_2, P_3) + \cdots \qquad \ldots ①$$

Here $P_0 = 105$, $P_1 = 42.7$, $P_2 = 25.3$, $P_3 = 16.7$ $f(P_0) = 2$,

$f(P_0, P_1) = -0.032$, $f(P_0, P_1, P_2) = 0.001$, $f(P_0, P_1, P_2, P_3) = -0.00005$,

$f(P_0, P_1, P_2, P_3, P_4) = 0000007$

Substituting these values in ①, we get

$$V = f(P) \equiv 2 + (P - 105)(-0.032) + (P - 105)(P - 42.7)(0.001)$$
$$+ (P - 105)(P - 42.7)(P - 25.3)(-0.00005)$$
$$+ (P - 105)(P - 42.7)(P - 25.3)(P - 16.7)(0.000007)$$

$$\therefore \frac{dV}{dP} = -0.032 + (2P - 147.7)(.001) + (3P^2 - 505.4P + 1992)(-.000005) +$$
$$[(3P^2 - 505.4P + 1992)(P - 16.7) + (P - 105)(P - 42.7)(P - 25.3)(1)](0.000007)$$

$$\left.\frac{dV}{dP}\right]_{P=105} = -0.032 + 0.0623 + 0.9 - 11.1258 = -10.1955$$

It is evident that volume decreases with increase in pressure.

Example 6 Find $y'(1)$ for the following data points of a polynomial $y = f(x)$

x	0	2	4	6	8
y	4	8	15	7	6

Solution: To find $y'(1)$ or $\left.\frac{dy}{dx}\right]_{x=1}$, constructing forward difference table for $y = f(x)$.

x	$y = f(x)$	Δ	Δ^2	Δ^3	Δ^4
0	4				
		4			
2	8		3		
		7		−18	
4	15		−15		40
		−8		22	
6	7		7		
		−1			
8	6				

Newton's forward interpolation formula given by:

$$y \equiv y_0 + p\Delta y_0 + \frac{p(p-1)}{2!}\Delta^2 y_0 + \frac{p(p-1)(p-2)}{3!}\Delta^3 y_0 + \frac{p(p-1)(p-2)(p-3)}{4!}\Delta^4 y_0 + \cdots$$

$$p = \frac{x - x_0}{h} \quad \ldots \text{①}$$

$$\Rightarrow \frac{dy}{dp} = \Delta y_0 + \frac{2p-1}{2!}\Delta^2 y_0 + \frac{3p^2-6p+2}{3!}\Delta^3 y_0 + \frac{4p^3-18p^2+22p-6}{4!}\Delta^4 y_0 + \cdots \quad \ldots \text{②}$$

Also $\frac{dp}{dx} = \frac{1}{h}$ ... ③

Again $\frac{dy}{dx} = \frac{dy}{dp}\frac{dp}{dx}$ ... ④

Using ② and ③ in ④, we get

$$\frac{dy}{dx} = \frac{1}{h}\left[\Delta y_0 + \frac{2p-1}{2!}\Delta^2 y_0 + \frac{3p^2-6p+2}{3!}\Delta^3 y_0 + \frac{4p^3-18p^2+22p-6}{4!}\Delta^4 y_0 + \cdots\right] \quad \ldots \text{⑤}$$

Now for $x = 1, x_0 = 0, h = 2, \therefore p = \frac{1-0}{2} = 0.5$

Also from table $\Delta y_0 = 4, \Delta^2 y_0 = 3, \Delta^3 y_0 = -18, \Delta^4 y_0 = 40$

Substituting these values in ⑤, we get

$$\left.\frac{dy}{dx}\right]_{x=1} = \frac{1}{2}\left[4 + \frac{1-1}{2}(3) + \frac{3(0.5)^2-6(0.5)+2}{6}(-18) + \frac{4(0.5)^3-18(0.5)^2+22(0.5)-6}{24}(40)\right]$$

$$\Rightarrow y'(1) = \left.\frac{dy}{dx}\right]_{x=1} = \frac{1}{2}[4 + 0 + 0.75 + 1.667] = \frac{1}{2}[6.417] = 3.2085$$

Note: Here formula for computing derivatives cannot be applied directly as $p \neq 0$ at $x = 1$,

i.e point at which derivative has to be computed does not exist in the table and has to be interpolated first.

12.2.5 Maxima and Minima of a Tabulated Function:

Newton's forward interpolation formula for the function $y = f(x)$ is given by

$$y \equiv y_0 + p\Delta y_0 + \frac{p(p-1)}{2!}\Delta^2 y_0 + \frac{p(p-1)(p-2)}{3!}\Delta^3 y_0 + \frac{p(p-1)(p-2)(p-3)}{4!}\Delta^4 y_0 + \cdots,$$

$$p = \frac{x-x_0}{h} \qquad \ldots ①$$

Differentiating ① with respect to p

$$\frac{dy}{dp} = \Delta y_0 + \frac{2p-1}{2!}\Delta^2 y_0 + \frac{3p^2-6p+2}{3!}\Delta^3 y_0 + \frac{4p^3-18p^2+22p-6}{4!}\Delta^4 y_0 + \cdots \qquad \ldots ②$$

For finding maxima/minima of a function $y = f(x)$, $\frac{dy}{dx} = \frac{dy}{dp}\frac{dp}{dx} = 0$

$$\frac{dp}{dx} = \frac{1}{h} \neq 0, \quad \therefore \frac{dy}{dp} = 0 \qquad \ldots ③$$

Neglecting 4th and higher order differences in equation ② and substituting in ③,

we get a quadratic equation of the form $A + Bp + Cp^2 = 0$, where A, B, C are constants. Solving for p and substituting in $x = x_0 + ph$, we get points of maxima/minima for the function $y = f(x)$.

> Newton's forward method is apt for finding extreme values of a tabulated data, wherever their location may be, by index p assuming values $|p| \geq 1$, if the extreme value is not in vicinity of the point (x_0, y_0). Yet we may also use Newton's backward or Stirling's central differences formulae to locate extreme values, if desired.

Example 7 From the following data, find maximum and minimum values of y.

x	0	2	4	6
$f(x)$	2	0	-50	-196

Solution: Constructing forward difference table for the function $y = f(x)$

x	y	Δ	Δ^2	Δ^3
0	2			
		-2		
2	0		-48	
		-50		-48
4	-50		-96	
		-146		
6	-196			

Newton's forward interpolation formula for the function $y = f(x)$ is given by

$$y \equiv y_0 + p\Delta y_0 + \frac{p(p-1)}{2!}\Delta^2 y_0 + \frac{p(p-1)(p-2)}{3!}\Delta^3 y_0 + \cdots, \quad p = \frac{x-x_0}{h} \qquad \ldots ①$$

Taking $x_0 = 0$, $y_0 = 2$, $\Delta y_0 = -2$, $\Delta^2 y_0 = -48$, $\Delta^3 y_0 = -48$

Substituting these values in ①, we get

$y \equiv 2 + p(-2) + \frac{p(p-1)}{2}(-48) + \frac{p(p-1)(p-2)}{6}(-48)$

$\Rightarrow y \equiv 2 - 2p - 24(p^2 - p) - 8(p^3 - 3p^2 + 2p)$

$\Rightarrow y \equiv -8p^3 + 6p + 2$ ...②

Differentiating ② with respect to p, we get

$\frac{dy}{dp} = -24p^2 + 6$

For y to be maximum, $\frac{dy}{dp} = 0$

$\Rightarrow -24p^2 + 6 = 0$

$\Rightarrow p = 0.5, -0.5$

Substituting in ②, maximum and minimum values of y are 4 and 0 respectively.

Example 8 From the following table, find x for which y is maximum.

x	3	4	5	6	7	8
$f(x)$	0.205	0.240	0.259	0.262	0.250	0.224

Also find maximum value of y.

Solution: Constructing forward difference table for the function $y = f(x)$, up to third differences

x	$y = f(x)$	Δ	Δ^2	Δ^3
3	0.205			
		0.035		
4	0.240		-0.016	
		0.019		0
5	0.259		-0.016	
		0.003		0.001
6	0.262		-0.015	
		-0.012		0.001
7	0.250		-0.014	
		-0.026		
8	0.224			

Newton's forward interpolation formula for the function $y = f(x)$ is given by

$y \equiv y_0 + p\Delta y_0 + \frac{p(p-1)}{2!}\Delta^2 y_0 + \frac{p(p-1)(p-2)}{3!}\Delta^3 y_0 + \cdots$, $p = \frac{x-x_0}{h}$ ...①

Taking $x_0 = 3$, $y_0 = 0.205$, $\Delta y_0 = 0.035$, $\Delta^2 y_0 = -0.016$, $\Delta^3 y_0 = 0$

Substituting these values in ①, we get

$y \equiv (0.205) + p(0.035) + \frac{p(p-1)}{2}(-0.016) + 0$ ... ②

Differentiating with respect to p, we get

$\frac{dy}{dp} = 0.035 + \frac{2p-1}{2}(-0.016) = 0.035 - (0.008)(2p-1)$

For y to be maximum, $\frac{dy}{dp} = 0$

$\Rightarrow 0.035 - (0.008)(2p-1) = 0$

$\Rightarrow p = 2.6875$

Also $p = \frac{x-x_0}{h}$ or $x = x_0 + ph$

$\Rightarrow x = 3 + 2.6875(1) = 5.6875$

∴ y is maximum when $x = 5.6875$ or $p = 2.6875$

Substituting in ②, maximum value of y is given by

$y \equiv (0.205) + (2.6875)(0.035) + \frac{(2.6875)(2.6875-1)}{2}(-0.016) = 0.2628$

12.3 Numerical Integration

Numerical Integration is the process of computing the value of definite integral $\int_a^b y \, dx$, when the integrand function $y = f(x)$ is given as discrete set of points (x_i, y_i), $i = 0, 1, 2, 3, \ldots, n$. As in case of numerical differentiation, here also the integrand $y = f(x)$ is first replaced with an interpolating polynomial, and then it is integrated to compute the value of the definite integral. This gives us 'quadrature formula' for numerical integration.

Newton Cotes Quadrature Formula

For an explicitly known function $y = f(x)$, let y take values $y_0, y_1, y_2, \ldots y_n$ for x taking values $x_0, x_1, x_2, \ldots x_n$, where x_i's are equispaced.

To evaluate $I = \int_a^b f(x)dx$, divide the interval (a, b) into n equal parts, each of width h, such that $a = x_0 < x_1 < x_2 < \cdots < x_n = b$

Clearly $x_n = x_0 + nh$ Then $I = \int_a^b f(x)dx = \int_{x_0}^{x_0+nh} f(x)dx$, $nh = b - a$

By Newton's forward interpolation formula:

$f(x) \equiv y_0 + p\Delta y_0 + \frac{p(p-1)}{2!}\Delta^2 y_0 + \frac{p(p-1)(p-2)}{3!}\Delta^3 y_0 + \cdots$, $x = x_0 + ph$

∴ $I = \int_{x_0}^{x_0+nh} f(x)dx = h\int_0^n f(x_0 + ph)dp$

∵ $x = x_0 + ph \Rightarrow dx = hdp$, also when $x = x_0, = 0$, $x = x_0 + nh, p = n$

$$\Rightarrow I \equiv h \int_0^n \left[y_0 + p\Delta y_0 + \frac{p(p-1)}{2!}\Delta^2 y_0 + \frac{p(p-1)(p-2)}{3!}\Delta^3 y_0 + \cdots \right] dp$$

$$\equiv h \left[py_0 + \frac{p^2}{2}\Delta y_0 + \frac{1}{2!}\left(\frac{p^3}{3} - \frac{p^2}{2}\right)\Delta^2 y_0 + \frac{1}{3!}\left(\frac{p^4}{4} - p^3 + p^2\right)\Delta^3 y_0 + \cdots \right]_0^n$$

$$\Rightarrow I \equiv nh \left[y_0 + \frac{n}{2}\Delta y_0 + \left(\frac{n^2}{3} - \frac{n}{2}\right)\frac{\Delta^2 y_0}{2!} + \left(\frac{n^3}{4} - n^2 + n\right)\frac{\Delta^3 y_0}{3!} + \cdots \right] \quad \ldots ①$$

This is known as Newton's Cote's quadrature formula.

Different quadrature formulae are derived by taking $n = 1, 2, 3, \ldots$ in equation ①.

12.3.1 Numerical Integration Using Trapezoidal Rule

If we put $n = 1$ in Newton's Cote's quadrature formula given by ①, we get the curve through (x_0, y_0) and (x_1, y_1) as a straight line and being linear equation in x, 2nd and higher order differences are zero.

$$\therefore ① \Rightarrow I = \int_{x_0}^{x_0+h} f(x)dx = h\left[y_0 + \frac{1}{2}\Delta y_0\right] = h\left[y_0 + \frac{1}{2}(y_1 - y_0)\right] = \frac{h}{2}(y_0 + y_1)$$

Similarly $\int_{x_0+h}^{x_0+2h} f(x)dx = \frac{h}{2}(y_1 + y_2)$

$\vdots$

$$\int_{x_0+(n-1)h}^{x_0+nh} f(x)dx = \frac{h}{2}(y_{n-1} + y_n)$$

Adding areas of all these intervals, we get:

$$\int_{x_0}^{x_0+nh} f(x)dx = \frac{h}{2}(y_0 + y_1) + \frac{h}{2}(y_1 + y_2) + \cdots + \frac{h}{2}(y_{n-1} + y_n)$$

$$\Rightarrow \int_a^b f(x)dx = \frac{h}{2}[y_0 + 2(y_1 + y_2 + \cdots + y_{n-1}) + y_n]$$

This is known as trapezoidal rule to evaluate $\int_a^b f(x)dx$, where the function $y = f(x)$ is given as discrete set of points (x_i, y_i), $i = 0, 1, 2, 3, \ldots, n$.

Geometrical Significance of Trapezoidal Rule

In trapezoidal rule, the curve $y = f(x)$ is replaced by n piecewise straight lines joining the points (x_0, y_0) and (x_1, y_1) ; (x_1, y_1) and (x_2, y_2); ...; (x_{n-1}, y_{n-1}) and (x_n, y_n).

The area under the curve $y = f(x)$, between the ordinates $x = x_0$; $x = x_n$ and above $x-$ axis is approximately equal to the sum of areas of n trapezoids obtained within the enclosed region, shown by shaded portion of adjoining figure.

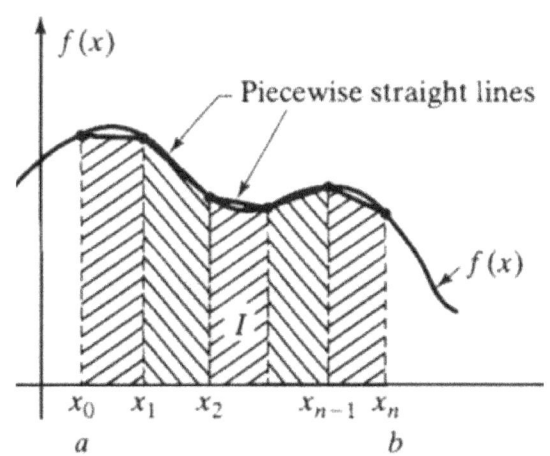

Error in Trapezoidal Rule

Area of first trapezoid in the interval $[x_0, x_1]$ is given by

$$A_1 = \frac{h}{2}(y_0 + y_1) \quad \ldots ①$$

Also expanding $y = f(x)$ by Taylor's series about $x = x_0$

$$y = f(x) = y_0 + (x - x_0)y_0' + \frac{1}{2!}(x - x_0)^2 y_0'' + \cdots \qquad \ldots ②$$

Putting $x = x_1 = x_0 + h$, we get

$$\Rightarrow y_1 = y_0 + hy_0' + \frac{h^2}{2!} y_0'' + \cdots \qquad \ldots ③$$

$$\because at\ x = x_1, y = y_1$$

Substituting ③ in ①, we get

$$A_1 = \frac{h}{2}\left(y_0 + y_0 + hy_0' + \frac{h^2}{2!} y_0'' + \cdots \right)$$

$$\Rightarrow A_1 = \frac{h}{2}\left(2y_0 + hy_0' + \frac{h^2}{2!} y_0'' + \cdots \right) = hy_0 + \frac{h^2}{2} y_0' + \frac{h^3}{2 \cdot 2!} y_0'' + \cdots$$

Again taking area of single strip, under the curve $y = f(x)$, between the ordinates $x = x_0$ and $x = x_1$ and above $x -$ axis

$$A_1' = \int_{x_0}^{x_0+h} y\, dx = \int_{x_0}^{x_0+h}\left[y_0 + (x - x_0)y_0' + \frac{1}{2!}(x - x_0)^2 y_0'' + \cdots \right] dx \quad \text{using } ②$$

$$= \left[y_0 x + \frac{(x-x_0)^2}{2!} y_0' + \frac{(x-x_0)^3}{3} \frac{y_0''}{2!} + \cdots \right]_{x_0}^{x_0+h}$$

$$= hy_0 + \frac{h^2}{2!} y_0' + \frac{h^3}{3!} y_0'' + \cdots$$

∴ Error in the interval $[x_0, x_1]$ is given by

$$A_1' - A_1 = \left[hy_0 + \frac{h^2}{2!} y_0' + \frac{h^3}{3!} y_0'' + \cdots \right] - \left[hy_0 + \frac{h^2}{2} y_0' + \frac{h^3}{2 \cdot 2!} y_0'' + \cdots \right]$$

$$= \left(\frac{1}{6} - \frac{1}{4}\right) h^3 y_0'' + \cdots = -\frac{h^3}{12} y_0'' + \cdots$$

Thus lowest order error in interval $[x_0, x_1] \approx -\frac{h^3}{12} y_0''$

Similarly lowest error in interval $[x_1, x_2] \approx -\frac{h^3}{12} y_1''$

⋮

∴ Lowest error in interval $[x_{n-1}, x_n] \approx -\frac{h^3}{12} y_{n-1}''$

Total error $E = -\frac{h^3}{12} [y_0'' + y_1'' + \cdots + y_{n-1}'']$

If $y_x'' = \text{Maximum}[y_0'', y_1'', \ldots, y_{n-1}'']$

Then Maximum $E \leq -\frac{nh^3}{12} y_x''$

$$= -\frac{nhh^2}{12} y_x'' = -\frac{(b-a)h^2}{12} y_x'' \qquad \because nh = (b-a)$$

∴ $E \leq kh^2$, where $k = -\frac{(b-a)}{12} y_x''$

Hence the error in trapezoidal rule is of order h^2, where h is the height of the interval.

12.3.2 Numerical Integration Using Simpson's One-Third Rule

If we put $n = 2$ in Newton's Cote's quadrature formula given by ①, we get the curve through the points (x_0, y_0), (x_1, y_1), (x_2, y_2) as a parabolic figure and being quadratic equation in x, 3rd and higher order differences are zero.

$$\therefore ① \Rightarrow I = \int_{x_0}^{x_0+2h} f(x)dx = 2h\left[y_0 + \frac{2}{2}\Delta y_0 + \left(\frac{4}{3} - \frac{2}{2}\right)\frac{\Delta^2 y_0}{2!}\right]$$

$$= 2h\left[y_0 + (y_1 - y_0) + \frac{1}{6}(y_2 - 2y_1 + y_0)\right]$$

$$= \frac{h}{3}(y_0 + 4y_1 + y_2)$$

Similarly $\int_{x_0+2h}^{x_0+4h} f(x)dx = \frac{h}{3}(y_2 + 4y_3 + y_4)$

$$\vdots$$

$$\int_{x_0+(n-2)h}^{x_0+nh} f(x)dx = \frac{h}{3}(y_{n-2} + 4y_{n-1} + y_n)$$

Adding areas of all these intervals, we get:

$$\int_{x_0}^{x_0+nh} f(x)dx = \frac{h}{3}(y_0 + 4y_1 + y_2) + \frac{h}{3}(y_2 + 4y_3 + y_4) + \cdots + \frac{h}{3}(y_{n-2} + 4y_{n-1} + y_n)$$

$$\therefore \int_a^b f(x)dx = \frac{h}{3}[(y_0 + y_n) + 4(y_1 + y_3 + \cdots + y_{n-1}) + 2(y_2 + y_4 + \cdots + y_{n-2})]$$

This is known as Simpson's one-third rule to evaluate $\int_a^b f(x)dx$, where the function $y = f(x)$ is given as discrete set of points (x_i, y_i), $i = 0, 1, 2, 3, \ldots, n$.

Geometrical Significance of Simpson's One-Third Rule

In Simpson's one-third rule, the curve $y = f(x)$ is replaced by arcs of 2nd degree parabolas with vertical axis as shown in given figure. Simpson's one-third rule requires the given interval to be divided into even number of sub-intervals, since we are finding areas of two strips at a time.

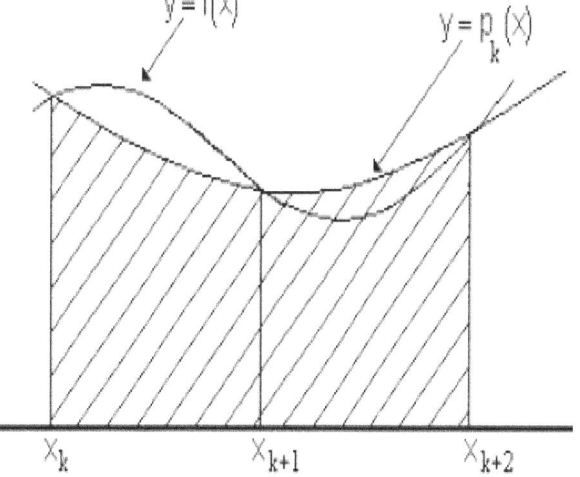

Error in Simpson's One-Third Rule

Area of first two strips in the interval $[x_0, x_2]$ is given by

$$A_2 = \frac{h}{3}(y_0 + 4y_1 + y_2) \quad \ldots ①$$

Also expanding $y = f(x)$ by Taylor's series about $x = x_0$

$$y = f(x) = y_0 + (x - x_0)y_0' + \frac{1}{2!}(x - x_0)^2 y_0'' + \frac{1}{3!}(x - x_0)^3 y_0''' \ldots \quad \ldots ②$$

Putting $x = x_1 = x_0 + h$ in ②, we get

$$y_1 = y_0 + hy_0' + \frac{h^2}{2!}y_0'' + \frac{h^3}{3!}y_0''' + \cdots \quad \ldots ③$$

$$\because \text{at } x = x_1, y = y_1$$

Putting $x = x_2 = x_0 + 2h$ in ②, we get

$$y_2 = y_0 + 2hy_0' + \frac{4h^2}{2!}y_0'' + \frac{8h^2}{3!}y_0''' + \cdots \qquad \ldots ④$$

$$\because \text{ at } x = x_2, y = y_2$$

Substituting ③ and ④ in ①, A_2 may be written as

$$\frac{h}{3}\left(y_0 + 4\left(y_0 + hy_0' + \frac{h^2}{2!}y_0'' + \frac{h^3}{3!}y_0'''\right) + \left(y_0 + 2hy_0' + \frac{4h^2}{2!}y_0'' + \frac{8h^2}{3!}y_0'''\right) + \cdots\right)$$

$$\Rightarrow A_2 = \frac{h}{3}\left(6y_0 + 6hy_0' + \frac{8h^2}{2!}y_0'' + \frac{12h^3}{3!}y_0''' + \cdots\right)$$

$$= 2hy_0 + 2h^2y_0' + \frac{4h^3}{3}y_0'' + \frac{2h^4}{3}y_0''' + \frac{5h^5}{18}y_0^{iv} + \cdots$$

Again taking actual area of two strip, under the curve $y = f(x)$, between the ordinates $x = x_0$ and $x = x_2$ and above $x-$axis

$$A_2' = \int_{x_0}^{x_0+2h} ydx = \int_{x_0}^{x_0+2h}\left[y_0 + (x-x_0)y_0' + \frac{1}{2!}(x-x_0)^2 y_0'' + \cdots\right]dx \quad \text{using ②}$$

$$= \left[y_0 x + \frac{(x-x_0)^2}{2!}y_0' + \frac{(x-x_0)^3}{3}\frac{y_0''}{2!} + \cdots\right]_{x_0}^{x_0+2h}$$

$$= 2hy_0 + \frac{4h^2}{2!}y_0' + \frac{8h^3}{3!}y_0'' + \frac{16h^4}{4!}y_0''' + \frac{32h^5}{5!}y_0^{iv} + \cdots$$

$$= 2hy_0 + 2h^2y_0' + \frac{4h^3}{3}y_0'' + \frac{2h^4}{3}y_0''' + \frac{4h^5}{15}y_0^{iv} + \cdots$$

$\therefore$ Error in the interval $[x_0, x_2]$ is given by

$$A_2' - A_2 = \left[2hy_0 + 2h^2y_0' + \frac{4h^3}{3}y_0'' + \frac{2h^4}{3}y_0''' + \frac{4h^5}{15}y_0^{iv} + \cdots\right]$$

$$- \left[2hy_0 + 2h^2y_0' + \frac{4h^3}{3}y_0'' + \frac{2h^4}{3}y_0''' + \frac{5h^5}{18}y_0^{iv} + \cdots\right]$$

$$= \left(\frac{4}{15} - \frac{5}{18}\right)h^5 y_0^{iv} + \cdots = -\frac{h^5}{90}y_0^{iv} + \cdots$$

Thus lowest order error in interval $[x_0, x_2] \approx -\frac{h^5}{90}y_0^{iv}$

Similarly lowest order error in interval $[x_1, x_2] \approx -\frac{h^5}{90}y_1^{iv}$

$\vdots$

Thus lowest order error in interval $[x_{n-1}, x_n] \approx -\frac{h^5}{90}y_{n-1}^{iv}$

Total error $E = -\frac{h^5}{90}\left[y_0^{iv} + y_1^{iv} + \cdots + y_{n-1}^{iv}\right]$

If $y_x^{iv} = \text{Maximum}\left[y_0^{iv}, y_1^{iv}, \ldots, y_{n-1}^{iv}\right]$

Then Maximum $E \leq -\frac{nh^5}{180}y_x^{iv}$ $\qquad \because$ number of intervals is $\frac{n}{2}$

$$= -\frac{nhh^4}{180}y_x^{iv} = -\frac{(b-a)h^4}{180}y_x^{iv} \qquad \because nh = (b-a)$$

$\therefore E \leq ph^4$, where $p = -\frac{(b-a)}{180}y_x^{iv}$

Hence the error in Simpson's one-third rule is of order h^4, where h is the height of the interval.

12.3.3 Numerical Integration Using Simpson's Three-Eighth Rule

If we put $n = 3$ in Newton's Cote's quadrature formula given by ①, we get the curve through the points (x_0, y_0), (x_1, y_1), (x_2, y_2), (x_3, y_3) as a cubic polynomial and hence 4th and higher order differences are zero.

∴Newton's Cote's quadrature formula reduces to:

$$I = \int_{x_0}^{x_0+3h} f(x)dx = 3h\left[y_0 + \frac{3}{2}\Delta y_0 + \left(\frac{9}{3} - \frac{3}{2}\right)\frac{\Delta^2 y_0}{2!} + \left(\frac{27}{4} - 9 + 3\right)\frac{\Delta^3 y_0}{3!}\right]$$

$$= 3h\left[y_0 + \frac{3}{2}(y_1 - y_0) + \frac{3}{4}(y_2 - 2y_1 + y_0) + \frac{1}{8}(y_3 - 3y_2 + 3y_1 - y_0)\right]$$

$$= \frac{3h}{8}[8y_0 + 12(y_1 - y_0) + 6(y_2 - 2y_1 + y_0) + (y_3 - 3y_2 + 3y_1 - y_0)]$$

$$= \frac{3h}{8}[(8 - 12 + 6 - 1)y_0 + (12 - 12 + 3)y_1 + (6 - 3)y_2 + y_3]$$

$$= \frac{3h}{8}[y_0 + 3y_1 + 3y_2 + y_3]$$

Similarly $\int_{x_0+3h}^{x_0+6h} f(x)dx = \frac{3h}{8}[y_3 + 3y_4 + 3y_5 + y_6]$

⋮

$$\int_{x_0+(n-3)h}^{x_0+nh} f(x)dx = \frac{3h}{8}[y_{n-3} + 3y_{n-2} + 3y_{n-1} + y_n]$$

Adding areas of all these intervals, we get:

$$\int_{x_0}^{x_0+nh} f(x)dx = \frac{3h}{8}[y_0 + 3y_1 + 3y_2 + y_3] + \frac{3h}{8}[y_3 + 3y_4 + 3y_5 + y_6]$$

$$+ \cdots + \frac{3h}{8}[y_{n-3} + 3y_{n-2} + 3y_{n-1} + y_n]$$

$$\therefore \int_a^b f(x)dx = \frac{3h}{8}[(y_0 + y_n) + 3(y_1 + y_2 + y_4 + y_5 + \cdots + y_{n-2} + y_{n-1})$$

$$+ 2(y_3 + y_6 + \cdots + y_{n-3})]$$

This is known as Simpson's three-eighths rule to evaluate $\int_a^b f(x)dx$, where the function $y = f(x)$ is given as discrete set of points (x_i, y_i), $i = 0, 1, 2, 3, \ldots, n$.

Geometrical Significance of Simpson's Three-Eighth Rule

The Simpson's 3/8 rule is similar to the 1/3 rule except that curve $y = f(x)$ is replaced by arcs of 3rd degree polynomial curve, as shown in given figure. It is used when it is required to take three segments at a time. Thus number of intervals must be a multiple of three.

It can be derived that the lowest order error in interval $[x_0, x_3] \approx -\frac{3h^5}{80} y_0^{iv}$

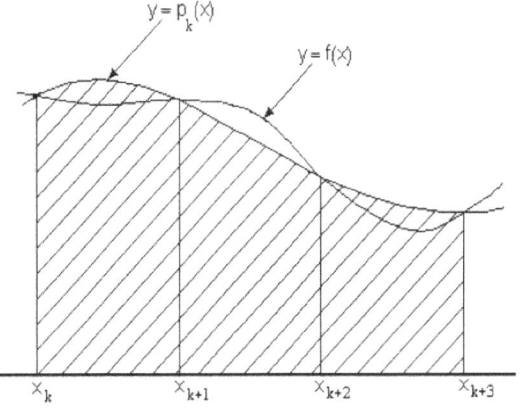

Maximum $E \leq -\frac{nh^5}{80} y_x^{iv}$ ∵ number of intervals is $\frac{n}{3}$

$$= -\frac{nhh^4}{80} y_x^{iv} = -\frac{(b-a)h^4}{80} y_x^{iv}$$

$$\because nh = (b-a)$$

$\therefore E \leq qh^4$, where $q = -\frac{(b-a)}{80} y_x^{iv}$

Hence error in Simpson's 3/8 rule is of order h^4, where h is the height of the interval.

12.3.4 Applications of Numerical Integration

Numerical integration has numerous practical applications in the field of calculus. Simpson's $\frac{1}{3}$ rule due to its ease in application and higher accuracy is a preferred method in various application areas as given below:

- Area bounded by a curve $y = f(x)$ between the ordinates $x = a$ and $x = b$, above $x -$ axis is given by $A = \int_a^b y \, dx$.
- Volume of solid formed by revolving the curve $y = f(x)$ between the ordinates $x = a$ and $x = b$ along $x -$ axis is given by $V = \int_a^b \pi y^2 \, dx$.
- Length of an arc of the curve $y = f(x)$ between the ordinates $x = a$ and $x = b$ and $x -$ axis is given by $\int_a^b \sqrt{1 + \left(\frac{dy}{dx}\right)^2} \, dx$
- To find velocity when acceleration at different times is given in tabular form.
- To find displacement when velocity is given as a function of time in discrete form.

Remarks:

> Simpson's rules ideally returns more accurate results compared to trapezoidal rule provided h is small, less than one essentially.
> Simpson's $\frac{1}{3}$ rule requires odd number of points (even number of sub-intervals) for application.
> Simpson's $\frac{3}{8}$ rule requires number of sub-intervals to be multiple of three.

Example 9 Evaluate $\int_0^1 \frac{1}{1+x^2} dx$ using

i. Trapezoidal rule taking $h = \frac{1}{5}$
ii. Simpson's $\frac{1}{3}$ rule taking $h = \frac{1}{4}$
iii. Simpson's $\frac{3}{8}$ rule taking $h = \frac{1}{6}$

Solution: i. To solve $\int_0^1 \frac{1}{1+x^2} dx$ using trapezoidal rule

Taking $h = \frac{1}{5} = 0.2$, $n = \frac{b-a}{h} = \frac{1-0}{0.2} = 5$

∴ Dividing the interval $(0,1)$ into 5 equal parts for the function $f(x) = \frac{1}{1+x^2}$

x	0	0.2	0.4	0.6	0.8	1
$y = f(x)$	1	0.96	0.86	0.74	0.61	0.5

By trapezoidal rule $\int_0^1 \frac{1}{1+x^2} dx = \frac{h}{2}[y_0 + 2(y_1 + y_2 + y_3 + y_4) + y_5]$

$= \frac{0.2}{2}[1 + 2(0.96 + 0.86 + 0.74 + 0.61) + 0.5]$

$\therefore \int_0^1 \frac{1}{1+x^2} dx = 0.784$ using trapezoidal rule.

ii. To solve $\int_0^1 \frac{1}{1+x^2} dx$ using Simpson's $\frac{1}{3}$ rule

Taking $h = \frac{1}{4} = 0.25$, $n = \frac{b-a}{h} = \frac{1-0}{0.25} = 4$

∴ Dividing the interval (0,1) into 4 equal parts for the function $f(x) = \frac{1}{1+x^2}$

x	0	0.25	0.5	0.75	1
$y = f(x)$	1	0.94	0.8	0.64	0.5

By Simpson's $\frac{1}{3}$ rule $\int_0^1 \frac{1}{1+x^2} dx = \frac{h}{3}[(y_0 + y_4) + 4(y_1 + y_3) + 2(y_2)]$

$= \frac{0.25}{3}[(1 + 0.5) + 4(0.94 + 0.64) + 2(0.8)]$

$\therefore \int_0^1 \frac{1}{1+x^2} dx = 0.7850$ using Simpson's $\frac{1}{3}$ rule

iii. To solve $\int_0^1 \frac{1}{1+x^2} dx$ using Simpson's $\frac{3}{8}$ rule

Taking $h = \frac{1}{6}$, $n = \frac{b-a}{h} = \frac{1-0}{\frac{1}{6}} = 6$

∴ Dividing the interval (0,1) into 6 equal parts for the function $f(x) = \frac{1}{1+x^2}$

x	0	$\frac{1}{6}$	$\frac{2}{6}$	$\frac{3}{6}$	$\frac{4}{6}$	$\frac{5}{6}$	1
$y = f(x)$	1	0.97	0.9	0.8	0.69	0.59	0.5

By Simpson's $\frac{3}{8}$ rule $\int_0^1 \frac{1}{1+x^2} dx = \frac{3h}{8}[(y_0 + y_6) + 3(y_1 + y_2 + y_4 + y_5) + 2(y_3)]$

$= \frac{3}{8} \cdot \frac{1}{6}[(1 + 0.5) + 3(0.97 + 0.9 + 0.69 + 0.59) + 2(0.8)]$

$\therefore \int_0^1 \frac{1}{1+x^2} dx = 0.7844$ using Simpson's $\frac{3}{8}$ rule

Example10 Evaluate $\int_0^{\frac{\pi}{2}} \sin x \, dx$ using

i. Trapezoidal rule   ii. Simpson's $\frac{1}{3}$ rule   iii. Simpson's $\frac{3}{8}$ rule, taking $n = 6$

Solution: Taking $n = 6$, $h = \frac{b-a}{n} = \frac{\frac{\pi}{2}-0}{6} = \frac{\pi}{12}$

∴ Dividing the interval $\left(0, \frac{\pi}{2}\right)$ into 6 equal parts for the function $f(x) = \sin x$

x	0	$\frac{\pi}{12}$	$\frac{2\pi}{12}$	$\frac{3\pi}{12}$	$\frac{4\pi}{12}$	$\frac{5\pi}{12}$	$\frac{\pi}{2}$
$y = f(x)$	0	0.2588	0.5	0.7071	0.866	0.9659	1

i. To solve $\int_0^{\frac{\pi}{2}} \sin x \, dx$ using trapezoidal rule

$\int_0^{\frac{\pi}{2}} \sin x \, dx = \frac{h}{2}[y_0 + 2(y_1 + y_2 + y_3 + y_4 + y_5) + y_6]$

$= \frac{\pi}{12} \cdot \frac{1}{2}[0 + 2(0.2588 + 0.5 + 0.7071 + 0.866 + 0.9659) + 1]$

∴ $\int_0^{\frac{\pi}{2}} \sin x \, dx = 0.9943$ using trapezoidal rule.

ii. To solve $\int_0^{\frac{\pi}{2}} \sin x \, dx$ using Simpson's $\frac{1}{3}$ rule

$\int_0^{\frac{\pi}{2}} \sin x \, dx = \frac{h}{3}[(y_0 + y_6) + 4(y_1 + y_3 + y_5) + 2(y_2 + y_4)]$

$= \frac{\pi}{12} \cdot \frac{1}{3}[(0 + 1) + 4(.2588 + .7071 + .9659) + 2(0.5 + .866)]$

∴ $\int_0^{\frac{\pi}{2}} \sin x \, dx = 1.000004$ using Simpson's $\frac{1}{3}$ rule

iii. To solve $\int_0^{\frac{\pi}{2}} \sin x \, dx$ using Simpson's $\frac{3}{8}$ rule

$\int_0^{\frac{\pi}{2}} \sin x \, dx = \frac{3h}{8}[(y_0 + y_6) + 3(y_1 + y_2 + y_4 + y_5) + 2(y_3)]$

$= \frac{\pi}{12} \cdot \frac{3}{8}[(0 + 1) + 3(0.2588 + .5 + .866 + .9659) + 2(0.7071)]$

∴ $\int_0^{\frac{\pi}{2}} \sin x \, dx = 1.00004$ using Simpson's $\frac{3}{8}$ rule

Example 11 Evaluate $\int_{0.5}^{1.3} \frac{1}{1+\log x} dx$ using

i. Trapezoidal rule   ii. Simpson's $\frac{1}{3}$ rule   iii. Simpson's $\frac{3}{8}$ rule, taking $n = 8$

Solution: Taking $n = 8$, $h = \frac{1.3 - 0.5}{8} = 0.1$

∴ Dividing the interval $(0.5, 1.3)$ into 8 equal parts for $f(x) = \frac{1}{1+\log x}$

x	0.5	0.6	0.7	0.8	0.9	1	1.1	1.2	1.3
$f(x)$	3.26	2.04	1.55	1.29	1.12	1	0.91	0.89	0.79

i. To solve $\int_{0.5}^{1.3} \frac{1}{1+\log x} dx$ using trapezoidal rule

$\int_{0.5}^{1.3} \frac{1}{1+\log x} dx = \frac{h}{2}[y_0 + 2(y_1 + y_2 + \cdots + y_7) + y_8]$

$= \frac{0.1}{2}[3.26 + 2(2.04 + 1.55 + 1.29 + 1.12 + 1 + .91 + .89) + .79]$

$= 0.05[21.65] = 1.0825$

∴ $\int_{0.5}^{1.3} \frac{1}{1+\log x} dx = 1.0825$ using trapezoidal rule.

ii. To solve $\int_{0.5}^{1.3} \frac{1}{1+\log x} dx$ using Simpson's $\frac{1}{3}$ rule

$\int_{0.5}^{1.3} \frac{1}{1+\log x} dx = \frac{h}{3}[(y_0 + y_8) + 4(y_1 + y_3 + y_5 + y_7) + 2(y_2 + y_4 + y_6)]$

$= \frac{0.1}{3}[(3.26 + .79) + 4(2.04 + 1.29 + 1 + .89) + 2(1.55 + 1.12 + .91)]$

$= \frac{0.1}{3}[32.09] = 1.070$

$\therefore \int_{0.5}^{1.3} \frac{1}{1+\log x} dx = 1.070$ using Simpson's $\frac{1}{3}$ rule

iii. Simpson's $\frac{3}{8}$ rule is not applicable, as n is not a multiple of 3

Example 12 Evaluate $\int_0^{\frac{\pi}{2}} e^{\cos x} dx$ up to 4 decimal places, using trapezoidal rule.

Solution: Taking $n = 3$, i.e. $h = \frac{\frac{\pi}{2}-0}{3} = \frac{\pi}{6}$

x	0	$\frac{\pi}{6}$	$\frac{\pi}{3}$	$\frac{\pi}{2}$
$y = f(x)$	2.71828	2.37744	1.64872	1

By trapezoidal rule

$\int_0^{\frac{\pi}{2}} e^{\cos x} dx = \frac{h}{2}[y_0 + 2(y_1 + y_2) + y_3]$

$= \frac{\pi}{6} \cdot \frac{1}{2}[2.71828 + 2(2.37744 + 1.64872) + 1]$

$= \frac{\pi}{12}[11.7706] = 3.0815$

$\therefore \int_0^{\frac{\pi}{2}} e^{\cos x} dx = 3.0815$ using trapezoidal rule

Example 13 From the following table, find the area bounded by the curve and x − axis, between the ordinates $x = 7.47$ to $x = 7.52$.

x	7.47	7.48	7.49	7.50	7.51	7.52
$y = f(x)$	1.93	1.95	1.98	2.01	2.03	2.06

Solution: As $n = 5$, Simpson's $\frac{1}{3}$ rule Simpson's $\frac{3}{8}$ rules are not applicable. Applying trapezoidal rule with $h = 0.01$

$\int_{7.47}^{7.52} f(x) dx = \frac{.01}{2}[1.93 + 2(1.95 + 1.98 + 2.01 + 2.03) + 2.06]$

$= 0.005[19.93] = 0.09965$ square units

Example 14 The velocity v of an airplane which starts from rest is given at fixed intervals of time t as shown:

t (minutes)	2	4	6	8	10	12	14	16	18	20
$v = f(t)$ (km/minutes)	8	17	24	28	30	20	12	6	2	0

Estimate the approximate distance covered in 20 minutes.

Solution: Since the airplane starts from rest, its initial velocity is zero. So the time/velocity relationship may be tabulated as:

t (minutes)	0	2	4	6	8	10	12	14	16	18	20
$v = f(t)$ (km/minutes)	0	8	17	24	28	30	20	12	6	2	0

Let S be the distance covered at any instant of time t,

Then $v = \frac{dS}{dt}$ or $dS = v\, dt$

∴ Distance covered in 20 minutes is given by:

$$S = \int_0^{20} dS = \int_0^{20} v\, dt = \frac{h}{3}[(v_0 + v_{10}) + 4(v_1 + v_3 + v_5 + v_7 + v_9)$$
$$2(v_2 + v_4 + v_6 + v_8)]$$

Applying Simpson's $\frac{1}{3}$ rule with $h = 2$, as $n = 10$

$$\Rightarrow S = \frac{2}{3}[(0 + 0) + 4(8 + 24 + 30 + 12 + 2) + 2(17 + 28 + 20 + 6)]$$

∴ $S = 297.33$ km

Example 15 A solid of revolution is formed by rotating about x − axis, the area between x − axis, the line $x = 0$ and a curve through the points with the following coordinates:

x	0	0.25	0.50	0.75	1
$y = f(x)$	1	0.5846	0.5586	0.5085	0.7328

Estimate the volume of solid formed, giving the answer up to 3 decimal places.

Solution: Volume of solid formed by revolving the curve $y = f(x)$ between the ordinates $x = a$ and $y = b$ along x − axis is given by $V = \int_a^b \pi y^2 dx$

By Simpson's $\frac{1}{3}$ rule with $h = 0.25$, as $n = 4$

∴ $V = \pi \int_0^1 y^2 dx = \frac{\pi h}{3}[y_0^2 + y_4^2) + 4(y_1^2 + y_3^2) + 2(y_2^2)]$

$\Rightarrow V = \frac{\pi}{12}\{[1^2 + (0.7328)^2] + 4[(0.5846)^2 + (0.5085)^2] + 2[(0.5586)^2]\}$

$= \frac{\pi}{12}[1.5370 + 4(0.60033) + 2(0.31203)] = 1.944$

Exercise 12

1. For the following data, approximate $\frac{dy}{dx}$, at $x = 2$

x	0	2	3
y	2	-2	-1

2. Following table gives the angle θ (radians) through which a rod is rotating in a plane for different times t (seconds).

t (seconds)	0	0.2	0.4	0.6	0.8	1.0	1.2
θ (radians)	0	0.12	0.49	1.12	2.02	3.2	4.67

Calculate the angular velocity and angular acceleration of the rod, when $t = 0.6$ seconds.

3. From the data given below, find the value of x, for which y is maximum.

x	3	4	5	6	7	8
y	0.205	0.24	0.259	0.262	0.25	0.224

Also find maximum y.

4. Find value of cos 1.747, using the values given in the table below:

x	1.7	1.74	1.78	1.82	1.86
sin x	0.9916	0.9857	0.9781	0.9691	0.9584

5. Derive Newton-Cote's quadrature formula for evaluating $\int_a^b f(x)\,dx$, where the interval $[a,b]$ is divided into n subintervals of height h.

6. Approximate $\int_0^6 \frac{1}{1+x^2}\,dx$ taking $h=1$, using:

   i. Trapezoidal rule   ii. Simpsons 1/3 rule   iii. Simpsons 3/8 rule and compare the results with direct integration.

7. A river is 80 feet wide. The depth d in feet at a distance x feet from one bank is given in the following table:

x(feet)	0	10	20	30	40	50	60	70	80
d(feet)	0	4	7	9	12	15	14	8	3

Find the approximate area of cross-section of the river taking $h = 10$ using trapezoidal rule.

8. The velocity v of an airplane, which starts from rest, is given at fixed intervals of time as shown below:

v(km)	2	4	6	8	10	12	14	16	18	20
t(minutes)	10	18	25	29	32	20	11	5	2	0

Find the approximate distance covered in 20 minutes.

9. Compute $\int_{0.2}^{1.4} (\sin x - \log_e x + e^x)\,dx$ using Simpson's $\frac{3}{8}$ rule and compare with direct result.

Answers

1. 0
2. 3.817 radians/seconds, 6.75 radians per seconds square
3. 5.6875, 0.2628
4. -0.176
6. i 1.4108   ii. 1.366174   iii. 1.357082, value by direct integration is 1.4056
7. 705 square feet
8. 309.33 km
9. 4.05116, value by direct integration is 4.05094

Chapter 13: Numerical Solutions of Ordinary Differential Equations

Numerical Solutions
Of
Ordinary Differential Equations

13.1 Introduction

An ordinary differential equation is a mathematical equation that relates one or more functions of an independent variable with its derivatives. Differential equations are of extreme importance to scientists and engineers as they are inevitable tools for mathematical modeling of any problem involving rate of change. Sometimes we encounter situations where these equations are not amenable to analytic solutions. They can either be solved using mathematical software or by using numerical techniques discussed in coming sections.

Many practical applications lead to second or higher order systems of ordinary differential equations, numerical methods for higher order initial value problems are entirely based on their reformulation as first order systems. Numerical solutions of ordinary differential equations require initial values as they are based on finite-dimensional approximations.

The first-order differential equation and the given initial value constitute a first-order initial value problem given as: $\frac{dy}{dx} = f(x, y)$; $y(x_0) = y_0$, whose numerical solution may be given using any of the following methodologies:

I. **Single-Step methods**
  (a) Taylor series method
  (b) Picard's method
  (c) Euler's method
  (d) Modified Euler's method
  (e) Runge-Kutta method
II. **Multi-steps methods**
  (a) Milne's Predictor corrector method
  (b) Adams-Bashforth method

All these methods will be discussed in detail in the coming sections.

13.2 Single Step Methods to Solve Initial Value Problems

Single-step methods such as Taylor series and Picard's method refer to initial value and its derivative to determine the current solution. Some methods such as Runge-Kutta take some intermediate steps to obtain a higher order method, but then they discard all previous information before taking a second step.

13.2.1 Taylor Series Method

Taylor's series expansion of a function $y(x)$ about $x = x_0$ is given by

$$y(x) = y_0 + (x - x_0)y_0' + \frac{1}{2!}(x - x_0)^2 y_0'' + \frac{1}{3!}(x - x_0)^3 y_0''' + \cdots \quad \cdots ①$$

To approximate $y(x)$ numerically for the initial value problem given by

$\frac{dy}{dx} = f(x, y)$; $y(x_0) = y_0$, we substitute the values of y_0 and its successive derivatives in Taylor's series given by ①. Working methodology is illustrated in the examples given below.

Example1 Solve the differential equation $\frac{dy}{dx} = x + y$; $y(0) = 1$, at $x = 0.2$, 0.4 correct to 3 decimal places, using Taylor's series method. Also compare the numerical solution obtained with the analytic solution.

Solution: Taylor's series expansion of $y(x)$ about $x = 0$ is given by:

$$y(x) = y_0 + (x - 0)y_0' + \frac{1}{2!}(x - 0)^2 y_0'' + \frac{1}{3!}(x - 0)^3 y_0''' + \frac{1}{4!}(x - 0)^4 y_0^{iv} + \cdots ①$$

$$\text{Given} \quad \frac{dy}{dx} = x + y \quad ; \quad y_0 = 1$$

$$\text{or} \quad y' = x + y \quad ; \quad y_0' = 1$$

$$\Rightarrow y'' = 1 + y' \quad ; \quad y_0'' = 2$$

$$y''' = y'' \quad ; \quad y_0''' = 2$$

$$y^{iv} = y''' \quad ; \quad y_0^{iv} = 2$$

$$\vdots$$

Substituting these values in ①, we get

$$y(x) = 1 + x(1) + \frac{1}{2!}x^2(2) + \frac{1}{3!}x^3(2) + \frac{1}{4!}x^4(2) + \cdots$$

or $\quad y(x) = 1 + x + x^2 + \frac{x^3}{3} + \frac{x^4}{12} + \cdots$

i. $y(0.2) = 1 + 0.2 + 0.04 + \frac{0.008}{3} + \frac{0.0016}{12} + \cdots$

$= 1 + 0.2 + 0.04 + 0.002667 + 0.00013 + \cdots$

The fifth term in this series is $0.00013 < 0.0005$

Hence value of $y(0.2)$ correct to 3 decimal places may be obtained by adding first four terms.

$\therefore y(0.2) \approx 1.24280 \approx 1.243$

ii. $y(0.4) = 1 + 0.4 + 0.16 + \frac{0.064}{3} + \frac{0.0256}{12} + \frac{0.01024}{60} + \cdots$

$= 1 + 0.4 + 0.16 + 0.02133 + 0.00213 + 0.00017 + \cdots$

The sixth term in this series is $0.00017 < 0.0005$

Hence value of $y(0.4)$ correct to 3 decimal places may be obtained by adding first five terms.

$\therefore y(0.4) \approx 1.58346 \approx 1.583$ correct to three decimal places.

Again to find exact solution of $\frac{dy}{dx} - y = x$, which is a linear differential equation

Integrating Factor (I.F.) $= e^{\int -dx} = e^{-x}$

Solution is given by $ye^{-x} = \int xe^{-x} dx$

$$\Rightarrow ye^{-x} = -xe^{-x} - e^{-x} + c$$
$$\Rightarrow y = -x - 1 + ce^x$$

Given that $y(0) = 1 \Rightarrow 1 = 0 - 1 + c \quad \therefore c = 2$

$\Rightarrow y = -x - 1 + 2e^x$

$y(0.2) \approx 1.243$ and $y(0.4) \approx 1.584$ correct to three decimal places.

Example 2 Solve the differential equation $\frac{dy}{dx} = 4y$; $(0) = 1$, at $x = 0.1$ using Taylor's series method correct to three decimal places.

Solution: Taylor's series of $y(x)$ about $x = 0$, is given by

$$y(x) = y_0 + (x-0)y_0' + \frac{1}{2!}(x-0)^2 y_0'' + \frac{1}{3!}(x-0)^3 y_0''' + \frac{1}{4!}(x-0)^4 y_0^{iv} + \cdots \text{①}$$

Given $\frac{dy}{dx} = 4y$; $y_0 = 1$

or $y' = 4y$; $y_0' = 4$

$\Rightarrow y'' = 4y'$; $y_0'' = 16$

$y''' = 4y''$; $y_0''' = 64$

$y^{iv} = 4y'''$; $y_0^{iv} = 256$

$\vdots$

Substituting these values in ①, we get

$$y(x) = 1 + x(4) + \frac{1}{2!}x^2(16) + \frac{1}{3!}x^3(64) + \frac{1}{4!}x^4(256) + \cdots$$

or $y(x) = 1 + 4x + \frac{16x^2}{2!} + \frac{64x^3}{3!} + \frac{256x^4}{4!} + \frac{256x^4}{5!} \cdots$

$\Rightarrow y(x) = 1 + 4x + 8x^2 + \frac{32}{3}x^3 + \frac{32}{3}x^4 + \cdots$

$y(0.1) = 1 + 4(0.1) + 8(0.1)^2 + \frac{32}{3}(0.1)^3 + \frac{32}{3}(0.1)^4 + \frac{128}{15}(0.1)^5 \ldots$

$\Rightarrow y(0.1) = 1 + 0.4 + 0.08 + 0.01067 + 0.00107 + 0.00009$

$y(0.1) \approx 1.49183 \approx 1.492$ correct to three decimal places

Again to find analytical solution of $\frac{dy}{dx} = 4y \Rightarrow \frac{dy}{y} = 4dx$

This is a variable separable equation, whose solution is given by:

$\log y = 4x + \log c$

$\Rightarrow y = ce^{4x}$

Given that $y(0) = 1 \quad \therefore c = 1$

$\Rightarrow y = e^{4x}$

$y(0.1) \approx 1.491824 \approx 1.492$ correct to three decimal places

Example 3 Using Taylor's series method, solve the differential equation

$\frac{dy}{dx} = y + 3e^x$; $(0) = 1$, at $x = 0.2$

Also compare the result with the exact solution.

Solution: Taylor's series expansion of $y(x)$ about $x = 0$ is given by:

$$y(x) = y_0 + (x-0)y_0' + \frac{1}{2!}(x-0)^2 y_0'' + \frac{1}{3!}(x-0)^3 y_0''' + \frac{1}{4!}(x-0)^4 y_0^{iv} + \cdots \quad \text{①}$$

Given $\frac{dy}{dx} = y + 3e^x$; $y_0 = 1$

or $y' = y + 3e^x$; $y_0' = 4$

$\Rightarrow y'' = y' + 3e^x$; $y_0'' = 7$

$y''' = y'' + 3e^x$; $y_0''' = 10$

$y^{iv} = y''' + 3e^x$; $y_0^{iv} = 13$

$y^v = y^{iv} + 3e^x$; $y_0^v = 16$

$\vdots$

Substituting these values in ①, we get

$$y(x) = 1 + x(4) + \frac{1}{2!}x^2(7) + \frac{1}{3!}x^3(10) + \frac{1}{4!}x^4(13) + \frac{1}{5!}x^5(16) + \cdots$$

or $y(x) = 1 + 4x + \frac{7}{2}x^2 + \frac{5}{3}x^3 + \frac{13}{24}x^4 + \frac{2}{15}x^5 + \cdots$

i. $y(0.2) = 1 + 4(0.2) + \frac{7}{2}(0.2)^2 + \frac{5}{3}(0.2)^3 + \frac{13}{24}(0.2)^4 + \frac{2}{15}(0.2)^5 + \cdots$

$= 1 + 0.8 + 0.14 + 0.01333 + 0.00087 + 0.00004 + \cdots$

The sixth term in this series is $0.00004 < 0.0005$

Hence value of $y(0.2)$ correct to 3 decimal places may be obtained by adding first five terms.

$\therefore y(0.2) \approx 1.9542 \approx 1.954$

Again to find exact solution of $\frac{dy}{dx} - y = 3e^x$, which is a linear equation

Integrating Factor (I.F.) $= e^{\int -dx} = e^{-x}$

Solution is given by $ye^{-x} = 3 \int e^x e^{-x} dx$

$\Rightarrow ye^{-x} = 3x + c$

$\Rightarrow y = (3x + c)e^x$

Given that $y(0) = 1 \Rightarrow c = 1$

$\Rightarrow y = (3x + 1)e^x$

$y(0.2) \approx 1.954244 \approx 1.954$ correct to three decimal places

13.2.2 Picard's Method of Successive Approximations

Consider the initial value problem given by $\frac{dy}{dx} = f(x, y)$; $y(x_0) = y_0$

$\Rightarrow dy = f(x, y)dx$

Integrating, we get

$\int_{y_0}^{y} dy = \int_{x_0}^{x} f(x,y)dx$

$\Rightarrow y - y_0 = \int_{x_0}^{x} f(x,y)dx$

$\Rightarrow y = y_0 + \int_{x_0}^{x} f(x,y)dx$

To obtain the first approximation, replacing y by y_0 on R.H.S.

$\Rightarrow \quad y_1 = y_0 + \int_{x_0}^{x} f(x,y_0)dx$

Similarly $y_2 = y_0 + \int_{x_0}^{x} f(x,y_1)dx$

$\vdots$

$y_n = y_0 + \int_{x_0}^{x} f(x,y_{n-1})dx$, where $y(x_0) = y_0$

Remark: Picard's method can be applied only to limited types of problems, which can be integrated successively.

Example4 Using Picard's method, solve the initial value problem $\frac{dy}{dx} = x + y$; $y(0) = 1$, up to 3 approximations.

Solution: Given $f(x,y) = x + y$, $\quad x_0 = 0$, $y_0 = 1$

Using Picard's approximation

$$y = y_0 + \int_{x_0}^{x} f(x,y)dx$$

1st approximation: $y_1 = y_0 + \int_{x_0}^{x} f(x,y_0)dx$

$\Rightarrow y_1 = 1 + \int_0^x (x+1)dx$

$= 1 + \left[\frac{x^2}{2} + x\right]_0^x = 1 + x + \frac{x^2}{2}$

2nd approximation: $y_2 = y_0 + \int_{x_0}^{x} f(x,y_1)dx$

$\Rightarrow y_2 = 1 + \int_0^x (x + y_1)dx$

$= 1 + \int_0^x \left(x + \left(1 + x + \frac{x^2}{2}\right)\right)dx$

$= 1 + x + x^2 + \frac{x^3}{6}$

3rd approximation: $y_3 = y_0 + \int_{x_0}^{x} f(x,y_2)dx$

$\Rightarrow y_3 = 1 + \int_0^x (x + y_2)dx$

$= 1 + \int_0^x \left(x + \left(1 + x + x^2 + \frac{x^3}{6}\right)\right)dx$

$= 1 + x + x^2 + \frac{x^3}{3} + \frac{x^4}{24}$

Example5 Using Picard's method, obtain the solution of $\frac{dy}{dx} = x(1 + x^3 y)$; $y(0) = 3$, at $x = 0.1$.

Solution: Given $f(x, y) = x(1 + x^3 y)$, $x_0 = 0$, $y_0 = 3$

Using Picard's approximation

$$y = y_0 + \int_{x_0}^{x} f(x, y) dx$$

1st approximation: $y_1 = y_0 + \int_{x_0}^{x} f(x, y_0) dx$

$$\Rightarrow y_1 = 3 + \int_0^x x(1 + x^3 y) \, dx$$

$$= 3 + \frac{x^2}{2} + \frac{3x^5}{5}$$

2nd approximation: $y_2 = y_0 + \int_{x_0}^{x} f(x, y_1) dx$

$$\Rightarrow y_2 = 3 + \int_0^x x \left[1 + x^3 \left(3 + \frac{x^2}{2} + \frac{3x^5}{5}\right)\right] dx$$

$$= 3 + \frac{x^2}{2} + \frac{3x^5}{5} + \frac{x^7}{14} + \frac{3x^{10}}{50}$$

Clearly y_1 and y_2 are coincident up to 3 terms, $\therefore$ let $y = 3 + \frac{x^2}{2} + \frac{3x^5}{5}$

Also $y(0.1) = 3 + \frac{(0.1)^2}{2} + \frac{3(0.1)^5}{5} = 3.00501$

Example6 Using Picard's method, find $y(1.1)$ for the initial value problem

$$\frac{dy}{dx} = xy; \quad y(1) = 2.$$

Solution: Given $f(x, y) = xy$, $x_0 = 1$, $y_0 = 2$

Using Picard's approximation

$$y = y_0 + \int_{x_0}^{x} f(x, y) dx$$

1st approximation: $y_1 = y_0 + \int_{x_0}^{x} f(x, y_0) dx$

$$\Rightarrow y_1 = 2 + \int_1^x x(2) dx$$

$$= 2 + [x^2]_1^x = 1 + x^2$$

2nd approximation: $y_2 = y_0 + \int_{x_0}^{x} f(x, y_1) dx$

$$\Rightarrow y_2 = 2 + \int_1^x (x. y_1) dx$$

$$= 2 + \int_1^x (x(1 + x^2)) dx = \frac{5}{4} + \frac{x^2}{2} + \frac{x^4}{4}$$

3rd approximation: $y_3 = y_0 + \int_{x_0}^{x} f(x, y_2) dx$

$$\Rightarrow y_3 = 2 + \int_{1}^{x} (x \cdot y_2) dx$$

$$= 2 + \int_{1}^{x} x \left(\frac{5}{4} + \frac{x^2}{2} + \frac{x^4}{4} \right) dx$$

$$= \frac{29}{24} + \frac{5x^2}{8} + \frac{x^4}{8} + \frac{x^6}{24}$$

$$\therefore y(1.1) = \frac{29}{24} + \frac{5(1.1)^2}{8} + \frac{(1.1)^4}{8} + \frac{(1.1)^6}{24} = 2.2214$$

13.2.3 Euler's Method

Euler's Method provides us with a numerical solution of the initial value problem $\frac{dy}{dx} = f(x, y); \ y(x_0) = y_0 \ \cdots \ ①$, by joining multiple small line segments $A_0 A_1$, $A_1 A_2$, $A_2 A_3, \cdots$, making an approximation of the actual curve, as shown in the adjoining figure.

Thus if $[x_0, x_1]$ is the small interval, where $x_1 = x_0 + h$, we approximate the curve by the tangent drawn to curve at the poin A_0, having coordinates (x_0, y_0), whose equation is given by

$y - y_0 = m(x - x_0)$, where m is slope of tangent at the point (x_0, y_0)

Also $m = \frac{dy}{dx}\Big|_{(x_0, y_0)} = f(x_0, y_0)$ from ①

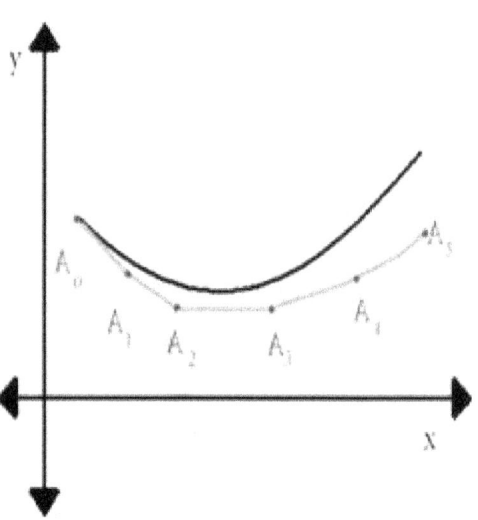

$\Rightarrow y = y_0 + f(x_0, y_0)(x - x_0)$

$\Rightarrow y_1 = y_0 + f(x_0, y_0)(x_1 - x_0) \quad \because y(x_1) = y_1$

$\Rightarrow y_1 = y_0 + hf(x_0, y_0) \quad \because x_1 - x_0 = h$

Similarly for range $[x_1, x_2]$

$y_2 = y_1 + hf(x_1, y_1)$

⋮

$y_n = y_{n-1} + hf(x_{n-1}, y_{n-1})$

It is evident from the given figure that h has to be kept small to avoid the approximations diverging away from the curve. As a result, this method is very slow and needs to be improved.

Example7 Using Euler's method, Compute $y(0.12)$ for the initial value problem:

$\frac{dy}{dx} = x^3 + y; \ y(0) = 1$, taking $h = 0.02$.

Solution: Given $f(x, y) = x^3 + y$, $x_0 = 0$, $y_0 = 1$, $x_n = x_{n-1} + h$, $h = 0.02$

$\therefore x_1 = 0.02, \ x_2 = 0.04, \ x_3 = 0.06, \ x_4 = 0.08, \ x_5 = 0.1$

Using Euler's method $y_n = y_{n-1} + hf(x_{n-1}, y_{n-1})$

$\Rightarrow y_n = y_{n-1} + h(x_{n-1}^3 + y_{n-1}) \quad \cdots ①$

Putting $n = 1$ in ①, $y_1 = y(0.02) = y_0 + h(x_0^3 + y_0)$

∴ $y_1 = 1 + 0.02(0 + 1) = 1.02$

Putting $n = 2$ in ①, $y_2 = y(0.04) = y_1 + h(x_1^3 + y_1)$

∴ $y_2 = 1.02 + 0.02((0.02)^3 + 1.02) = 1.04040016$

Putting $n = 3$ in ①, $y_3 = y(0.06) = y_2 + h(x_2^3 + y_2)$

∴ $y_3 = 1.04040016 + 0.02((0.04)^3 + 1.04040016) = 1.061209443$

Putting $n = 4$ in ①, $y_4 = y(0.08) = y_3 + h(x_3^3 + y_3)$

∴ $y_4 = 1.061209443 + 0.02((0.06)^3 + 1.061209443) = 1.082437952$

Putting $n = 5$ in ①, $y_5 = y(0.1) = y_4 + h(x_4^3 + y_4)$

∴ $y_5 = 1.082437952 + 0.02((0.08)^3 + 1.082437952) = 1.104096951$

Putting $n = 6$ in ①, $y_6 = y(0.12) = y_5 + h(x_5^3 + y_5)$

∴ $y_6 = 1.104096951 + 0.02((0.1)^3 + 1.104096951) = 1.126198890$

Thus at $x = 0.12$, $y = 1.126198890$ ⇒ $y(0.12) = 1.126198890$

Example 8 Using Euler's method, solve $\frac{dy}{dx} = \frac{x-y}{2}$; $y(0) = 1$, over the interval $[0,2]$, taking the step size $\frac{1}{2}$

Solution: Given $f(x,y) = \frac{x-y}{2}$, $x_0 = 0$, $y_0 = 1$, $x_n = x_{n-1} + h$, $h = \frac{1}{2}$

∴ $x_1 = \frac{1}{2} = 0.5$, $x_2 = 1$, $x_3 = \frac{3}{2} = 1.5$, $x_4 = 2$

Using Euler's method $y_n = y_{n-1} + hf(x_{n-1}, y_{n-1})$

⇒ $y_n = y_{n-1} + \frac{h}{2}(x_{n-1} - y_{n-1})$

or $y_n = y_{n-1} + 0.25(x_{n-1} - y_{n-1})$ ... ①

Putting $n = 1$ in ①, $y_1 = y\left(\frac{1}{2}\right) = y_0 + 0.25(x_0 - y_0)$

∴ $y_1 = 1 + 0.25(0 - 1) = 0.75$

Putting $n = 2$ in ①, $y_2 = y(1) = y_1 + 0.25(x_1 - y_1)$

∴ $y_2 = 0.75 + 0.25(0.5 - 0.75) = 0.6875$

Putting $n = 3$ in ①, $y_3 = y\left(\frac{3}{2}\right) = y_2 + 0.25(x_2 - y_2)$

∴ $y_3 = 0.6875 + 0.25(1 - 0.6875) = 0.765625$

Putting $n = 4$ in ①, $y_4 = y(2) = y_3 + 0.25(x_3 - y_3)$

∴ $y_4 = 0.765625 + 0.25(1.5 - 0.765625) = 0.94921875$

13.2.4 Modified Euler's Method

Though Euler's method is quite easy to implement, but unless the step size h is very small, the truncation error will be large and the results will be inaccurate.

As per Modified Euler's method, a better approximation of y_1 is given by improving $f(x_0, y_0)$ obtained by Euler's method as shown:

$$y_1^{(1)} = y_0 + \frac{h}{2}[f(x_0, y_0) + f(x_1, y_1)]$$

$$y_1^{(2)} = y_0 + \frac{h}{2}[f(x_0, y_0) + f(x_1, y_1^{(1)})]$$

$$\vdots$$

Continue approximating y_1 until two consecutive values are coincident to a specific degree of accuracy.

$$\therefore y_1^{(k)} = y_0 + \frac{h}{2}[f(x_0, y_0) + f(x_1, y_1^{(k-1)})]$$

Repeat the procedure for y_2, y_3, y_4 ... to find y_n

Example 9 Use Modified Euler's method to obtain $y(0.2)$, $y(0.4)$ correct to 3 decimal places, given that $\frac{dy}{dx} = y - x^2$; $y(0) = 1$

Solution: Given $f(x, y) = y - x^2$, $x_0 = 0$, $y_0 = 1$

By Euler's method $y_n = y_{n-1} + hf(x_{n-1}, y_{n-1})$

i. To evaluate $y(0.2)$, $h = 0.2$, $x_1 = 0 + 0.2 = 0.2$

$y_1 = y(0.2) = y_0 + hf(x_0, y_0)$, $f(x_0, y_0) = y_0 - x_0^2 = 1 - 0 = 1$

$\therefore y_1 = 1 + 0.2(1) = 1.2$

$f(x_1, y_1) = y_1 - x_1^2 = 1.2 - (0.2)^2 = 1.16$

Now improving y_1 using Modified Euler's method

$$y_1^{(1)} = y_0 + \frac{h}{2}(f(x_0, y_0) + f(x_1, y_1))$$

$$\therefore y_1^{(1)} = 1 + \frac{0.2}{2}(1 + 1.16) = 1.216$$

$$f(x_1, y_1^{(1)}) = y_1^{(1)} - x_1^2 = 1.216 - (0.2)^2 = 1.176$$

$$y_1^{(2)} = y_0 + \frac{h}{2}[f(x_0, y_0) + f(x_1, y_1^{(1)})]$$

$$\therefore y_1^{(2)} = 1 + \frac{0.2}{2}(1 + 1.176) = 1.2176$$

$$f(x_1, y_1^{(2)}) = y_1^{(2)} - x_1^2 = 1.2176 - (0.2)^2 = 1.1776$$

$$y_1^{(3)} = y_0 + \frac{h}{2}[f(x_0, y_0) + f(x_1, y_1^{(2)})]$$

$$\therefore y_1^{(3)} = 1 + \frac{0.2}{2}(1 + 1.1776) = 1.21776 = y(0.2)$$

Thus by Modified Euler's method, we have improved $y(0.2)$ from 1.2 to 1.21776

ii. To evaluate $y(0.4)$, $h = 0.2$, $x_2 = 0.2 + 0.2 = 0.4$

$y_2 = y(0.4) = y_1 + hf(x_1, y_1)$,

$f(x_1, y_1) = y_1 - x_1^2 = 1.21776 - (0.2)^2 = 1.17776$

$$\therefore y_2 = 1.21776 + 0.2(1.17776) = 1.453312$$

$$f(x_2, y_2) = y_2 - x_2^2 = 1.453312 - (0.4)^2 = 1.293312$$

Now improving y_1 using Modified Euler's method

$$y_2^{(1)} = y_1 + \frac{h}{2}(f(x_1, y_1) + f(x_2, y_2))$$

$$\therefore y_2^{(1)} = 1.21776 + \frac{0.2}{2}(1.17776 + 1.293312) = 1.4648672$$

$$f(x_2, y_2^{(1)}) = y_2^{(1)} - x_2^2 = 1.4648672 - (0.4)^2 = 1.3048672$$

$$y_2^{(2)} = y_1 + \frac{h}{2}\left(f(x_1, y_1) + f(x_2, y_2^{(1)})\right)$$

$$\therefore y_2^{(2)} = 1.21776 + \frac{0.2}{2}(1.17776 + 1.3048672) = 1.46602272$$

$$f(x_2, y_2^{(2)}) = y_2^{(2)} - x_2^2 = 1.46602272 - (0.4)^2 = 1.30602272$$

$$y_2^{(3)} = y_1 + \frac{h}{2}\left(f(x_1, y_1) + f(x_2, y_2^{(2)})\right)$$

$$\therefore y_2^{(3)} = 1.21776 + \frac{0.2}{2}(1.17776 + 1.30602272) = 1.466138272$$

Thus by Modified Euler's method, we have improved $y(0.4)$ from 1.453312 to 1.466138272 correct to 3 decimal places.

Example 10 Use Modified Euler's method to obtain $y(1.2)$ correct to 3 decimal places, given that $\frac{dy}{dx} = \ln(x + y)$; $y(1) = 2$

Solution: Given $f(x, y) = \ln(x + y)$, $x_0 = 1$, $y_0 = 2$

By Euler's method $y_n = y_{n-1} + hf(x_{n-1}, y_{n-1})$

To evaluate $y(1.2)$, $h = 0.2$, $x_1 = 1 + 0.2 = 1.2$

$$y_1 = y(1.2) = y_0 + hf(x_0, y_0)$$

$$f(x_0, y_0) = \ln(x_0 + y_0) = \ln(1 + 2) = 1.09861$$

$$\therefore y_1 = 2 + 0.2(1.09861) = 2.21972$$

$$f(x_1, y_1) = \ln(x_1 + y_1) = \ln(1 + 2.21972) = 1.16929$$

Now improving y_1 using Modified Euler's method

$$y_1^{(1)} = y_0 + \frac{h}{2}(f(x_0, y_0) + f(x_1, y_1))$$

$$\therefore y_1^{(1)} = 2 + \frac{0.2}{2}(1.09861 + 1.16929) = 2.22679$$

$$f(x_1, y_1^{(1)}) = \ln(x_1 + y_1^{(1)}) = \ln(1 + 2.22679) = 1.17149$$

$$y_1^{(2)} = y_0 + \frac{h}{2}[f(x_0, y_0) + f(x_1, y_1^{(1)})]$$

$$\therefore y_1^{(2)} = 2 + \frac{0.2}{2}(1.09861 + 1.17149) = 2.22701$$

$$f(x_1, y_1^{(2)}) = \ln(x_1 + y_1^{(2)}) = \ln(1 + 2.22701) = 1.17156$$

$$y_1^{(3)} = y_0 + \frac{h}{2}[f(x_0, y_0) + f(x_1, y_1^{(2)})]$$

$$\therefore y_1^{(3)} = 2 + \frac{0.2}{2}(1.09861 + 1.17156) = 2.227017 = y(1.2)$$

Thus by Modified Euler's method, we have improved $y(1.2)$ from 2.21972 to 2.227017 correct to 4 decimal places

13.2.5 Runge- Kutta's Method

The Taylor's series method of solving differential equations numerically is impaired due to requirement of higher order derivatives. Euler's method does not obtain reasonable accuracy if the step height h is not kept small. Runge-Kutta method is preferment of the concepts used in Euler's and Modified Euler's methods. It is designed to give greater accuracy with the advantage of requiring only the function values at some selected points on the sub-interval. These methods agree with Taylor's series solution upto the order h^r where r is the order of the Runge-Kutta method. Consider the initial value problem

$$\frac{dy}{dx} = f(x, y); \quad y(x_0) = y_0 \quad \cdots \text{①}$$

Taylor's series expansion of a function $y(x)$ about $x = x_0$ is given by

$$y(x) = y_0 + (x - x_0)y_0' + \frac{1}{2!}(x - x_0)^2 y_0'' + \frac{1}{3!}(x - x_0)^3 y_0''' + \cdots$$

Now $y_1 = y(x_0 + h)$, $\therefore$ Putting $x = x_0 + h$ in Taylor's series, we get

$$y_1 = y(x_0 + h) = y_0 + hy_0' + \frac{h^2}{2!}y_0'' + \cdots \quad \cdots \text{②}$$

Also by Euler's method $y_1 = y_0 + hf(x_0, y_0) = y_0 + hy_0' \quad \cdots \text{③}$

From ② and ③, Euler's method is in consonant to Taylor's series expansion upto first 2 terms i.e. till the term containing h of order one. Euler's method itself is **first order Runge-Kutta method** and is having a local error of order 2.

Similarly it can be shown that Modified Euler's method coincides with Taylor's series expansion upto first 3 terms.

Modified Euler's method is given by $y_1 = y_0 + \frac{h}{2}[f(x_0, y_0) + f(x_1, y_1)]$

$\Rightarrow y_1 = y_0 + \frac{1}{2}[hf(x_0, y_0) + hf(x_1, y_1)]$

Now $x_1 = x_0 + h$ and $y_1 = y_0 + hf(x_0, y_0)$ by Euler's method

$\Rightarrow y_1 = y_0 + \frac{1}{2}[hf(x_0, y_0) + hf(x_0 + h, y_0 + hf(x_0, y_0))]$

$\Rightarrow y_1 = y_0 + \frac{1}{2}[K_1 + K_2]$

Where $K_1 = hf(x_0, y_0)$, $K_2 = hf(x_0 + h, y_0 + K_1)$

$\therefore$ Modified Euler's method itself is **second order Runge-Kutta method** having local error of order three as it is in consonant to Taylor's series expansion up to first three terms i.e. till the term containing h^2.

Similarly **third order Runge-Kutta method** tallies with Taylor's expansion upto first 4 terms i.e. till the term containing h^3 and is having a local error of order 3.

Solution by third order Runge-Kutta method is given by:

$y_1 = y_0 + \frac{1}{6}[K_1 + 4K_2 + K_3]$

where $K_1 = hf(x_0, y_0)$

$K_2 = hf\left(x_0 + \frac{h}{2}, y_0 + \frac{K_1}{2}\right)$,

$K_3 = hf(x_0 + h, y_0 + hf(x_0 + h, y_0 + K_1))$

On the similar lines, **Runge- Kutta's method of order four** is collateral with Taylor's series expansion upto first 5 terms i.e. till the term containing h^4 with a local error of order 5.

Numerical solution of initial value problem given by ①, using fourth order Runge-Kutta method is: $y_1 = y_0 + \frac{1}{6}[K_1 + 2K_2 + 2K_3 + K_4]$

where $K_1 = hf(x_0, y_0)$

$K_2 = hf\left(x_0 + \frac{h}{2}, y_0 + \frac{K_1}{2}\right)$

$K_3 = hf\left(x_0 + \frac{h}{2}, y_0 + \frac{K_2}{2}\right)$

$K_4 = hf(x_0 + h, y_0 + K_3)$

Fourth order Runge- Kutta's method (commonly known as Runge- Kutta method)**,** provides most accurate result and is widely used to approximate initial value problems.

Example11 Solve the differential equation $\frac{dy}{dx} = y - x$; $y(0) = 1$, at $x = 0.1$, using Runge-Kutta method. Also compare the numerical solution obtained with the exact solution.

Solution: Given $f(x, y) = x + y$, $x_0 = 0$, $y_0 = 1$, $h = 0.1$

Runge-Kutta method of 4th order is given by

$y_1 = y_0 + \frac{1}{6}[K_1 + 2K_2 + 2K_3 + K_4]$ ... ①

$K_1 = hf(x_0, y_0) = h(y_0 - x_0) = 0.1(1 - 0) = 0.1$

$K_2 = hf\left(x_0 + \frac{h}{2}, y_0 + \frac{K_1}{2}\right) = 0.1\left(\left(1 + \frac{0.1}{2}\right) - \left(0 + \frac{0.1}{2}\right)\right) = 0.1$

$K_3 = hf\left(x_0 + \frac{h}{2}, y_0 + \frac{K_2}{2}\right) = 0.1\left(\left(1 + \frac{0.1}{2}\right) - \left(0 + \frac{0.1}{2}\right)\right) = 0.1$

$K_4 = hf(x_0 + h, y_0 + K_3) = 0.1((1 + 0.1) - (0 + 0.1)) = 0.1$

Substituting values of K_1, K_2, K_3, K_4 in ①, we get the solution as:

$y_1 = 1 + \frac{1}{6}[0.1 + 2(0.1) + 2(0.1) + 0.1] = 1.1$

Again to find exact solution of the initial value problem

$\frac{dy}{dx} - y = -x$, which is a linear differential equation

Integrating Factor (I.F.) = $e^{\int -dx} = e^{-x}$

Solution is given by $ye^{-x} = -\int xe^{-x}dx$

$$\Rightarrow ye^{-x} = xe^{-x} + e^{-x} + c$$
$$\Rightarrow y = x + 1 + ce^x$$

Given that $y(0) = 1 \Rightarrow 1 = 0 + 1 + c \quad \therefore c = 0$

$$\Rightarrow y = x + 1$$

$\therefore y(0.1) = 0.1 + 1 = 1.1$

Example12 Solve the differential equation $\frac{dy}{dx} = \ln(x+y); \; y(0) = 2$, at $x = 0.3$, using Runge-Kutta method of 4th order by dividing into two steps of $h = 0.15$ each. Compare the results with one step solution.

Solution: i. Given $f(x,y) = \ln(x+y), \quad x_0 = 0, \; y_0 = 2, h = 0.15$

Runge-Kutta method of 4th order is given by

$$y_1 = y_0 + \frac{1}{6}[K_1 + 2K_2 + 2K_3 + K_4] \quad \cdots \text{①}$$

$K_1 = hf(x_0, y_0) = 0.15 \ln(x_0 + y_0) = 0.15 \ln(0+2) = 0.10397$

$K_2 = hf\left(x_0 + \frac{h}{2}, y_0 + \frac{K_1}{2}\right) = 0.15 \ln\left(0 + \frac{0.15}{2} + 2 + \frac{0.10397}{2}\right) = 0.11321$

$K_3 = hf\left(x_0 + \frac{h}{2}, y_0 + \frac{K_2}{2}\right) = 0.15 \ln\left(0 + \frac{0.15}{2} + 2 + \frac{0.11321}{2}\right) = 0.11353$

$K_4 = hf(x_0 + h, y_0 + K_3) = 0.15 \ln(0 + 0.15 + 2 + 0.11353) = 0.12254$

Substituting values of K_1, K_2, K_3, K_4 in ①, we get the solution as:

$$y_1 = y(0.15) = 2 + \frac{1}{6}[0.10397 + 2(0.11321) + 2(0.11353) + 0.12254]$$
$$= 2.11333$$

$\therefore x_1 = 0.15, \; y_1 = 2.11333, h = 0.15$

Again $y_2 = y_1 + \frac{1}{6}[K_1 + 2K_2 + 2K_3 + K_4] \quad \cdots \text{②}$

$K_1 = hf(x_1, y_1) = 0.15 \ln(x_1 + y_1) = .15 \ln(.15 + 2.11333) = .12253$

$K_2 = hf\left(x_1 + \frac{h}{2}, y_1 + \frac{K_1}{2}\right) = .15 \ln\left(.15 + \frac{.15}{2} + 2.11333 + \frac{.12253}{2}\right) = .13129$

$K_3 = hf\left(x_1 + \frac{h}{2}, y_1 + \frac{K_2}{2}\right) = .15 \ln\left(.15 + \frac{.15}{2} + 2.11333 + \frac{.13129}{2}\right) = .13157$

$K_4 = hf(x_1 + h, y_1 + K_3) = .15 \ln(.15 + .15 + 2.11333 + .13157) = .14011$

Substituting values of K_1, K_2, K_3, K_4 in ②, we get the solution as:

$$y_2 = 2.11333 + \frac{1}{6}[.12253 + 2(.13129) + 2(.13157) + .14011]$$

$\therefore y(0.3) = 2.24472$

ii. Solving in single step of $h = 0.3$

Given $f(x,y) = \ln(x+y)$, $x_0 = 0$, $y_0 = 2$, $h = 0.3$

Runge-Kutta method of 4th order is given by

$$y_1 = y_0 + \frac{1}{6}[K_1 + 2K_2 + 2K_3 + K_4] \qquad \cdots \text{①}$$

$K_1 = hf(x_0, y_0) = 0.3 \ln(x_0 + y_0) = 0.3 \ln(0+2) = 0.20794$

$K_2 = hf\left(x_0 + \frac{h}{2}, y_0 + \frac{K_1}{2}\right) = 0.3 \ln\left(0 + \frac{0.3}{2} + 2 + \frac{0.20794}{2}\right) = 0.24381$

$K_3 = hf\left(x_0 + \frac{h}{2}, y_0 + \frac{K_2}{2}\right) = 0.3 \ln\left(0 + \frac{0.3}{2} + 2 + \frac{0.24381}{2}\right) = 0.24619$

$K_4 = hf(x_0 + h, y_0 + K_3) = 0.3 \ln(0 + 0.3 + 2 + 0.24619) = 0.28038$

Substituting values of K_1, K_2, K_3, K_4 in ①, we get the solution as:

$y_1 = 2 + \frac{1}{6}[0.20794 + 2(0.24381) + 2(0.24619) + 0.28038] = 2.24472$

Example13 Solve the differential equation $\frac{dy}{dx} = x^2 + y^2$; $y(0) = 2$, at $x = 0.1$, using Runge-Kutta method.

Solution: Given $f(x,y) = x^2 + y^2$, $x_0 = 0$, $y_0 = 2$, $h = 0.1$

Runge-Kutta method of 4th order is given by

$$y_1 = y_0 + \frac{1}{6}[K_1 + 2K_2 + 2K_3 + K_4] \qquad \cdots \text{①}$$

$K_1 = hf(x_0, y_0) = h(x_0^2 + y_0^2) = 0.1(0 + 4) = 0.4$

$K_2 = hf\left(x_0 + \frac{h}{2}, y_0 + \frac{K_1}{2}\right) = 0.1\left(\left(0 + \frac{0.1}{2}\right)^2 + \left(2 + \frac{0.4}{2}\right)^2\right) = 0.48425$

$K_3 = hf\left(x_0 + \frac{h}{2}, y_0 + \frac{K_2}{2}\right) = 0.1\left(\left(0 + \frac{0.1}{2}\right)^2 + \left(2 + \frac{0.48425}{2}\right)^2\right) = 0.50296$

$K_4 = hf(x_0 + h, y_0 + K_3) = 0.1((0 + 0.1)^2 + (2 + 0.50296)^2) = 0.62748$

Substituting values of K_1, K_2, K_3, K_4 in ①, we get the solution as:

$y_1 = 2 + \frac{1}{6}[0.4 + 2(0.48425) + 2(0.50296) + 0.62748] = 2.50032$

13.3 Multistep Methods

Single steps methods (Euler's and Picard's) discussed in previous section, refer to immediate preceding value and its derivative to determine the current value. Even **Modified Euler's method** is single-step method as it uses just one preceding value to approximate the current value. Methods such as **Runge-Kutta** take some intermediate steps, to obtain the present value, but none of the methods uses other than immediate preceding value to estimate present value. Multistep methods attempt to gain efficiency by using collective proceeding values rather than discarding them. Most multistep methods make a rough estimation of current values using multiple previous values, known as **predictor** and thereby make corrections in predictor value using present and nearby preceding values, known as **corrector**.

13.3.1 Milne's Predictor Corrector Method

Milne's Predictor corrector method is a multistep method, since it uses last four preceding values to estimate current value (predictor) and then amends the current value (corrector) using present and nearby preceding values.

Consider the initial value problem given by $\frac{dy}{dx} = f(x,y)$; $y(x_0) = y_0$ $\cdots$ ①

Let us illustrate step by step methodology to compute the current value using preceding estimates.

Step1: To determine predictor value

i. Determine first thee values y_1, y_2, y_3 corresponding to $x_1 = x_0 + h$, $x_2 = x_0 + 2h$, $x_3 = x_0 + 3h$, using any of the single steps methods, preferably Runge-Kutta because of its simplicity and accuracy.

ii. To determine predictor value of y_4 at $x_4 = x_0 + 4h$, denoted by y_4^p,

$$\text{Let} \quad y_4^p = y_0 + \int_{x_0}^{x_0+4h} f(x,y)dx$$

By Newton's forward interpolation formula

$$f(x,y) \equiv f_0 + n\Delta f_0 + \frac{n(n-1)}{2!}\Delta^2 f_0 + \frac{n(n-1)(n-2)}{3!}\Delta^3 f_0 + \cdots, \quad x = x_0 + nh$$

$$\Rightarrow y_4^p = y_0 + \int_{x_0}^{x_0+4h}\left(f_0 + n\Delta f_0 + \frac{n(n-1)}{2!}\Delta^2 f_0 + \frac{n(n-1)(n-2)}{3!}\Delta^3 f_0 + \cdots\right) dx$$

Putting $x = x_0 + nh$, $dx = hdn$,

$$\Rightarrow y_4^p = y_0 + h\int_0^4 \left(f_0 + n\Delta f_0 + \frac{n(n-1)}{2!}\Delta^2 f_0 + \frac{n(n-1)(n-2)}{3!}\Delta^3 f_0 + \cdots\right) dn$$

$$= y_0 + h\left[nf_0 + \frac{n^2}{2}\Delta f_0 + \left(\frac{n^3}{6} - \frac{n^2}{4}\right)\Delta^2 f_0 + \frac{1}{6}\left(\frac{n^4}{4} - n^3 + n^2\right)\Delta^3 f_0 + \cdots\right]_0^4$$

Neglecting 4th and higher order terms, we get

$$y_4^p = y_0 + h\left[4f_0 + 8\Delta f_0 + \frac{20}{3}\Delta^2 f_0 + \frac{8}{3}\Delta^3 f_0\right]$$

$$= y_0 + h\left[4f_0 + 8(E-1)f_0 + \frac{20}{3}(E^2 - 2E + 1)f_0 + \frac{8}{3}(E^3 - 3E^2 + 3E - 1)f_0\right]$$

$$= y_0 + h\left[4f_0 + 8(f_1 - f_0) + \frac{20}{3}(f_2 - 2f_1 + f_0) + \frac{8}{3}(f_3 - 3f_2 + 3f_1 - f_0)\right]$$

$$\Rightarrow y_4^p = y_0 + \frac{4h}{3}(2f_1 - f_2 + 2f_3) \quad \text{(Milne's Predictor formula)}$$

This formula can be used to predict the value of y_4 when y_1, y_2, y_3 are known.

iii. Compute the predictor function $f_4^p = f(x_4, y_4^p)$

Step2: To determine corrector value

Corrector formula is obtained by using Newton's forward interpolation formula in an intermediate corrector value y_2^c, which can be generalized to obtain y_4^c.

i. Let $y_2^{(1)} = y_0 + \int_{x_0}^{x_0+2h} f(x,y)dx$

$$\Rightarrow y_2^{(1)} = y_0 + \int_{x_0}^{x_0+2h}\left(f_0 + n\Delta f_0 + \frac{n(n-1)}{2!}\Delta^2 f_0 + \cdots\right) dx$$

Putting $x = x_0 + nh$, $dx = hdn$,

$$\Rightarrow y_2^{(1)} = y_0 + h \int_0^2 \left(f_0 + n\Delta f_0 + \frac{n(n-1)}{2!}\Delta^2 f_0 + \cdots \right) dn$$

$$= y_0 + h \left[nf_0 + \frac{n^2}{2}\Delta f_0 + \left(\frac{n^3}{6} - \frac{n^2}{4}\right)\Delta^2 f_0 + \cdots \right]_0^2$$

Neglecting 3rd and higher order terms, we get

$$y_2^{(1)} = y_0 + h\left[2f_0 + 2\Delta f_0 + \frac{1}{3}\Delta^2 f_0 \right]$$

$$= y_0 + h\left[2f_0 + 2(E-1)f_0 + \frac{1}{3}(E^2 - 2E + 1)f_0 \right]$$

$$= y_0 + h\left[2f_0 + 2(f_1 - f_0) + \frac{1}{3}(f_2 - 2f_1 + f_0) \right]$$

$$\therefore y_2^{(1)} = y_0 + \frac{h}{3}(f_0 + 4f_1 + f_2)$$

Similarly $y_3^{(1)} = y_1 + \frac{h}{3}(f_1 + 4f_2 + f_3)$

$$y_4^{(1)} = y_2 + \frac{h}{3}(f_2 + 4f_3 + f_4^p), f_4^p \text{ is the predicted value of } f_4$$

ii. Compute the corrected function $f_4^{(1)} = f(x_4, y_4^{(1)})$

iii. Continue correcting y_4 until two consecutive values are coincident to a specific degree of accuracy to find

$$y_4^c = y_2 + \frac{h}{3}(f_2 + 4f_3 + f_4^{c-1}) \quad \textbf{(Milne's Corrector formula)}$$

iv. Compute $f_4 = f(x_4, y_4^c)$

Step3: To determine next predictor and corrector values

i. Use the predicted value $y_5^p = y_1 + \frac{4h}{3}(2f_2 - f_3 + 2f_4)$

ii. Compute the predictor function $f_5^p = f(x_5, y_5^p)$

iii. Apply corrector values to stabilize the corrector functions
$y_5^{(1)} = y_3 + \frac{h}{3}(f_3 + 4f_4 + f_5^p) \ldots$ upto $y_5^c = y_3 + \frac{h}{3}(f_3 + 4f_4 + f_5^{c-1})$ and compute the corrected function $f_5 = f(x_5, y_5^c)$.

iv. Repeat the procedure for y_6, y_7, ... to find y_n

Example14 If $y(0.1) = 1.1169$, $y(0.2) = 1.2773$, $y(0.3) = 1.5040$ for the initial value problem $\frac{dy}{dx} = xy + y^2$; $y(0) = 1$, compute $y(0.4)$ using Milne's method.

Solution: Given $f(x,y) = xy + y^2$, $h = 0.1$

Also $x_0 = 0$, $y_0 = 1 \Rightarrow f_0 = f(x_0, y_0) = x_0 y_0 + y_0^2 = 1$

$x_1 = 0.1$, $y_1 = 1.1169 \Rightarrow f_1 = 0.1(1.1169) + (1.1169)^2 = 1.3592$

$x_2 = 0.2$, $y_2 = 1.2773 \Rightarrow f_2 = 0.2(1.2773) + (1.2773)^2 = 1.8870$

$x_3 = 0.3$, $y_3 = 1.5040 \Rightarrow f_3 = 0.3(1.5049) + (1.5049)^2 = 2.7132$

To determine predictor value y_4^p:

Predictor value of y_4 is given by $y_4^p = y_0 + \frac{4h}{3}(2f_1 - f_2 + 2f_3)$

$\Rightarrow y_4^p = 1 + \frac{4(0.1)}{3}(2(1.3592) - (1.8870) + 2(2.7132)) = 1.8344$

$\therefore f_4^p = f(x_4, y_4^p) = 0.4(1.8344) + (1.8344)^2 = 4.0988$

To find corrector value y_4^c by improving y_4^p

$y_4^{(1)} = y_2 + \frac{h}{3}(f_2 + 4f_3 + f_4^p)$

$= 1.2773 + \frac{0.1}{3}(1.8870 + 4(2.7132) + 4.0988) = 1.8386$

$f_4^{(1)} = f(x_4, y_4^{(1)}) = 0.4(1.8386) + (1.8386)^2 = 4.1159$

$y_4^{(2)} = y_2 + \frac{h}{3}(f_2 + 4f_3 + f_4^{(1)})$

$= 1.2773 + \frac{0.1}{3}(1.8870 + 4(2.7132) + 4.1159) = 1.8392$

$f_4^{(2)} = f(x_4, y_4^{(2)}) = 0.4(1.8392) + (1.8392)^2 = 4.1183$

$y_4^{(3)} = y_2 + \frac{h}{3}(f_2 + 4f_3 + f_4^{(2)})$

$= 1.2773 + \frac{0.1}{3}(1.8870 + 4(2.7132) + 4.1183) = 1.8392$

$y_4^{(2)}$ and $y_4^{(3)}$ coincide, $\therefore y_4^c = 1.8392$

Hence $y(0.4) = 1.8392$

Example 15 Solve the initial value problem $\frac{dy}{dx} = x - y^2$; $y(0) = 0$, $h = 0.2$

in the range $[0,1]$ using Milne's Predictor Corrector method.

Solution: Given $f(x, y) = x - y^2$, $x_0 = 0$, $y_0 = 0$

$h = 0.2 \therefore x_1 = 0.2$, $x_2 = 0.4$, $x_3 = 0.6$, $x_4 = 0.8$, $x_5 = 1$

$f_0 = f(x_0, y_0) = x_0 - y_0^2 = 0$

To determine predictor value y_4^p:

Using Picard's approximation to estimate y_1, y_2, y_3

1st approximation ($x_1 = 0.2$)

$y_1 = y_0 + \int_{x_0}^{x} f(x, y_0) dx$

$\Rightarrow y_1 = 0 + \int_0^x (x - y_0^2) dx = \int_0^x x\, dx = \frac{x^2}{2}$

$\therefore y_1 = \frac{(0.2)^2}{2} = 0.02$

2nd approximation ($x_2 = 0.4$)

$y_2 = y_0 + \int_{x_0}^{x} f(x, y_1) dx$

$$\Rightarrow y_2 = 0 + \int_0^x (x - y_1^2)\, dx = \int_0^x \left(x - \left(\frac{x^2}{2}\right)^2\right) dx = \frac{x^2}{2} - \frac{x^5}{20}$$

$$\therefore y_2 = \frac{(0.4)^2}{2} - \frac{(0.4)^5}{20} = 0.079488$$

3rd approximation ($x_3 = 0.6$)

$$y_3 = y_0 + \int_{x_0}^x f(x, y_2)\, dx$$

$$\Rightarrow y_3 = 0 + \int_0^x (x - y_2^2)\, dx = \int_0^x \left(x - \left(\frac{x^2}{2} - \frac{x^5}{20}\right)^2\right) dx$$

$$= \frac{x^2}{2} - \frac{x^5}{20} + \frac{x^8}{160} - \frac{x^{11}}{4400}$$

$$\therefore y_3 = \frac{(0.6)^2}{2} - \frac{(0.6)^5}{20} + \frac{(0.6)^8}{160} - \frac{(0.6)^{11}}{4400} = 0.17622$$

Also $f_1 = f(x_1, y_1) = x_1 - y_1^2 = 0.2 - (0.02)^2 = 0.1996$

$f_2 = f(x_2, y_2) = x_2 - y_2^2 = 0.4 - (0.079488)^2 = 0.3937$

$f_3 = f(x_3, y_3) = x_3 - y_3^2 = 0.6 - (0.17622)^2 = 0.5689$

Predictor value of y_4 is given by $y_4^p = y_0 + \frac{4h}{3}(2f_1 - f_2 + 2f_3)$

$$\Rightarrow y_4^p = 0 + \frac{4(0.2)}{3}\big(2(0.1996) - (0.3937) + 2(0.5689)\big) = 0.3049$$

$$\therefore f_4^p = f(x_4, y_4^p) = x_4 - y_4^{p^2} = 0.8 - (0.3049)^2 = 0.7070$$

To find corrector value y_4^c by improving y_4^p

$$y_4^{(1)} = y_2 + \frac{h}{3}(f_2 + 4f_3 + f_4^p)$$

$$= 0.079488 + \frac{0.2}{3}(0.3937 + 4(0.5689) + 0.7070) = 0.3046$$

$$f_4^{(1)} = f(x_4, y_4^{(1)}) = x_4 - \left(y_4^{(1)}\right)^2 = 0.8 - (0.3046)^2 = 0.7072$$

$$y_4^{(2)} = y_2 + \frac{h}{3}(f_2 + 4f_3 + f_4^{(1)})$$

$$= 0.079488 + \frac{0.2}{3}(0.3937 + 4(0.5689) + 0.7072) = 0.3046$$

$y_4^{(1)}$ and $y_4^{(2)}$ coincide, $\therefore y_4^c = 0.3046$ and $f_4 = f(x_4, y_4^c) = 0.7072$

To determine predictor value y_5^p at $x_5 = 1.0$

Predictor is given by $y_5^p = y_1 + \frac{4h}{3}(2f_2 - f_3 + 2f_4)$

$$\Rightarrow y_5^p = 0.02 + \frac{4(0.2)}{3}\big(2(0.3937) - 0.5689 + 2(0.7070)\big) = 0.4553$$

$$f_5^p = f(x_5, y_5^p) = x_5 - \left(y_5^p\right)^2 = 1.0 - (0.4553)^2 = 0.7927$$

To find corrector value y_5^c by improving y_5^p

$$y_5^{(1)} = y_3 + \frac{h}{3}(f_3 + 4f_4 + f_5^p)$$

$\Rightarrow y_5^{(1)} = 0.17622 + \frac{0.2}{3}(0.5689 + 4(0.7070) + 0.7927) = 0.4555$

$f_5^{(1)} = x_5 - \left(y_5^{(1)}\right)^2 = 1.0 - (0.4555)^2 = 0.7925$

$y_5^{(2)} = y_3 + \frac{h}{3}\left(f_3 + 4f_4 + f_5^{(1)}\right)$

$= 0.17622 + \frac{0.2}{3}(0.5689 + 4(0.7070) + 0.7925) = 0.4555$

$y_5^{(1)}$ and $y_5^{(2)}$ coincide, $\therefore y_5 = 0.4555$

Example 16 Using Runge- Kutta method, find y for $x = 1.1, 1.2, 1.3$ and hence find $y(1.4)$ using Milne's Predictor Corrector method for the initial value problem

$\frac{dy}{dx} = x\sqrt{y};\ y(1) = 1.$

Solution: Given $f(x,y) = x\sqrt{y},\quad x_0 = 1,\ y_0 = 1,\ h = 0.1$

$\therefore x_1 = 1.1,\ x_2 = 1.2,\ x_3 = 1.3,\ x_4 = 1.4$

Runge-Kutta method of 4th order is given by

$y_1 = y_0 + \frac{1}{6}[K_1 + 2K_2 + 2K_3 + K_4] \quad \cdots ①$

$K_1 = hf(x_0, y_0) = h(x_0\sqrt{y_0}) = 0.1(1\sqrt{1}) = 0.1$

$K_2 = hf\left(x_0 + \frac{h}{2}, y_0 + \frac{K_1}{2}\right) = 0.1\left(\left(1 + \frac{0.1}{2}\right)\sqrt{\left(1 + \frac{0.1}{2}\right)}\right) = 0.10759$

$K_3 = hf\left(x_0 + \frac{h}{2}, y_0 + \frac{K_2}{2}\right) = 0.1\left(\left(1 + \frac{0.1}{2}\right)\sqrt{\left(1 + \frac{0.10759}{2}\right)}\right) = 0.10779$

$K_4 = hf(x_0 + h, y_0 + K_3) = 0.1\left((1 + 0.1)\sqrt{(1 + 0.10779)}\right) = 0.11578$

Substituting values of K_1, K_2, K_3, K_4 in ①, we get the solution as:

$y_1 = 1 + \frac{1}{6}[0.1 + 2(0.10759) + 2(0.10779) + 0.11578] = 1.10775$

$\therefore x_1 = 1.1,\ y_1 = 1.10775$

Again $y_2 = y_1 + \frac{1}{6}[K_1 + 2K_2 + 2K_3 + K_4] \quad \cdots ②$

$K_1 = hf(x_1, y_1) = h(x_1\sqrt{y_1}) = .1(1.1\sqrt{1.10775}) = 0.11577$

$K_2 = hf\left(x_1 + \frac{h}{2}, y_1 + \frac{K_1}{2}\right) = .1\left(\left(1.1 + \frac{.1}{2}\right)\sqrt{\left(1.10775 + \frac{0.11577}{2}\right)}\right) = 0.12416$

$K_3 = hf\left(x_1 + \frac{h}{2}, y_1 + \frac{K_2}{2}\right) = .1\left(\left(1.1 + \frac{.1}{2}\right)\sqrt{\left(1.10775 + \frac{0.12416}{2}\right)}\right) = 0.12438$

$K_4 = hf(x_1 + h, y_1 + K_3) = .1\left((1.1 + .1)\sqrt{(1.10775 + 0.12438)}\right) = 0.13320$

Substituting values of K_1, K_2, K_3, K_4 in ②, we get the solution as:

$$y_2 = 1.10775 + \frac{1}{6}[.11577 + 2(.12416) + 2(.12438) + 0.13320] = 1.23209$$

$\therefore x_2 = 1.2$, $y_2 = 1.23209$

Again $y_3 = y_2 + \frac{1}{6}[K_1 + 2K_2 + 2K_3 + K_4]$ \quad\quad\quad ...③

$K_1 = hf(x_2, y_2) = h(x_2\sqrt{y_2}) = .1(1.2\sqrt{1.23209}) = 0.13320$

$K_2 = hf\left(x_2 + \frac{h}{2}, y_2 + \frac{K_1}{2}\right) = .1\left((1.2 + \frac{.1}{2})\sqrt{(1.23209 + \frac{0.1332}{2})}\right) = 0.14245$

$K_3 = hf\left(x_2 + \frac{h}{2}, y_2 + \frac{K_2}{2}\right) = .1\left((1.2 + \frac{.1}{2})\sqrt{(1.23209 + \frac{0.14245}{2})}\right) = 0.1427$

$K_4 = hf(x_2 + h, y_2 + K_3) = .1\left((1.2 + .1)\sqrt{(1.23209 + 0.1427)}\right) = 0.15242$

Substituting values of K_1, K_2, K_3, K_4 in ③, we get the solution as:

$$y_3 = 1.23209 + \frac{1}{6}[.1332 + 2(.14245) + 2(.1427) + .15242] = 1.37474$$

$\therefore x_3 = 1.3$, $y_3 = 1.37474$

Also $f_1 = f(x_1, y_1) = x_1\sqrt{y_1} = (1.1)\sqrt{1.10775} = 1.15775$

$\quad\quad f_2 = f(x_2, y_2) = x_2\sqrt{y_2} = (1.2)\sqrt{1.23209} = 1.33199$

$\quad\quad f_3 = f(x_3, y_3) = x_3\sqrt{y_3} = (1.3)\sqrt{1.37474} = 1.52424$

Now by **Milne's Predictor Corrector method**

Predictor value of y_4 is given by $y_4^p = y_0 + \frac{4h}{3}(2f_1 - f_2 + 2f_3)$

$\Rightarrow y_4^p = 1 + \frac{4(0.1)}{3}(2(1.15775) - (1.33199) + 2(1.52424)) = 1.5376$

$\therefore f_4^p = f(x_4, y_4^p) = x_4\sqrt{y_4} = 1.4\sqrt{1.5376} = 1.736$

To find corrector value y_4^c by improving y_4^p

$y_4^{(1)} = y_2 + \frac{h}{3}(f_2 + 4f_3 + f_4^p)$

$\quad\quad = 1.23209 + \frac{0.1}{3}(1.33199 + 4(1.52424) + 1.736) = 1.53759$

$f_4^{(1)} = f(x_4, y_4^{(1)}) = x_4\sqrt{y_4^{(1)}} = 1.4\sqrt{1.53759} = 1.73599$

$y_4^{(2)} = y_2 + \frac{h}{3}(f_2 + 4f_3 + f_4^{(1)})$

$\quad\quad = 1.23209 + \frac{0.1}{3}(1.33199 + 4(1.52424) + 1.73599) = 1.53759$

$y_4^{(1)}$ and $y_4^{(2)}$ coincide, $\therefore y_4 = 1.53759$ i.e. $y(1.4) = 1.53759$

13.3.2 Adams- Bashforth Method

Adams procedure is also a multistep method using Newton's backward interpolation formula to compute $f(x, y)$ illustrated below:

Consider the initial value problem given by $\frac{dy}{dx} = f(x, y)$; $y(x_0) = y_0$ ... ①

Step1: To determine predictor value

i. Determine y_{-1}, y_{-2}, y_{-3} corresponding to $x_{-1} = x_0 - h$, $x_{-2} = x_0 - 2h$, $x_3 = x_0 - 3h$, using any of the single steps methods.

ii. To determine predictor value of y_4 at $x_4 = x_0 + 4h$, denoted by y_4^p,

Let $\quad y_1^p = y_0 + \int_{x_0}^{x_0+h} f(x, y) dx$

Newton's backward interpolation formula can be written as:

$$f(x, y) \equiv f_0 + n\nabla f_0 + \frac{n(n+1)}{2!}\nabla^2 f_0 + \frac{n(n+1)(n+2)}{3!}\nabla^3 f_0 + \cdots, n = \frac{x - x_n}{h}$$

$$\Rightarrow y_1^p = y_0 + \int_{x_0}^{x_0+h} \left(f_0 + n\nabla f_0 + \frac{n(n+1)}{2!}\nabla^2 f_0 + \frac{n(n+1)(n+2)}{3!}\nabla^3 f_0 + \cdots\right) dx$$

Putting $x = x_0 + nh$, $dx = h\,dn$,

$$\Rightarrow y_1^p = y_0 + h \int_0^1 \left(f_0 + n\nabla f_0 + \frac{n(n+1)}{2!}\nabla^2 f_0 + \frac{n(n+1)(n+2)}{3!}\nabla^3 f_0 + \cdots\right) dn$$

$$= y_0 + h\left[nf_0 + \frac{n^2}{2}\nabla f_0 + \left(\frac{n^3}{6} + \frac{n^2}{4}\right)\nabla^2 f_0 + \frac{1}{6}\left(\frac{n^4}{4} + n^3 + n^2\right)\nabla^3 f_0 + \cdots\right]_0^1$$

Neglecting 4th and higher order terms, we get

$$y_1^p = y_0 + h\left[f_0 + \frac{1}{2}\nabla f_0 + \frac{5}{12}\nabla^2 f_0 + \frac{3}{8}\nabla^3 f_0\right]$$

$$= y_0 + h\left[f_0 + \frac{1}{2}(1 - E^{-1})f_0 + \frac{5}{12}(1 - 2E^{-1} + E^{-2})f_0 + \frac{3}{8}(1 - 3E^{-1} + 3E^{-2} - E^{-3})f_0\right]$$

$$= y_0 + h\left[f_0 + \frac{1}{2}(f_0 - f_{-1}) + \frac{5}{12}(f_0 - 2f_{-1} + f_{-2}) + \frac{3}{8}(f_0 - 3f_{-1} + 3f_{-2} - f_{-3})\right]$$

$$\Rightarrow y_1^p = y_0 + \frac{h}{24}(55f_0 - 59f_{-1} + 37f_{-2} - 9f_{-3}) \text{ (Adams Predictor formula)}$$

This formula can be used to predict the value of y_1 when y_{-1}, y_{-2}, y_{-3} are known.

iii. Compute the predictor function $f_1^p = f(x_1, y_1^p)$

Step2: To determine corrector value

Corrector formula is obtained by using Newton's backward interpolation formula to obtain y_1^c

i. Let $y_1^{(1)} = y_0 + \int_{x_0}^{x_1} f(x, y) dx$

$$\Rightarrow y_1^{(1)} = y_0 + \int_{x_0}^{x_0+2h} \left(f_1 + n\nabla f_1 + \frac{n(n+1)}{2!}\nabla^2 f_1 + \frac{n(n+1)(n+2)}{3!}\nabla^3 f_1 + \cdots\right) dx$$

Putting $x = x_1 + nh$, $dx = h\,dn$,

$$\Rightarrow y_1^{(1)} = y_0 + h \int_{-1}^0 \left(f_1 + n\nabla f_1 + \frac{n(n+1)}{2!}\nabla^2 f_1 + \frac{n(n+1)(n+2)}{3!}\nabla^3 f_1 + \cdots\right) dn$$

Neglecting 4th and higher order terms and solving, we get the corrector formula as:

$y_1^{(1)} = y_0 + \frac{h}{24}(9f_1^p + 19f_0 - 5f_{-1} + f_{-2})$, f_1^p is the predicted value of f_1

ii. Compute the corrected function $f_1^{(1)} = f(x_1, y_1^{(1)})$

iii. Continue correcting y_1 until two consecutive values are coincident to a specific degree of accuracy to find

$y_1^{(c)} = y_0 + \frac{h}{24}(9f_1^{-c} + 19f_0 - 5f_{-1} + f_{-2})$ **(Adams Corrector formula)**

iv. Compute $f_1 = f(x_1, y_1^c)$

Step 3: To determine next predictor and corrector values

i. Use the predicted value $y_2^p = y_1 + \frac{h}{24}(55f_1 - 59f_0 + 37f_{-1} - 9f_{-2})$

ii. Compute the predictor function $f_2^p = f(x_2, y_2^p)$

iii. Apply corrector values to stabilize the corrector functions

$y_2^{(1)} = y_1 + \frac{h}{24}(9f_2^p + 19f_1 - 5f_0 + f_{-1})$ upto

$y_2^c = y_1 + \frac{h}{24}(9f_2^{c-1} + 19f_1 - 5f_0 + f_{-1})$

iv. Compute the function $f_2 = f(x_2, y_2^c)$

v. Repeat the procedure for y_3, y_4, ... to find y_n

Example 17 If $y(0.1) = 1.3916$, $y(0.2) = 1.7725$, $y(0.3) = 2.1509$ for the initial value problem $\frac{dy}{dx} = 6e^x - 2y$; $y(0) = 1$, compute $y(0.4)$ using Adams method.

Solution: Given $f(x, y) = 6e^x - 2y$, $h = 0.1$,

Also $y(0) = 1 \Rightarrow f(0,1) = 6e^0 - 2(1) = 4$

$y(0.1) = 1.3916 \Rightarrow f(0.1, 1.3916) = 6e^{0.1} - 2(1.3916) = 3.8478$

$y(0.2) = 1.7725 \Rightarrow f(0.2, 1.7725) = 6e^{0.2} - 2(1.7725) = 3.7834$

$y(0.3) = 2.1509 \Rightarrow f(0.3, 2.1509) = 6e^{0.3} - 2(2.1509) = 3.7974$

To compute $y(0.4)$ using Adams method, taking

$x_0 = 0.3$, $y_0 = 2.1509$ $\Rightarrow f_0 = f(x_0, y_0) = 3.7974$

$x_{-1} = 0.2$, $y_{-1} = 1.7725$ $\Rightarrow f_{-1} = f(x_{-1}, y_{-1}) = 3.7834$

$x_{-2} = 0.1$, $y_{-2} = 1.3916$ $\Rightarrow f_{-2} = f(x_{-2}, y_{-2}) = 3.8478$

$x_{-3} = 0$, $y_{-3} = 1$ $\Rightarrow f_{-3} = f(x_{-3}, y_{-3}) = 4$

To determine predictor value y_1^p:

$y_1^p = y_0 + \frac{h}{24}(55f_0 - 59f_{-1} + 37f_{-2} - 9f_{-3})$

$\Rightarrow y_1^p = 2.1509 + \frac{0.1}{24}(55(3.7974) - 59(3.7834) + 37(3.8478) - 9(4))$

$= 2.5342$

$\therefore f_1^p = f(x_1, y_1^p) = 6e^{0.4} - 2(2.5342) = 3.8825$

To find corrector value y_1^c by improving y_1^p

$y_1^{(1)} = y_0 + \frac{h}{24}(9f_1^p + 19f_0 - 5f_{-1} + f_{-2})$

$= 2.1509 + \frac{0.1}{24}(9(3.8825) + 19(3.7974) - 5(3.7834) + (3.8478)) = 2.5343$

$f_1^{(1)} = f(x_1, y_1^{(1)}) = 6e^{0.4} - 2(2.5343) = 3.8823$

$y_1^{(2)} = y_0 + \frac{h}{24}(9f_1^{(1)} + 19f_0 - 5f_{-1} + f_{-2})$

$= 2.1509 + \frac{0.1}{24}(9(3.8823) + 19(3.7974) - 5(3.7834) + (3.8478)) = 2.5343$

$y_1^{(1)}$ and $y_1^{(2)}$ coincide, $\therefore y_1^c = 2.5343 = y(0.4)$

Example 18 For the initial value problem $\frac{dy}{dx} = x^2(1+y)$; $y(1) = 1$,

find $y(1.4)$ using Adams-Bashforth method.

Solution: We know that Runge-Kutta (or any single step method) derives current value from immediate previous value. Given $f(x,y) = x^2(1+y)$, $y(1) = 1$

Taking $h = 0.1$ $x_{-3} = 1$, $y_{-3} = 1$ and computing y_{-2}, y_{-1}, y_0 corresponding to $x_{-2} = 1.1$, $x_{-1} = 1.2$, $x_0 = 1.3$ using Runge-Kutta method.

Runge-Kutta method of 4th order is given by

$y_{-2} = y_{-3} + \frac{1}{6}[K_1 + 2K_2 + 2K_3 + K_4]$ ... ①

$K_1 = hf(x_{-3}, y_{-3}) = h((x_{-3})^2(1 + y_{-3})) = 0.1(1(1+1)) = 0.2$

$K_2 = hf\left(x_{-3} + \frac{h}{2}, y_{-3} + \frac{K_1}{2}\right) = 0.1\left(\left(1 + \frac{0.1}{2}\right)^2 \left(1 + 1 + \frac{0.2}{2}\right)\right) = 0.2315$

$K_3 = hf\left(x_{-3} + \frac{h}{2}, y_{-3} + \frac{K_2}{2}\right) = 0.1\left(\left(1 + \frac{0.1}{2}\right)^2 \left(1 + 1 + \frac{0.2315}{2}\right)\right) = 0.2333$

$K_4 = hf(x_{-3} + h, y_{-3} + K_3) = 0.1((1+0.1)^2(1+1+0.2333)) = 0.2702$

Substituting values of K_1, K_2, K_3, K_4 in ①, we get the solution as:

$y_{-2} = 1 + \frac{1}{6}[0.2 + 2(0.2315) + 2(0.2333) + 0.2702] = 1.2333$

$\therefore x_{-2} = 1.1$, $y_{-2} = 1.2333$

Again $y_{-1} = y_{-2} + \frac{1}{6}[K_1 + 2K_2 + 2K_3 + K_4]$ ... ②

$K_1 = hf(x_{-2}, y_{-2}) = h((x_{-2})^2(1 + y_{-2})) = .1((1.1)^2(1 + 1.2333)) = 0.2702$

$K_2 = hf\left(x_{-2} + \frac{h}{2}, y_{-2} + \frac{K_1}{2}\right) = 0.1\left(\left(1.1 + \frac{0.1}{2}\right)^2 \left(1 + 1.2333 + \frac{0.2702}{2}\right)\right) = 0.3132$

$K_3 = hf\left(x_{-2} + \frac{h}{2}, y_{-2} + \frac{K_2}{2}\right) = 0.1\left(\left(1.1 + \frac{0.1}{2}\right)^2 \left(1 + 1.2333 + \frac{0.3132}{2}\right)\right) = 0.3161$

$K_4 = hf(x_{-2} + h, y_{-2} + K_3) = 0.1((1.1+0.1)^2(1 + 1.2333 + 0.3161)) = 0.3671$

Substituting values of K_1, K_2, K_3, K_4 in ②, we get the solution as:

$$y_{-1} = 1.2333 + \frac{1}{6}[0.2702 + 2(0.3132) + 2(0.3161) + 0.3671] = 1.5493$$

$$\therefore x_{-1} = 1.2, y_{-1} = 1.5493$$

Again $y_0 = y_{-1} + \frac{1}{6}[K_1 + 2K_2 + 2K_3 + K_4]$ \quad\quad ...③

$$K_1 = hf(x_{-1}, y_{-1}) = h((x_{-1})^2(1 + y_{-1})) = 0.1((1.2)^2(1 + 1.5493)) = 0.3671$$

$$K_2 = hf\left(x_{-1} + \frac{h}{2}, y_{-1} + \frac{K_1}{2}\right) = 0.1\left(\left(1.2 + \frac{0.1}{2}\right)^2\left(1 + 1.5493 + \frac{0.3671}{2}\right)\right) = 0.4270$$

$$K_3 = hf\left(x_{-1} + \frac{h}{2}, y_{-1} + \frac{K_2}{2}\right) = 0.1\left(\left(1.2 + \frac{0.1}{2}\right)^2\left(1 + 1.5493 + \frac{0.4270}{2}\right)\right) = 0.4317$$

$$K_4 = hf(x_{-1} + h, y_{-1} + K_3) = 0.1((1.2 + 0.1)^2(1 + 1.5493 + 0.4317)) = 0.5038$$

Substituting values of K_1, K_2, K_3, K_4 in ②, we get the solution as:

$$y_0 = 1.5493 + \frac{1}{6}[0.3671 + 2(0.4270) + 2(0.4317) + 0.5038] = 1.9807$$

$$\therefore x_0 = 1.3, y_0 = 1.9807$$

Now to compute $y_1 = y(0.4)$ using Adams method, taking

$$f_{-3} = f(x_{-3}, y_{-3}) = (x_{-3})^2(1 + y_{-3}) = (1)^2(1 + 1) = 2.0$$
$$f_{-2} = f(x_{-2}, y_{-2}) = (x_{-2})^2(1 + y_{-2}) = (1.1)^2(1 + 1.2333) = 2.7023$$
$$f_{-1} = f(x_{-1}, y_{-1}) = (x_{-1})^2(1 + y_{-1}) = (1.2)^2(1 + 1.5493) = 3.671$$
$$f_0 = f(x_0, y_0) \quad = (x_0)^2(1 + y_0) \quad = (1.3)^2(1 + 1.9807) = 5.0374$$

To determine predictor value y_1^p:

$$y_1^p = y_0 + \frac{h}{24}(55f_0 - 59f_{-1} + 37f_{-2} - 9f_{-3})$$

$$\Rightarrow y_1^p = 1.9807 + \frac{1}{24}(55(5.0374) - 59(3.671) + 37(2.7023) - 9(2)) = 2.5743$$

$$\therefore f_1^p = f(x_1, y_1^p) = (1.4)^2(1 + 2.5743) = 7.0056$$

To find corrector value y_1^c by improving y_1^p

$$y_1^{(1)} = y_0 + \frac{h}{24}(9f_1^p + 19f_0 - 5f_{-1} + f_{-2})$$

$$= 1.9807 + \frac{0.1}{24}(9(7.0056) + 19(5.0374) - 5(3.671) + (2.7023)) = 2.5770$$

$$f_1^{(1)} = f(x_1, y_1^{(1)}) = (1.4)^2(1 + 2.5770) = 7.0109$$

$$y_1^{(2)} = y_0 + \frac{h}{24}(9f_1^{(1)} + 19f_0 - 5f_{-1} + f_{-2})$$

$$= 1.9807 + \frac{0.1}{24}(9(7.0109) + 19(5.0374) - 5(3.671) + (2.7023)) = 2.5772$$

$$f_1^{(2)} = f(x_1, y_1^{(2)}) = (1.4)^2(1 + 2.5772) = 7.0113$$

$$y_1^{(3)} = y_0 + \frac{h}{24}\left(9f_1^{(2)} + 19f_0 - 5f_{-1} + f_{-2}\right)$$
$$= 1.9807 + \frac{0.1}{24}(9(7.0113) + 19(5.0374) - 5(3.671) + (2.7023)) = 2.5772$$

$y_1^{(2)}$ and $y_1^{(3)}$ coincide, $\therefore y_1^c = 2.5772 = y(0.4)$

13.4 Simultaneous Differential Equations of First Order

Consider the first simultaneous order differential equations:

$$\frac{dy}{dx} = f(x,y,z), \quad y(x_0) = y_0 \qquad \cdots \text{①}$$

$$\frac{dz}{dx} = g(x,y,z), \quad z(x_0) = z_0 \qquad \cdots \text{②}$$

System of equations given by ① and ② can be solved numerically using any of the single step methods such as Taylor series, Picard's or Runge Kutta.

To solve the given system of equations using Taylor's series method

Expanding $y(x)$ and $z(x)$ by Taylor's series about $x = x_0$, we get

$$y_1 = y_0 + (x - x_0)y_0' + \frac{1}{2!}(x - x_0)^2 y_0'' + \frac{1}{3!}(x - x_0)^3 y_0''' + \cdots$$

$$z_1 = z_0 + (x - x_0)z_0' + \frac{1}{2!}(x - x_0)^2 z_0'' + \frac{1}{3!}(x - x_0)^3 z_0''' + \cdots$$

Differentiating ① and ② successively, we can find values of $y_0', z_0', y_0'', z_0'', \ldots$ and then they are substituted in Taylor's series. Subsequent approximations $y_2, z_2, y_3, z_3, \cdots$ can be evaluated in the similar way.

To solve the given system of equations using Picard's method

By Picard's method, first and subsequent approximations can be found, as shown:

$$y_1 = y_0 + \int f(x, y_0, z_0)dx, \quad z_1 = z_0 + \int g(x, y_0, z_0)dx$$
$$y_2 = y_0 + \int f(x, y_1, z_1)dx, \quad z_1 = z_0 + \int g(x, y_1, z_1)dx$$
$$y_3 = y_0 + \int f(x, y_2, z_2)dx, \quad z_1 = z_0 + \int g(x, y_2, z_2)dx$$
$$\vdots$$

To solve the given system of equations using Runge-Kutta method

For solving given system using Runge-Kutta method, start with the initial value (x_0, y_0, z_0), y_1 and z_1 can be computed as:

$$y_1 = y_0 + \frac{1}{6}[K_1 + 2K_2 + 2K_3 + K_4], \quad z_1 = z_0 + \frac{1}{6}[L_1 + 2L_2 + 2L_3 + L_4]$$

Where

$K_1 = hf(x_0, y_0, z_0)$ $\qquad L_1 = hg(x_0, y_0, z_0)$

$K_2 = hf\left(x_0 + \frac{h}{2}, y_0 + \frac{K_1}{2}, z_0 + \frac{L_1}{2}\right)$ $\qquad L_2 = hg\left(x_0 + \frac{h}{2}, y_0 + \frac{K_1}{2}, z_0 + \frac{L_1}{2}\right)$

$K_3 = hf\left(x_0 + \frac{h}{2}, y_0 + \frac{K_2}{2}, z_0 + \frac{L_2}{2}\right)$ $\qquad L_3 = hg\left(x_0 + \frac{h}{2}, y_0 + \frac{K_2}{2}, z_0 + \frac{L_2}{2}\right)$

$K_4 = hf(x_0 + h, y_0 + K_3, z_0 + L_3)$ $\qquad L_4 = hg(x_0 + h, y_0 + K_3, z_0 + L_3)$

Subsequent approximations $y_2, z_2, y_3, z_3, \cdots$ can be evaluated in the similar way.

Example 19 Solve the given system of simultaneous equations:

$$\frac{dy}{dx} = 1 + xz \ , \ \frac{dz}{dx} = -xy \text{ for } x = 0.2, \text{ given } x = y = 0, z = 1, \text{ using}$$

i. Picard's method *ii.* Runge-Kutta method *iii.* Taylor series method

Solution: Given $f(x, y, z) = 1 + xz, \quad g(x, y, z) = -xy$

$x_0 = 0, \ y_0 = 0, \ z_0 = 1 \text{ and } h = 0.2$

i. Using Picard's approximation

$$y = y_0 + \int_{x_0}^{x} f(x, y, z) dx \ , \quad z = z_0 + \int_{x_0}^{x} g(x, y, z) dx$$

1st approximation:

$$y_1 = y_0 + \int_{x_0}^{x} f(x, y_0, z_0) dx \qquad\qquad z_1 = z_0 + \int_{x_0}^{x} g(x, y_0, z_0) dx$$

$$= 0 + \int_{0}^{x} (1 + xz_0) dx \qquad\qquad\qquad = 1 - \int_{0}^{x} xy_0 dx$$

$$= \int_{0}^{x} (1 + x) dx \qquad\qquad\qquad\qquad = 1 - \int_{0}^{x} 0 \, dx$$

$$\Rightarrow y_1 = x + \frac{x^2}{2} \qquad\qquad\qquad\qquad\qquad \Rightarrow z_1 = 1$$

2nd approximation:

$$y_2 = y_0 + \int_{x_0}^{x} f(x, y_1, z_1) dx \qquad\qquad z_2 = z_0 + \int_{0}^{x} g(x, y_1, z_1) dx$$

$$= 0 + \int_{0}^{x} (1 + xz_1) dx \qquad\qquad\qquad = 1 - \int_{0}^{x} xy_1 dx$$

$$= 0 + \int_{0}^{x} (1 + x) dx \qquad\qquad\qquad = 1 - \int_{0}^{x} \left(x^2 + \frac{x^3}{2}\right) dx$$

$$\Rightarrow y_2 = x + \frac{x^2}{2} \qquad\qquad\qquad\qquad\qquad \Rightarrow z_2 = 1 - \frac{x^3}{3} - \frac{x^4}{8}$$

3rd approximation:

$$y_3 = y_0 + \int_{x_0}^{x} f(x, y_2, z_2) dx \qquad\qquad z_2 = z_0 + \int_{x_0}^{x} g(x, y_2, z_2) dx$$

$$= 0 + \int_{0}^{x} (1 + xz_2) dx \qquad\qquad\qquad = 1 - \int_{0}^{x} xy_2 dx$$

$$= \int_{0}^{x} \left(1 + x - \frac{x^4}{3} - \frac{x^5}{8}\right) dx \qquad\qquad = 1 - \int_{x_0}^{x} \left(x^2 + \frac{x^3}{2}\right) dx$$

$$= x + \frac{x^2}{2} - \frac{x^5}{15} - \frac{x^6}{48} \qquad\qquad\qquad = 1 - \frac{x^3}{3} - \frac{x^4}{8}$$

$$\therefore y(0.2) = (0.2) + \frac{(0.2)^2}{2} - \frac{(0.2)^5}{15} - \frac{(0.2)^6}{48} \qquad z(0.2) = 1 - \frac{(0.2)^3}{3} - \frac{(0.2)^4}{8}$$

$$= 0.21998 , \qquad\qquad\qquad\qquad\qquad\qquad\qquad = 0.9971$$

ii. By Runge-Kutta method for simultaneous equations:

$$y_1 = y_0 + \frac{1}{6}[K_1 + 2K_2 + 2K_3 + K_4] , \ z_1 = z_0 + \frac{1}{6}[L_1 + 2L_2 + 2L_3 + L_4]$$

$$\cdots \textcircled{1}$$

$K_1 = hf(x_0, y_0, z_0) = h(1 + x_0 z_0) = 0.2(1 + (0)(1)) = 0.2$

$L_1 = hg(x_0, y_0, z_0) = h(-x_0 z_0) = 0.2(-(0)(1)) = 0$

$K_2 = hf\left(x_0 + \frac{h}{2}, y_0 + \frac{K_1}{2}, z_0 + \frac{L_1}{2}\right) = 0.2\left(1 + \left(0 + \frac{0.2}{2}\right)\left(1 + \frac{0}{2}\right)\right) = 0.22$

$L_2 = hg\left(x_0 + \frac{h}{2}, y_0 + \frac{K_1}{2}, z_0 + \frac{L_1}{2}\right) = 0.2\left(-\left(0 + \frac{0.2}{2}\right)\left(0 + \frac{0.2}{2}\right)\right) = -0.002$

$K_3 = hf\left(x_0 + \frac{h}{2}, y_0 + \frac{K_2}{2}, z_0 + \frac{L_2}{2}\right) = 0.2\left(1 + \left(0 + \frac{0.2}{2}\right)\left(1 - \frac{0.002}{2}\right)\right) = 0.22$

$L_3 = hg\left(x_0 + \frac{h}{2}, y_0 + \frac{K_2}{2}, z_0 + \frac{L_2}{2}\right) = 0.2\left(-\left(0 + \frac{0.2}{2}\right)\left(0 + \frac{0.22}{2}\right)\right) = -0.0022$

$K_4 = hf(x_0 + h, y_0 + K_3, z_0 + L_3) = 0.2(1 + (0.2)(1 - .0022)) = 0.2399$

$L_4 = hg(x_0 + h, y_0 + K_3, z_0 + L_3) = 0.2(-(0.2)(0.22)) = -0.0088$

Substituting values of K's and L's ①, we get the solution as:

$y_1 = 0 + \frac{1}{6}[0.2 + 2(0.22) + 2(0.22) + 0.2399] = 0.21998$

$z_1 = 1 + \frac{1}{6}[0 + 2(-0.002) + 2(-0.0022) + -0.0088] = 0.9971$

iii. Expanding $y(x)$ and $z(x)$ in Taylor's series about $x = 0$, upto 4th order

$y(x) = y_0 + (x - 0)y_0' + \frac{1}{2!}(x - 0)^2 y_0'' + \frac{1}{3!}(x - 0)^3 y_0''' + \frac{1}{4!}(x - 0)^4 y_0^{iv}$

$z(x) = z_0 + (x - 0)z_0' + \frac{1}{2!}(x - 0)^2 z_0'' + \frac{1}{3!}(x - 0)^3 z_0''' + \frac{1}{4!}(x - 0)^4 z_0^{iv}$

$\frac{dy}{dx} = 1 + xz$,	$y_0 = 0$	$\frac{dz}{dx} = -xy$,	$z_0 = 1$
$y' = 1 + xz$	$y_0' = 1$	$z' = -xy$	$z_0' = 0$
$y'' = xz' + z$	$y_0'' = 1$	$z'' = -xy' - y$	$z_0'' = 0$
$y''' = xz'' + 2z'$	$y_0''' = 0$	$z''' = -xy'' - 2y'$	$z_0''' = -2$
$y^{iv} = xz''' + 3z''$	$y_0^{iv} = 0$	$z^{iv} = -xy''' - 3y''$	$z_0^{iv} = -3$

Substituting the values of $y_0, y_0', y_0'', y_0''', y_0^{iv}$ and $x_0, x_0', x_0'', x_0''', x_0^{iv}$ in Taylor's series expansion, we get

$y(x) = 0 + x(1) + \frac{1}{2}x^2(1) + \frac{1}{6}x^3(0) + \frac{1}{24}x^4(0)$

$z(x) = 1 + x(0) + \frac{1}{2}x^2(0) + \frac{1}{6}x^3(-2) + \frac{1}{24}x^4(-3)$

$\therefore y(0.2) = (0.2)(1) + \frac{1}{2}(0.2)^2(1) = 0.22$

$z(0.2) = 1 + \frac{1}{6}(0.2)^3(-2) + \frac{1}{24}(0.2)^4(-3) = 0.9971$

Example 20 Solve the simultaneous equations:

$$\frac{dx}{dt} = -2x - 3y \quad, \quad \frac{dy}{dt} = 2e^{2t} - 3x - 2y \text{ for } t = 0.1,$$

Given that at $t = 0$ $x = 1$ and $y = 2$, using

    i. Runge-Kutta method   ii. Taylor series method

Solution: Given $f(t, x, y) = -2x - 3y, \quad g(t, x, y) = 2e^{2t} - 3x - 2y$

$t_0 = 0, \ x_0 = 1, \ y_0 = 2$ and $h = 0.1$

Here x & y are dependent variables and t is the independent variable.

i. By Runge-Kutta method for simultaneous equations:

$$x_1 = x_0 + \frac{1}{6}[K_1 + 2K_2 + 2K_3 + K_4], \quad y_1 = y_0 + \frac{1}{6}[L_1 + 2L_2 + 2L_3 + L_4]$$

$$\cdots ①$$

$K_1 = hf(t_0, x_0, y_0) = h(-2x_0 - 3y_0) = 0.1(-2(1) - 3(2)) = -0.8$

$L_1 = hg(t_0, x_0, y_0) = h(2e^{2t_0} - 3x_0 - 2y_0) = .1(2e^0 - 3(1) - 2(2)) = -0.5$

$K_2 = hf\left(t_0 + \frac{h}{2}, x_0 + \frac{K_1}{2}, y_0 + \frac{L_1}{2}\right) = 0.1\left(-2\left(1 - \frac{.8}{2}\right) - 3\left(2 - \frac{.5}{2}\right)\right) = -0.645$

$L_2 = hg\left(t_0 + \frac{h}{2}, x_0 + \frac{K_1}{2}, y_0 + \frac{L_1}{2}\right)$

$= 0.1\left(2e^{2\left(\frac{0.1}{2}\right)} - 3\left(1 - \frac{0.8}{2}\right) - 2\left(2 - \frac{0.5}{2}\right)\right) = -0.309$

$K_3 = hf\left(t_0 + \frac{h}{2}, x_0 + \frac{K_2}{2}, y_0 + \frac{L_2}{2}\right)$

$= 0.1\left(-2\left(1 - \frac{0.645}{2}\right) - 3\left(2 - \frac{0.309}{2}\right)\right) = -0.6892$

$L_3 = hg\left(t_0 + \frac{h}{2}, x_0 + \frac{K_2}{2}, y_0 + \frac{L_2}{2}\right)$

$= 0.1\left(2e^{2\left(\frac{.1}{2}\right)} - 3\left(1 - \frac{.645}{2}\right) - 2\left(2 - \frac{.309}{2}\right)\right) = -0.3513$

$K_4 = hf(t_0 + h, x_0 + K_3, y_0 + L_3)$

$= 0.1(-2(1 - 0.6892) - 3(2 - 0.3513)) = -0.5568$

$L_4 = hg(t_0 + h, x_0 + K_3, y_0 + L_3)$

$= 0.1\left(2e^{2\left(\frac{0.1}{2}\right)} - 3(1 - 0.6892) - 2(2 - 0.3513)\right) = -0.2019$

Substituting values of K's and L's ①, we get the solution as:

$x_1 = x(0.1) = 1 + \frac{1}{6}[-0.8 + 2(-0.645) + 2(-0.6892) - 0.5568] = 0.3291$

$y_1 = y(0.1) = 2 + \frac{1}{6}[-0.5 + 2(-0.309) + 2(-0.3513) - 0.2019] = 1.6629$

ii. Expanding $y(t)$ and $z(t)$ by Taylor's series about $t = 0$, upto 4th order

$x(t) = x_0 + (t-0)x_0' + \frac{1}{2!}(t-0)^2 x_0'' + \frac{1}{3!}(t-0)^3 x_0''' + \frac{1}{4!}(t-0)^4 x_0^{iv}$

$y(t) = y_0 + (t-0)y_0' + \frac{1}{2!}(t-0)^2 y_0'' + \frac{1}{3!}(t-0)^3 y_0''' + \frac{1}{4!}(t-0)^4 y_0^{iv}$

$\frac{dx}{dt} = -2x - 3y \qquad\qquad\qquad \frac{dy}{dt} = 2e^{2t} - 3x - 2y$

$x_0 = 1 \qquad\qquad\qquad\qquad\qquad y_0 = 2$

$\begin{array}{llll} x' = -2x - 3y & x_0' = -8 & y' = 2e^{2t} - 3x - 2y & y_0' = -5 \\ x'' = -2x' - 3y' & x_0'' = 31 & y'' = 4e^{2t} - 3x' - 2y' & y_0'' = 38 \\ x''' = -2x'' - 3y'' & x_0''' = -176 & y''' = 8e^{2t} - 3x'' - 2y'' & y_0''' = -161 \\ x^{iv} = -2x''' - 3y''' & x_0^{iv} = 835 & y^{iv} = 4e^{2t} - 3x''' - 2y''' & y_0^{iv} = 866 \end{array}$

Substituting the values of $x_0\ x_0', x_0'', x_0''', x_0^{iv}$ and $y_0\ y_0', y_0'', y_0''', y_0^{iv}$ in Taylor's series expansion, we get

$x(t) = 1 + t(-8) + \frac{1}{2}t^2(31) + \frac{1}{6}t^3(-176) + \frac{1}{24}t^4(835)$

$y(t) = 2 + t(-5) + \frac{1}{2}t^2(38) + \frac{1}{6}t^3(-161) + \frac{1}{24}t^4(866)$

$\therefore x(0.1) = 1 + (0.1)(-8) + \frac{1}{2}(0.1)^2(31) + \frac{1}{6}(0.1)^3(-176) + \frac{1}{24}(0.1)^4(835) = 0.3291$

$y(0.1) = 2 + (0.1)(-5) + \frac{1}{2}(0.1)^2(38) + \frac{1}{6}(0.1)^3(-161) + \frac{1}{24}(0.1)^4(866) = 1.6668$

13.5 Second Order Differential Equations

Consider the 2nd order differential equation $\frac{d^2y}{dx^2} = f\left(x, y, \frac{dy}{dx}\right)$ ... ①

Rewriting ① as $\frac{d}{dx}\left(\frac{dy}{dx}\right) = f\left(x, y, \frac{dy}{dx}\right)$

$\Rightarrow \frac{dz}{dx} = f(x, y, z) \qquad$ By putting $\frac{dy}{dx} = z$

$\therefore$ The 2nd order differential equation given by ① can be remodeled as a system of simultaneous equations given by: $\frac{dy}{dx} = z, \frac{dz}{dx} = f(x, y, z)$

Example 21 Solve the differential equation $y'' = xy'^2 - y^2$ for $x = 0.2$, using Runge-Kutta method, given that $x = 0, y = 1, y' = 0$

Solution: Given $\frac{d^2y}{dx^2} = x\left(\frac{dy}{dx}\right)^2 - y^2$

$\Rightarrow \frac{d}{dx}\left(\frac{dy}{dx}\right) = x\left(\frac{dy}{dx}\right)^2 - y^2$

$\Rightarrow \frac{dy}{dx} = z, \frac{dz}{dx} = xz^2 - y^2 \qquad\qquad$ by putting $\frac{dy}{dx} = z$

Let $f(x, y, z) = z$, $g(x, y, z) = xz^2 - y^2$, initial conditions are $x_0 = 0$, $y_0 = 1$, $z_0 = 0$

By Runge-Kutta method for simultaneous equations:

$$y_1 = y_0 + \frac{1}{6}[K_1 + 2K_2 + 2K_3 + K_4], \quad z_1 = z_0 + \frac{1}{6}[L_1 + 2L_2 + 2L_3 + L_4]$$

$\quad\quad\quad\quad\quad\quad\quad\quad\quad\quad\quad\quad\quad\quad\quad\quad\quad\quad\quad\quad\quad\quad \cdots ①$

$K_1 = hf(x_0, y_0, z_0) = h(z_0) = 0.2(0) = 0$

$L_1 = hg(x_0, y_0, z_0) = h(x_0 z_0^2 - y_0^2) = 0.2(0 - 1) = -0.2$

$K_2 = hf\left(x_0 + \frac{h}{2}, y_0 + \frac{K_1}{2}, z_0 + \frac{L_1}{2}\right) = 0.2\left(0 - \frac{0.2}{2}\right) = -0.02$

$L_2 = hg\left(x_0 + \frac{h}{2}, y_0 + \frac{K_1}{2}, z_0 + \frac{L_1}{2}\right) = 0.2\left(\frac{0.2}{2}\left(-\frac{0.2}{2}\right)^2 - \left(1 + \frac{0}{2}\right)^2\right) = -0.1998$

$K_3 = hf\left(x_0 + \frac{h}{2}, y_0 + \frac{K_2}{2}, z_0 + \frac{L_2}{2}\right) = 0.2\left(0 - \frac{0.1998}{2}\right) = -0.02$

$L_3 = hg\left(x_0 + \frac{h}{2}, y_0 + \frac{K_2}{2}, z_0 + \frac{L_2}{2}\right) = .2\left(\frac{.2}{2}\left(-\frac{.1998}{2}\right)^2 - \left(1 - \frac{.02}{2}\right)^2\right) = -0.1958$

$K_4 = hf(x_0 + h, y_0 + K_3, z_0 + L_3) = 0.2(0 - 0.1958) = -0.0392$

$L_4 = hg(x_0 + h, y_0 + K_3, z_0 + L_3) = .2(.2(-.1958)^2 - (1 - .02)^2) = -0.1905$

Substituting values of K's and L's ①, we get the solution as:

$y_1 = y(0.2) = 1 + \frac{1}{6}[0 + 2(-0.02) + 2(-0.02) - 0.0392] = 0.9801$

$z_1 = z(0.2) = 0 + \frac{1}{6}[-0.2 + 2(-0.1998) + 2(-.1958) + -.1905] = -0.1970$

Example 22 Find $y(0.2)$ & $y'(0.2)$ for the initial value problem $y'' + xy' + y = 0$ given that $y(0) = 1$, $y'(0) = 0$.

Solution: Given $\frac{d^2y}{dx^2} = -x\frac{dy}{dx} - y$

$\Rightarrow \frac{d}{dx}\left(\frac{dy}{dx}\right) = -x\frac{dy}{dx} - y$

$\Rightarrow \frac{dy}{dx} = z, \quad \frac{dz}{dx} = -xz - y \quad\quad$ by putting $\frac{dy}{dx} = z$

Let $f(x, y, z) = z$, $g(x, y, z) = -xz - y$

Initial conditions are $x_0 = 0$, $y_0 = 1$, $z_0 = 0 \quad \because y' = z$

Using Picard's method to obtain $y(0.1)$, $y(0.2)$ and $y(0.3)$

$y = y_0 + \int_{x_0}^{x} f(x, y, z)dx \quad\quad\quad z = z_0 + \int_{x_0}^{x} g(x, y, z)dx$

1st approximation:

$y_1 = y_0 + \int_{x_0}^{x} f(x, y_0, z_0)dx \quad\quad z_1 = z_0 + \int_{x_0}^{x} g(x, y_0, z_0)dx$

$\quad = 1 + \int_0^x z_0 dx \quad\quad\quad\quad\quad\quad = 0 - \int_0^x (xz_0 + y_0)dx$

$\quad = 1 + \int_0^x 0\, dx \quad\quad\quad\quad\quad\quad = 0 - \int_0^x (0 + 1)dx$

$\Rightarrow y_1 = 1 \quad\quad\quad\quad\quad\quad\quad\quad\quad\quad \Rightarrow z_1 = -x$

2nd approximation:

$$y_2 = y_0 + \int_{x_0}^{x} f(x, y_1, z_1)dx \qquad\qquad z_2 = z_0 + \int_0^x g(x, y_1, z_1)dx$$

$$= 1 + \int_0^x z_1 dx \qquad\qquad = 0 - \int_0^x (xz_1 + y_1)dx$$

$$= 1 - \int_0^x x \, dx \qquad\qquad = 0 + \int_0^x (x^2 - 1)dx$$

$$\Rightarrow y_2 = 1 - \frac{x^2}{2} \qquad\qquad \Rightarrow z_2 = \frac{x^3}{3} - x$$

3rd approximation:

$$y_3 = y_0 + \int_{x_0}^{x} f(x, y_2, z_2)dx \qquad\qquad z_3 = z_0 + \int_0^x g(x, y_2, z_2)dx$$

$$= 1 + \int_0^x z_2 dx \qquad\qquad = 0 - \int_0^x (xz_2 + y_2)dx$$

$$= 1 + \int_0^x \left(\frac{x^3}{3} - x\right)dx \qquad\qquad = 0 - \int_0^x \left(\frac{x^4}{3} - \frac{x^2}{2} - 1\right)dx$$

$$\Rightarrow y_3 = 1 - \frac{x^2}{2} + \frac{x^4}{12} \qquad\qquad \Rightarrow z_3 = -x + \frac{x^3}{2} - \frac{x^5}{15}$$

Now $y(0.2) = 1 - \frac{(0.2)^2}{2} + \frac{(0.2)^4}{12} = 0.9801$

Also $y'(0.2) = z(0.2) = -0.2 + \frac{(0.2)^3}{2} - \frac{(0.2)^5}{15} = 0.196 \qquad \because y' = z$

13.6 Boundary Value Problems (BVPs):

Boundary value problems require determination of the function, subject to specified values at the two extremes (starting and end points) of the solution domain in contrast to initial value problems which require only one value (typically starting point). Boundary value problems often have multiple solutions or sometimes even fail to have a solution.

There are many numerical techniques to solve BVPs; we shall confine our studies to finite difference method, which is based on converting the boundary value problem with given constraints into an algebraic system of equations.

13.6.1 Finite Differences Method

Finite difference method provides a workable solution to 2nd and higher order boundary value problems with boundary conditions $y(x_0) = a$, $y(x_n) = b$, by providing solution at interior pivotal points taken at regular intervals.

We Divide the interval (x_0, x_n) into n equal sub-intervals of height h giving interior pivotal points $x_1 = x_0 + h$, $x_2 = a + 2h$, ..., $x_{n-1} = x_0 + (n-1)h$.

Now given that $y(x_0) = a$, $y(x_n) = b$, we need to evaluate $y_1, y_2, ..., y_{n-1}$.

Step by step procedure is illustrated below.

1. Using finite approximations to replace the derivatives by Taylor's series expansion of the function $y(x + h)$ and $y(x - h)$ given by:

$$y(x + h) = y(x) + h y'(x) + \frac{h^2}{2!} y''(x) + \frac{h^3}{3!} y'''(x) + \cdots \qquad \cdots ①$$

$$y(x - h) = y(x) - hy'(x) + \frac{h^2}{2!}y''(x) - \frac{h^3}{3!}y'''(x) + \cdots \qquad \cdots ②$$

Adding ① and ②, we get

$$y(x + h) + y(x - h) = 2\left[y(x) + \frac{h^2}{2!}y''(x) + \frac{h^4}{4!}y^{iv}(x) + \cdots\right] \qquad \cdots ③$$

Subtracting ② from ① we get

$$y(x + h) - y(x - h) = 2\left[hy'(x) + \frac{h^3}{3!}y'''(x) + \cdots\right] \qquad \cdots ④$$

Neglecting higher order terms, each derivative may be changed to finite differences using relations ③ and ④ as given below:

$$y'(x) = \frac{1}{2h}[y(x+h) - y(x-h)]$$

$$y''(x) = \frac{1}{h^2}[y(x+h) - 2y(x) + y(x-h)]$$

$\vdots$

2. These approximation are potent to change the given BVP into a set of algebraic equations, by changing each derivative into finite differences given as:

$$y'_i = \frac{1}{2h}[y_{i+1} - y_{i-1}]$$

$$y''_i = \frac{1}{h^2}[y_{i+1} - 2y_i + y_{i-1}]$$

$$y'''_i = \frac{1}{2h^3}[y_{i+2} - 2y_{i+1} + 2y_{i-1} - y_{i-2}]$$

$$y^{iv}_i = \frac{1}{h^4}[y_{i+2} - 4y_{i+1} + 6y_i - 4y_{i-1} + y_{i-2}]$$

$\vdots$

Here i can take values as per number of equations required for solving a particular set of algebraic equations.

3. Solve the set of algebraic equations to obtain the solution of the BVP.

Note: The accuracy of this method depends upon the number of intervals (n).

Higher the value of n, higher the accuracy, yet increasing the number of intervals increases the number of algebraic equations formed. This suggests keeping value of n to a reasonable threshold, to balance the accuracy and computational efforts.

Example 23 Solve the boundary value problem $y'' - x + y = 0$, $y(0) = y(1) = 0$

Solution: Given BVP is $y'' = x - y$, $y(0) = y(1) = 0$ $\qquad \cdots ①$

Dividing the interval $(0,1)$ into four sub-intervals taking $h = \frac{1}{4}$ such that pivotal points are $x_0 = 0$, $x_1 = \frac{1}{4}$, $x_2 = \frac{1}{2}$, $x_3 = \frac{3}{4}$, $x_4 = 1$. We need to evaluate y_1, y_2 and y_3, given that $y_0 = 0$, $y_4 = 0$.

The boundary value problem given by ① may be approximated in terms of finite differences as:

$$\frac{1}{h^2}[y_{i+1} - 2y_i + y_{i-1}] = x_i - y_i, \qquad i = 1,2,3$$

$\Rightarrow 16[y_{i+1} - 2y_i + y_{i-1}] = x_i - y_i \qquad \because h = \frac{1}{4}$

$\Rightarrow 16y_{i+1} - 31y_i + 16y_{i-1} = x_i$

$i = 1 \quad \Rightarrow 16y_2 - 31y_1 = \frac{1}{4}$

$i = 2 \quad \Rightarrow 16y_3 - 31y_2 + 16y_1 = \frac{1}{2}$

$i = 3 \quad \Rightarrow -31y_3 + 16y_2 = \frac{3}{4}$

Solving these three equations, we get

$y_1 = -\frac{2465}{55676} = -0.04427, \; y_2 = -\frac{63}{898} = -0.07016, \; y_3 = -\frac{3363}{55676} = -0.0604$

Example 24 Solve the boundary value problem $y'' + 2y' + y = x^2$,

$y(0) = 0.2, \; y(1) = 0.8$.

Solution: Given BVP is $y'' + 2y' = x^2 - y, \; y(0) = 0.2, \; y(1) = 0.8 \quad \cdots \; ①$

Dividing the interval $(0,1)$ into four sub-intervals taking $h = \frac{1}{4}$ such that pivotal points are $x_0 = 0$, $x_1 = 0.25$, $x_2 = 0.5$, $x_3 = 0.75$, $x_4 = 1$.

We need to evaluate y_1, y_2 and y_3, given that $y_0 = 0.2$ and $y_4 = 0.8$

The boundary value problem given by ① may be approximated in terms of finite differences as:

$\frac{1}{h^2}[y_{i+1} - 2y_i + y_{i-1}] + \frac{2}{2h}[y_{i+1} - y_{i-1}] = x_i^2 - y_i, \quad i = 1,2,3$

$\Rightarrow 16[y_{i+1} - 2y_i + y_{i-1}] + 4[y_{i+1} - y_{i-1}] = x_i^2 - y_i \qquad \because h = \frac{1}{4}$

$\Rightarrow 20y_{i+1} - 31y_i + 12y_{i-1} = x_i^2$

$i = 1 \quad \Rightarrow 20y_2 - 31y_1 = (0.25)^2 - 12(0.2) = -2.3375$

$i = 2 \quad \Rightarrow 20y_3 - 31y_2 + 12y_1 = (0.5)^2 = 0.25$

$i = 3 \quad \Rightarrow -31y_3 + 12y_2 = (0.75)^2 - 20(0.8) = -15.4375$

Solving these three equations, we get

$y_1 = \frac{616427}{1192880} = 0.5168, \; y_2 = \frac{6581}{9620} = 0.6841, \; y_3 = \frac{909923}{1192880} = 0.7628$

Example 25 Solve the boundary value problem $y^{iv} + 81y = 81x^2$, in the interval $(0,1)$ with given boundary value conditions: $y(0) = y(1) = y''(0) = y''(1) = 0$, taking $h = \frac{1}{3}$

Solution: Given BVP is $y^{iv} = 81x^2 - 81y, \; y(0) = y(1) = y''(0) = y''(1) = 0 \quad \cdots \; ①$

Dividing the interval $(0,1)$ into three sub-intervals taking $h = \frac{1}{3}$ such that pivotal points are

$x_0 = 0, \; x_1 = \frac{1}{3}, \; x_2 = \frac{2}{3}, \; x_3 = 1$

We need to evaluate y_1, y_2, given that $y_0 = y_3 = y_0'' = y_3'' = 0$

The boundary value problem given by ① may be approximated in terms of finite differences as:

$\frac{1}{h^4}[y_{i+2} - 4y_{i+1} + 6y_i - 4y_{i-1} + y_{i-2}] = 81x_i^2 - 81y_i$, $i = 1,2$

$\Rightarrow 81[y_{i+2} - 4y_{i+1} + 6y_i - 4y_{i-1} + y_{i-2}] = 81x_i^2 - 81y_i$ $\because h = \frac{1}{3}$

$\Rightarrow y_{i+2} - 4y_{i+1} + 7y_i - 4y_{i-1} + y_{i-2} = x_i^2$

Putting $i = 1,2$ and using the boundary value conditions $y_0 = y_3 = 0$, we get

$i = 1 \quad \Rightarrow -4y_2 + 7y_1 + y_{-1} = \left(\frac{1}{3}\right)^2$ ... ②

$i = 2 \quad \Rightarrow y_4 + 7y_2 - 4y_1 = \left(\frac{2}{3}\right)^2$ ... ③

Again $\quad y_i'' = \frac{1}{h^2}[y_{i+1} - 2y_i + y_{i-1}]$

$i = 0 \Rightarrow \quad y_0'' = 9[y_1 - 2y_0 + y_{-1}]$

$\Rightarrow \quad 0 = 9[y_1 - 2(0) + y_{-1}]$ $\because y_0 = y_0'' = 0$

$\Rightarrow \quad y_{-1} = -y_1$ ... ④

$i = 3 \Rightarrow \quad y_3'' = 9[y_4 - 2y_3 + y_2]$

$\Rightarrow \quad 0 = 9[y_4 - 0 + y_2]$ $\because y_3 = y_3'' = 0$

$\Rightarrow \quad y_4 = -y_2$ ... ⑤

Using ④ in ② and ⑤ in ③, we get

$6y_1 - 4y_2 = \frac{1}{9}$ and $-4y_1 + 6y_2 = \frac{4}{9}$

Solving these two equations, we get

$y_1 = \frac{11}{90} = 0.1222$, $y_2 = \frac{7}{45} = 0.1556$

Exercise 13

1. If $\frac{dy}{dx} = \frac{y-x}{y+x}$, $y(0) = 1$, find the value of y for $x = 0.1$ using Picard's method.
2. Use Taylor's series method to find approximate value of y at $x = 0.2$, given that $\frac{dy}{dx} = 2y + 3e^x$, $y(0) = 0$. Compare the results by solving the differential equation directly.
3. Given that $\frac{dy}{dx} = \frac{y-x}{y+x}$, $y(0) = 1$, find the value of y for $x = 0.1$ using Euler's method.
4. If $\frac{dy}{dx} = x + |\sqrt{y}|$, $y(0) = 1$, find the value of y for the range $0 \le x \le 0.6$ in steps of 0.2 using modified Euler's method.
5. Given that $\frac{dy}{dx} = \frac{y^2 - x^2}{y^2 + x^2}$, $y(0) = 1$, find the value of y for $x = 0.2$ using Runge-Kutta method of 4th order.

6. If $\frac{dy}{dx} = x - y^2$, $y(0) = 0$, Apply Milne's method find the value of y in the range $0 \leq x \leq 1$, taking $h = 0.2$.

7. For the initial value problem $\frac{dy}{dx} = x - y^2$, $y(0) = 1$, apply Adam's method to find the value of $y(0.4)$ by evaluating starting values using Runge-Kutta method of order 4 taking $h = 0.1$.

8. Find approximate values of y and z correct to 4 decimal places; corresponding to $x = 0.1$ for the simultaneous differential equations $\frac{dy}{dx} = x + z$, $\frac{dz}{dx} = x - y^2$; using Picard's method given that $y(0) = 2, z(0) = 1$.

9. Find $y(0.1), y(0.2)$ and $y(0.3)$ using Taylor's series method for the initial value problem $y'' + xy' + y = 0$, $y(0) = 1$, $y'(0) = 0$

10. Solve the boundary value problem $y'' - x - y = 0$, $y(0) = y(1) = 0$

Answers

1. **0.9828**
2. 0.8110, 0.8112
3. 1.0928
4. At $x = 0$; $y = 1.2$,   at $x = 0.2$; $y = 1.2309$
   at $x = 0.4$; $y = 1.5253$, at $x = 0.6$; $y = 1.8861$
5. 1.196
6. $y(0.2) = 0.02$, $y(0.4) = 0.0795$, $y(0.6) = 0.1762$, $y(0.8) = 0.3049$, $y(1) = 0.4555$
7. $y(0.4) = 0.7785$
8. $y(0.1) = 2.0845$, $z(0.1) = 0.5867$
9. $y(0.1) = 0.995$, $y(0.2) = 0.9802$, $y(0.3) = 0.956$
10. $y(0.25) = -0.03488$, $y(0.5) = -0.05632$, $y(0.75) = -0.05003$

Appendix

Scilab code to add two matrices

```
// Matrix Addition script file
clc
m=input("enter number of rows of the Matrix: ");
n=input("enter number of columns of the Matrix: ");
disp('enter the first Matrix')
for i=1:m
   for j=1:n
   A(i,j)=input('\');
     end
end
disp('enter the second Matrix')
for i=1:m
   for j=1:n
   B(i,j)=input('\');
     end
end
for i=1:m
   for j=1:n
   C(i,j)=A(i,j)+B(i,j);
     end
end
disp('The first matrix is')
disp(A)
disp('The Second matrix is')
disp(B)
disp('The sum of the two matrices is')
disp(C)
```

Scilab code to multiply two matrices

```
// matrix multiplication script file
clc
m=input("Enter number of rows of the first Matrix: ");
n=input("Enter number of columns of the first Matrix: ");
p=input("Enter number of rows of the second Matrix: ");
q=input("Enter number of columns of the second Matrix: ");
if n==p
```

```
  disp('Matrices are conformable for multiplication')
else
   disp('Matrices are not conformable for multiplication')
   break;
end
disp('enter the first Matrix')
for i=1:m
   for j=1:n
   A(i,j)=input('\');
      end
end
disp('enter the second Matrix')
for i=1:p
   for j=1:q
   B(i,j)=input('\');
      end
end
C=zeros(m,q);
for i=1:m
 for j=1:q
    for k=1:n
 C(i,j)=C(i,j)+A(i,k)*B(k,j);
      end
   end
end
disp('The first matrix is')
disp(A)
disp('The Second matrix is')
disp(B)
disp('The product of the two matrices is')
disp(C)
```

Scilab code to find matrix transpose

```
// matrix transpose script file
m=input("Enter number of rows of the Matrix: ");

n=input("Enter number of columns of the Matrix: ");
disp('Enter the Matrix')
for i=1:m
```

```
    for j=1:n
      A(i,j)=input('\');
       end
end
B=zeros(n,m);
for i=1:n
  for j=1:m
 B(i,j)=A(j,i)
   end
end
disp('Entered matrix is')
disp(A)
disp('Transposed matrix is')
disp(B)
```

Scilab code to find the inverse of a matrix

```
// Inverse of a 3 by 3 matrix using gauss jordan Method
 clc
disp('Enter a 3 by 3 matrix row-wise, make sure that diagonal elements are non -zeros')
for i=1:3
    for j=1:3
    A(i,j)=input('\');
     end
end
disp('Entered Matrix is')
disp(A)
if det(A)==0
disp('Matrix is singular, Inverse does not exist')
break;
end
//Taking the augmented matrix [A|I],
B=[A eye(3,3)]
disp('Augumented matrix is:')
disp(B)
//Making B(1,1)=1
B(1,:) = B(1,:)/B(1,1);
//Making B(2,1) and B(3,1)=0
B(2,:) = B(2,:) - B(2,1)*B(1,:);
```

```
B(3,:) = B(3,:) - B(3,1)*B(1,:);
//Making B(2,2)=1 and B(1,2), B(3,2)=0
B(2,:) = B(2,:)/B(2,2);
B(1,:) = B(1,:) - B(1,2)*B(2,:);
B(3,:) = B(3,:) - B(3,2)*B(2,:);
// Making B(3,3)=1 and B(1,3), B(2,3)=0
B(3,:) = B(3,:)/B(3,3);
B(1,:) = B(1,:) - B(1,3)*B(3,:);
B(2,:) = B(2,:) - B(2,3)*B(3,:);
disp('Augumented matrix after row operations is:')
disp(B)
B(:,1:3)=[ ]
disp('Inverse of the Matrix is')
disp(B)
```

Scilab code to find eigenvalues of a matrix

```
// Eigen Values
clc
disp('enter the Matrix')
for i=1:2
   for j=1:2
   A(i,j)=input('\');
    end
end
b=A(1,1)+A(2,2);
c=A(1,1)*A(2,2)-A(1,2)*A(2,1);
// characteristic equation is e^2-trace(A)+ det(A)=0
 disp('The characteristic equation is:')
disp(['  e^2 + ' string(-b) '*e + ' string(c) ' = 0'])
e1=(b+sqrt(b^2-4*c))/2;
e2=(b-sqrt(b^2-4*c))/2;
if A(1,2) ~= 0
  v1 = [A(1,2); e1-A(1,1)];
  v2 = [A(1,2); e2-A(1,1)];
elseif A(2,1) ~= 0
  v1 = [e1-A(2,2); A(2,1)];
  v2 = [e2-A(2,2); A(2,1)];
else
  v1 = [1; 0];
  v2 = [0; 1];
end
          disp('First Eigen value is:');
          disp(e1)
```

```
        disp('First Eigen vector is:');
        disp (v1)
        disp('Second Eigen value is:');
        disp(e2)
        disp('Second Eigen vector is:');
        disp (v2)
```

Scilab code to find mean, S.D. and first *r* moments about mean of a grouped data

```
//Program to find mean, S.D. and first r moments about mean of given grouped data
clc
n=input('Enter the no. of observations:');
disp('Enter the values of xi');
for i=1:n
   x(i)=input('\');
end;
disp('Enter the corresponding frequencies fi:')
sum=0;
for i=1:n
   f(i)=input('\');
   sum=sum+f(i);
end;
r=input('How many moments to be calculated:');
sum1=0
for i=1:n
   sum1=sum1+f(i)*x(i);
end
A=sum1/sum;    //Calculate the average
printf('Average=%f\n',A);
for j=1:r
   sum2=0;
   for i=1:n
      y(i)=f(i)*(x(i)-A)^j;
      sum2=sum2+y(i);
   end
 M(j)=(sum2/sum);   //Calculate the moments
 printf('Moment about mean M(%d)=%f\n',j,M(j));
end
sd=sqrt(M(2));    //Calculate the standard deviation
printf('Standard deviation=%f\n',sd);
```

Scilab code to find mean, S.D. and first *r* moments about mean of an ungrouped grouped data

// program to find mean, mode, median, moments, skewness and kurtosis of linear data
clc
function [ ]=moments(**A**)
// save file as moments.sce
B=gsort(**A**);
n = length(B);
meanA = sum(B)/n;
if pmodulo(n,2)==0
medianA =((B(n/2)+B(n/2 +1)))/2;
else medianA = B((n+1)/2);
end
C = diff(B)
//Y= diff(X) calculates differences between adjacent elements of X along the first array dimension whose size does not equal 1:
//If X is a vector of length m, then Y = diff(X) returns a vector of length m-1. The elements of Y are the differences between adjacent elements of X.
    //Y = [X(2)-X(1) X(3)-X(2) ... X(m)-X(m-1)]
D = find(C) *//D = find(C) finds the idices(positions), where value is non zero*
E = diff(D)
[m k] = max(E) *// maximum 'm' at kth position*
modeA = B(D(k)+1)
printf('Mean of the given data is : %f \n\n', meanA);
printf('Median of the given data is : %f \n\n', medianA);
printf('Mode of the given data is : %f \n\n', modeA);
printf('First moment about the mean(M1)= %f \n\n', 0);
for i=1:n
X(i)=**A**(i)-meanA;
 end
M2 = sum(X.*X)/n;
M3 = sum(X.*X.*X)/n;
M4 = sum(X.*X.*X.*X)/n;
printf('Second moment about the mean(M2)= %f \n\n', M2);
printf('Third moment about the mean(M3)= %f \n\n', M3);
printf('Fourth moment about the mean(M4)= %f \n\n', M4);
sd= sqrt (M2);
printf('Standard deviation: %f \n\n', sd);

```
Csk= (meanA - modeA)/sd;
printf('Coefficient of skewness: %f \n\n', Csk);
Sk= (M3)^2/(M2)^3;
printf('Skewness: %f \n\n', Sk);
Kur= M4/(M2)^2;
printf('Kurtosis: %f \n\n', Kur);
endfunction
```

Scilab code to find a line of regression

```
// Regression line
clc
n=input('Enter the number of terms:')
  printf(' Enter the values of xi')
  for i=1:n
  x(i)=input('\');
  end
  printf(' Enter the values of yi')
  for i=1:n
  y(i)=input('\');
  end
    sumx=0;sumy=0;sumxy=0;sumx2=0;
  for i=1:n
    sumx=sumx +x(i);
    sumx2=sumx2 +x(i)*x(i);
    sumy=sumy +y(i);
    sumxy=sumxy +x(i)*y(i);
  end
  a=((sumx2*sumy -sumx*sumxy)*1.0/(n*sumx2-sumx*sumx)*1.0);
  b=((n*sumxy-sumx*sumy)*1.0/(n*sumx2-sumx*sumx)*1.0);
  printf('The line is Y=%3.3f +%3.3f X',a,b)
```

Scilab code to fit a straight line to given set of data points

```
// Program of straight line fitting for given n pairs of values (x,y)
clc;clear;close;
n=input('Enter the no. of pairs of values (x,y):')
disp('Enter the values of x:')
for i=1:n
   x(i)=input(' ')
end
disp('Enter the corresponding values of y:')
```

```
for i=1:n
    y(i)=input(' ')
end
sumx=0;sumx2=0;sumy=0;sumxy=0
for i=1:n
    sumx=sumx+x(i);
    sumx2=sumx2+x(i)*x(i);
    sumy=sumy+y(i);
    sumxy=sumxy+x(i)*y(i);
end
A=[sumx n; sumx2 sumx];
B=[sumy;sumxy];
C=inv(A)*B
printf('The fitted line is y=(%g)x+(%g)',C(1,1),C(2,1))
```

Scilab code to fit a parabola to given set of data points

```
clc;clear;close;
n=input('Enter the no. of pairs of values (x,y):')
disp('Enter the valyes of x:')
for i=1:n
    x(i)=input(' ')
end
disp('Enter the corrersponding valyes of y:')
for i=1:n
    y(i)=input(' ')
end
sumx=0;sumx2=0;sumx3=0;sumx4=0;sumy=0;sumxy=0;sumx2y=0;
for i=1:n
    sumx=sumx+x(i);
    sumx2=sumx2+x(i)^2;
    sumx3=sumx3+x(i)^3;
    sumx4=sumx4+x(i)^4;
    sumy=sumy+y(i);
    sumxy=sumxy+x(i)*y(i);
    sumx2y=sumx2y+x(i)^2*y(i);
end
A=[n sumx sumx2; sumx sumx2 sumx3;sumx2 sumx3 sumx4];
B=[sumy;sumxy;sumx2y];
C=inv(A)*B
printf('The fitted parabola is y=(%g)+(%g)x+(%g)x^2',C(1,1),C(2,1),C(3,1))
```

Scilab code to find roots of an equation using Bisection method

```
//Bisection method
clc
deff('y=f(x)','y=x^3+x^2-3*x-3')
a=input("enter initial interval value: ");
b=input("enter final interval value: ");
fa = f(a);   //compute initial values of f(a) and f(b)
fb = f(b);
if  sign(fa) == sign(fb)    // sanity check: f(a) and f(b) must have different signs
disp('f must have different signs at the endpoints a and b')
error
end
e=input(" answer correct upto : ");
iter=0;
printf('Iteration\ta\t\tb\t\troot\t\tf(root)\n')
while abs(a-b)>2*e
   root=(a+b)/2
   printf('   %i\t\t%f\t%f\t%f\t%f\n',iter,a,b,root,f(root))
   if f(root)*f(a)>0
      a=root
      else
     b=root
end
iter=iter+1
end
printf('\n\nThe solution of given equation is %f after %i Iterations',root,iter-1)
```

Scilab code to find roots of an equation using N-R method

```
// newton raphson method x(n+1)= x(n)-f(x(n))/df(x(n))
clc
deff('y=f(x)','y=x^3+x^2-3*x-3')
deff('y=df(x)','y=3*x^2+2*x-3')
x(1)=input('Enter Initial Guess:');
e=input(" answer correct upto : ");
for i=1:100
x(i+1)=x(i)-((f(x(i))/df(x(i))));
err(i)=abs((x(i+1)-x(i))/x(i));
if err(i)<e
```

```
break;
end
end
printf('the solution is %f',x(i))
```

Scilab code to find roots of an equation using Regula Falsi method

```
// regula falsi method
clc
deff('y=f(x)','y=x^3+x^2-3*x-3')
a=input("enter initial interval value: ");
b=input("enter final interval value: ");
e=input(" answer correct upto : ");
for i=2:100
if f(b)>f(a)
   xn=b-((f(b)*(b-a))/(f(b)-f(a)));
else
   xn=a-((f(a)*(a-b))/(f(a)-f(b)));
end
if f(b)*f(xn)<0
   a=xn;
else
   b=xn;
end
if f(a)*f(xn)<0
   b=xn;
else
   a=xn;
end
xnew(1)=0;
xnew(i)=xn;
if abs((xnew(i)-xnew(i-1))/xnew(i))<e;
   break;
   end
end
printf('Solution using Regula Falsi method is %f',xnew(i))
```

Scilab Code for numerical integration using Simpson's 1/3 rule

```
// SIMPSON'S 1/3 RULE
clc
```

```
deff('y=f(x)','y=x/(x^3+10)');
x1=0;
x2=1;
n=4;
h=(x2-x1)/n;
x(1)=x1;
sum=f(x1);
for i=2:n
    x(i)=x(i-1)+h;
end
for j=2:2:n
    sum=sum+4*f(x(j));
end
for k=3:2:n
    sum=sum+2*f(x(k));
end
sum=sum+f(x2);
value=sum*h/3;
printf('\nThe value of the integral using SIMPSONS 1/3RD RULE is %f',value)
```

Scilab Code for numerical integration using Simpson's 3/8 rule

```
// Simpsom's 3/8 rule
clc
deff('y=f(x)','y=x/(x^3+10)');
x1=0;
x2=1;
n=4;
h=(x2-x1)/n;
x(1)=x1;
sum=f(x1);
for i=2:n
    x(i)=x(i-1)+h;
end
for j=2:3:n
    sum=sum+3*f(x(j));
end
for k=3:3:n
    sum=sum+3*f(x(k));
```

end
for l=4:3:n
   sum=sum+2*f(x(l));
end
sum=sum+f(x2);
value=sum*3*h/8;
printf('\nThe value of the integral SIMPSONS 3/8th RULE is %f ',value)

Scilab Code for solving initial value problems using Euler's method

// If f is a Scilab function, its calling sequence must be ydot = f(t,y)

//where t is a real scalar (the time) and y is a real vector (the state) and ydot is a real vector (the first order derivative dy/dt).

1. // Solution of Initial value problem dy/dt = y^2-y*sin(t)+cos(t), y(0) = 0

function ydot=fun(t, y)
   ydot=y^2-y*sin(t)+cos(t)
endfunction
y0=0;
t0=0;
t=0:0.1:%pi;
y = ode(y0,t0,t,fun);
plot(t,y)

2. // Solution of Initial value problem dy/dt = sin(y)/(1 + t^2), y(0) = 1

function ydot=fct(t, y)
    ydot = sin(y)/(1 + t^2)
    endfunction
    t=[0:.1:10];
    y = ode(1, 0, t, fct)
    clf; plot(t, y)

Scilab Code for solving initial value problem using Runge Kutta method

// RUNGE KUTTA METHOD
clc
function ydot=f(x, y)
    ydot =x+y^2
    endfunction
x1=0;
y1=1;
h=0.1;
x(1)=x1;
y(1) = y1;

```
for i=1:2
    k_1 = h*f(x(i),y(i));
    k_2 = h*f(x(i)+0.5*h,y(i)+0.5*k_1);
    k_3 = h*f((x(i)+0.5*h),(y(i)+0.5*k_2));
    k_4 = h*f((x(i)+h),(y(i)+k_3));
    k   = (1/6)*(k_1 +2*k_2 +2*k_3 +k_4);
    y(i+1)= y(i)+ k;
    printf('\n The value of y at x=%f is %f ', i*h,y(i+1))
    x(i+1)=x(1)+ i*h;
end
```

Table1: Area Under Standard Normal Curve

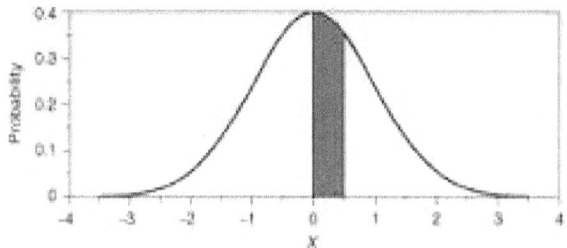

Area under the Normal Curve from 0 to X

X	0.00	0.01	0.02	0.03	0.04	0.05	0.06	0.07	0.08	0.09
0.0	0.00000	0.00399	0.00798	0.01197	0.01595	0.01994	0.02392	0.02790	0.03188	0.03586
0.1	0.03983	0.04380	0.04776	0.05172	0.05567	0.05962	0.06356	0.06749	0.07142	0.07535
0.2	0.07926	0.08317	0.08706	0.09095	0.09483	0.09871	0.10257	0.10642	0.11026	0.11409
0.3	0.11791	0.12172	0.12552	0.12930	0.13307	0.13683	0.14058	0.14431	0.14803	0.15173
0.4	0.15542	0.15910	0.16276	0.16640	0.17003	0.17364	0.17724	0.18082	0.18439	0.18793
0.5	0.19146	0.19497	0.19847	0.20194	0.20540	0.20884	0.21226	0.21566	0.21904	0.22240
0.6	0.22575	0.22907	0.23237	0.23565	0.23891	0.24215	0.24537	0.24857	0.25175	0.25490
0.7	0.25804	0.26115	0.26424	0.26730	0.27035	0.27337	0.27637	0.27935	0.28230	0.28524
0.8	0.28814	0.29103	0.29389	0.29673	0.29955	0.30234	0.30511	0.30785	0.31057	0.31327
0.9	0.31594	0.31859	0.32121	0.32381	0.32639	0.32894	0.33147	0.33398	0.33646	0.33891
1.0	0.34134	0.34375	0.34614	0.34849	0.35083	0.35314	0.35543	0.35769	0.35993	0.36214
1.1	0.36433	0.36650	0.36864	0.37076	0.37286	0.37493	0.37698	0.37900	0.38100	0.38298
1.2	0.38493	0.38686	0.38877	0.39065	0.39251	0.39435	0.39617	0.39796	0.39973	0.40147
1.3	0.40320	0.40490	0.40658	0.40824	0.40988	0.41149	0.41308	0.41466	0.41621	0.41774
1.4	0.41924	0.42073	0.42220	0.42364	0.42507	0.42647	0.42785	0.42922	0.43056	0.43189
1.5	0.43319	0.43448	0.43574	0.43699	0.43822	0.43943	0.44062	0.44179	0.44295	0.44408
1.6	0.44520	0.44630	0.44738	0.44845	0.44950	0.45053	0.45154	0.45254	0.45352	0.45449
1.7	0.45543	0.45637	0.45728	0.45818	0.45907	0.45994	0.46080	0.46164	0.46246	0.46327
1.8	0.46407	0.46485	0.46562	0.46638	0.46712	0.46784	0.46856	0.46926	0.46995	0.47062
1.9	0.47128	0.47193	0.47257	0.47320	0.47381	0.47441	0.47500	0.47558	0.47615	0.47670
2.0	0.47725	0.47778	0.47831	0.47882	0.47932	0.47982	0.48030	0.48077	0.48124	0.48169
2.1	0.48214	0.48257	0.48300	0.48341	0.48382	0.48422	0.48461	0.48500	0.48537	0.48574
2.2	0.48610	0.48645	0.48679	0.48713	0.48745	0.48778	0.48809	0.48840	0.48870	0.48899
2.3	0.48928	0.48956	0.48983	0.49010	0.49036	0.49061	0.49086	0.49111	0.49134	0.49158
2.4	0.49180	0.49202	0.49224	0.49245	0.49266	0.49286	0.49305	0.49324	0.49343	0.49361
2.5	0.49379	0.49396	0.49413	0.49430	0.49446	0.49461	0.49477	0.49492	0.49506	0.49520
2.6	0.49534	0.49547	0.49560	0.49573	0.49585	0.49598	0.49609	0.49621	0.49632	0.49643
2.7	0.49653	0.49664	0.49674	0.49683	0.49693	0.49702	0.49711	0.49720	0.49728	0.49736
2.8	0.49744	0.49752	0.49760	0.49767	0.49774	0.49781	0.49788	0.49795	0.49801	0.49807
2.9	0.49813	0.49819	0.49825	0.49831	0.49836	0.49841	0.49846	0.49851	0.49856	0.49861
3.0	0.49865	0.49869	0.49874	0.49878	0.49882	0.49886	0.49889	0.49893	0.49896	0.49900
3.1	0.49903	0.49906	0.49910	0.49913	0.49916	0.49918	0.49921	0.49924	0.49926	0.49929
3.2	0.49931	0.49934	0.49936	0.49938	0.49940	0.49942	0.49944	0.49946	0.49948	0.49950
3.3	0.49952	0.49953	0.49955	0.49957	0.49958	0.49960	0.49961	0.49962	0.49964	0.49965
3.4	0.49966	0.49968	0.49969	0.49970	0.49971	0.49972	0.49973	0.49974	0.49975	0.49976
3.5	0.49977	0.49978	0.49978	0.49979	0.49980	0.49981	0.49981	0.49982	0.49983	0.49983
3.6	0.49984	0.49985	0.49985	0.49986	0.49986	0.49987	0.49987	0.49988	0.49988	0.49989
3.7	0.49989	0.49990	0.49990	0.49990	0.49991	0.49991	0.49992	0.49992	0.49992	0.49992
3.8	0.49993	0.49993	0.49993	0.49994	0.49994	0.49994	0.49994	0.49995	0.49995	0.49995
3.9	0.49995	0.49995	0.49996	0.49996	0.49996	0.49996	0.49996	0.49996	0.49997	0.49997
4.0	0.49997	0.49997	0.49997	0.49997	0.49997	0.49997	0.49998	0.49998	0.49998	0.49998

Table2: Standard Normal Cumulative Probability Table

Cumulative probabilities for Z - values are given in the following table:

Z	.00	.01	.02	.03	.04	.05	.06	.07	.08	.09
0.0	.5000	.5040	.5080	.5120	.5160	.5199	.5239	.5279	.5319	.5359
0.1	.5398	.5438	.5478	.5517	.5557	.5596	.5636	.5675	.5714	.5753
0.2	.5793	.5832	.5871	.5910	.5948	.5987	.6026	.6064	.6103	.6141
0.3	.6179	.6217	.6255	.6293	.6331	.6368	.6406	.6443	.6480	.6517
0.4	.6554	.6591	.6628	.6664	.6700	.6736	.6772	.6808	.6844	.6879
0.5	.6915	.6950	.6985	.7019	.7054	.7088	.7123	.7157	.7190	.7224
0.6	.7257	.7291	.7324	.7357	.7389	.7422	.7454	.7486	.7517	.7549
0.7	.7580	.7611	.7642	.7673	.7704	.7734	.7764	.7794	.7823	.7852
0.8	.7881	.7910	.7939	.7967	.7995	.8023	.8051	.8078	.8106	.8133
0.9	.8159	.8186	.8212	.8238	.8264	.8289	.8315	.8340	.8365	.8389
1.0	.8413	.8438	.8461	.8485	.8508	.8531	.8554	.8577	.8599	.8621
1.1	.8643	.8665	.8686	.8708	.8729	.8749	.8770	.8790	.8810	.8830
1.2	.8849	.8869	.8888	.8907	.8925	.8944	.8962	.8980	.8997	.9015
1.3	.9032	.9049	.9066	.9082	.9099	.9115	.9131	.9147	.9162	.9177
1.4	.9192	.9207	.9222	.9236	.9251	.9265	.9279	.9292	.9306	.9319
1.5	.9332	.9345	.9357	.9370	.9382	.9394	.9406	.9418	.9429	.9441
1.6	.9452	.9463	.9474	.9484	.9495	.9505	.9515	.9525	.9535	.9545
1.7	.9554	.9564	.9573	.9582	.9591	.9599	.9608	.9616	.9625	.9633
1.8	.9641	.9649	.9656	.9664	.9671	.9678	.9686	.9693	.9699	.9706
1.9	.9713	.9719	.9726	.9732	.9738	.9744	.9750	.9756	.9761	.9767
2.0	.9772	.9778	.9783	.9788	.9793	.9798	.9803	.9808	.9812	.9817
2.1	.9821	.9826	.9830	.9834	.9838	.9842	.9846	.9850	.9854	.9857
2.2	.9861	.9864	.9868	.9871	.9875	.9878	.9881	.9884	.9887	.9890
2.3	.9893	.9896	.9898	.9901	.9904	.9906	.9909	.9911	.9913	.9916
2.4	.9918	.9920	.9922	.9925	.9927	.9929	.9931	.9932	.9934	.9936
2.5	.9938	.9940	.9941	.9943	.9945	.9946	.9948	.9949	.9951	.9952
2.6	.9953	.9955	.9956	.9957	.9959	.9960	.9961	.9962	.9963	.9964
2.7	.9965	.9966	.9967	.9968	.9969	.9970	.9971	.9972	.9973	.9974
2.8	.9974	.9975	.9976	.9977	.9977	.9978	.9979	.9979	.9980	.9981
2.9	.9981	.9982	.9982	.9983	.9984	.9984	.9985	.9985	.9986	.9986
3.0	.9987	.9987	.9987	.9988	.9988	.9989	.9989	.9989	.9990	.9990

Table3: *t*-table

cum. prob	$t_{.50}$	$t_{.75}$	$t_{.80}$	$t_{.85}$	$t_{.90}$	$t_{.95}$	$t_{.975}$	$t_{.99}$	$t_{.995}$	$t_{.999}$	$t_{.9995}$
one-tail	0.50	0.25	0.20	0.15	0.10	0.05	0.025	0.01	0.005	0.001	0.0005
two-tails	1.00	0.50	0.40	0.30	0.20	0.10	0.05	0.02	0.01	0.002	0.001
df											
1	0.000	1.000	1.376	1.963	3.078	6.314	12.71	31.82	63.66	318.31	636.62
2	0.000	0.816	1.061	1.386	1.886	2.920	4.303	6.965	9.925	22.327	31.599
3	0.000	0.765	0.978	1.250	1.638	2.353	3.182	4.541	5.841	10.215	12.924
4	0.000	0.741	0.941	1.190	1.533	2.132	2.776	3.747	4.604	7.173	8.610
5	0.000	0.727	0.920	1.156	1.476	2.015	2.571	3.365	4.032	5.893	6.869
6	0.000	0.718	0.906	1.134	1.440	1.943	2.447	3.143	3.707	5.208	5.959
7	0.000	0.711	0.896	1.119	1.415	1.895	2.365	2.998	3.499	4.785	5.408
8	0.000	0.706	0.889	1.108	1.397	1.860	2.306	2.896	3.355	4.501	5.041
9	0.000	0.703	0.883	1.100	1.383	1.833	2.262	2.821	3.250	4.297	4.781
10	0.000	0.700	0.879	1.093	1.372	1.812	2.228	2.764	3.169	4.144	4.587
11	0.000	0.697	0.876	1.088	1.363	1.796	2.201	2.718	3.106	4.025	4.437
12	0.000	0.695	0.873	1.083	1.356	1.782	2.179	2.681	3.055	3.930	4.318
13	0.000	0.694	0.870	1.079	1.350	1.771	2.160	2.650	3.012	3.852	4.221
14	0.000	0.692	0.868	1.076	1.345	1.761	2.145	2.624	2.977	3.787	4.140
15	0.000	0.691	0.866	1.074	1.341	1.753	2.131	2.602	2.947	3.733	4.073
16	0.000	0.690	0.865	1.071	1.337	1.746	2.120	2.583	2.921	3.686	4.015
17	0.000	0.689	0.863	1.069	1.333	1.740	2.110	2.567	2.898	3.646	3.965
18	0.000	0.688	0.862	1.067	1.330	1.734	2.101	2.552	2.878	3.610	3.922
19	0.000	0.688	0.861	1.066	1.328	1.729	2.093	2.539	2.861	3.579	3.883
20	0.000	0.687	0.860	1.064	1.325	1.725	2.086	2.528	2.845	3.552	3.850
21	0.000	0.686	0.859	1.063	1.323	1.721	2.080	2.518	2.831	3.527	3.819
22	0.000	0.686	0.858	1.061	1.321	1.717	2.074	2.508	2.819	3.505	3.792
23	0.000	0.685	0.858	1.060	1.319	1.714	2.069	2.500	2.807	3.485	3.768
24	0.000	0.685	0.857	1.059	1.318	1.711	2.064	2.492	2.797	3.467	3.745
25	0.000	0.684	0.856	1.058	1.316	1.708	2.060	2.485	2.787	3.450	3.725
26	0.000	0.684	0.856	1.058	1.315	1.706	2.056	2.479	2.779	3.435	3.707
27	0.000	0.684	0.855	1.057	1.314	1.703	2.052	2.473	2.771	3.421	3.690
28	0.000	0.683	0.855	1.056	1.313	1.701	2.048	2.467	2.763	3.408	3.674
29	0.000	0.683	0.854	1.055	1.311	1.699	2.045	2.462	2.756	3.396	3.659
30	0.000	0.683	0.854	1.055	1.310	1.697	2.042	2.457	2.750	3.385	3.646
40	0.000	0.681	0.851	1.050	1.303	1.684	2.021	2.423	2.704	3.307	3.551
60	0.000	0.679	0.848	1.045	1.296	1.671	2.000	2.390	2.660	3.232	3.460
80	0.000	0.678	0.846	1.043	1.292	1.664	1.990	2.374	2.639	3.195	3.416
100	0.000	0.677	0.845	1.042	1.290	1.660	1.984	2.364	2.626	3.174	3.390
1000	0.000	0.675	0.842	1.037	1.282	1.646	1.962	2.330	2.581	3.098	3.300
Z	0.000	0.674	0.842	1.036	1.282	1.645	1.960	2.326	2.576	3.090	3.291
	0%	50%	60%	70%	80%	90%	95%	98%	99%	99.8%	99.9%

Confidence Level

Table 4: Chi-Square (χ^2) Distribution Table

The shaded area is equal to α for $\chi^2 = \chi^2_\alpha$.

df	$\chi^2_{.995}$	$\chi^2_{.990}$	$\chi^2_{.975}$	$\chi^2_{.950}$	$\chi^2_{.900}$	$\chi^2_{.100}$	$\chi^2_{.050}$	$\chi^2_{.025}$	$\chi^2_{.010}$	$\chi^2_{.005}$
1	0.000	0.000	0.001	0.004	0.016	2.706	3.841	5.024	6.635	7.879
2	0.010	0.020	0.051	0.103	0.211	4.605	5.991	7.378	9.210	10.597
3	0.072	0.115	0.216	0.352	0.584	6.251	7.815	9.348	11.345	12.838
4	0.207	0.297	0.484	0.711	1.064	7.779	9.488	11.143	13.277	14.860
5	0.412	0.554	0.831	1.145	1.610	9.236	11.070	12.833	15.086	16.750
6	0.676	0.872	1.237	1.635	2.204	10.645	12.592	14.449	16.812	18.548
7	0.989	1.239	1.690	2.167	2.833	12.017	14.067	16.013	18.475	20.278
8	1.344	1.646	2.180	2.733	3.490	13.362	15.507	17.535	20.090	21.955
9	1.735	2.088	2.700	3.325	4.168	14.684	16.919	19.023	21.666	23.589
10	2.156	2.558	3.247	3.940	4.865	15.987	18.307	20.483	23.209	25.188
11	2.603	3.053	3.816	4.575	5.578	17.275	19.675	21.920	24.725	26.757
12	3.074	3.571	4.404	5.226	6.304	18.549	21.026	23.337	26.217	28.300
13	3.565	4.107	5.009	5.892	7.042	19.812	22.362	24.736	27.688	29.819
14	4.075	4.660	5.629	6.571	7.790	21.064	23.685	26.119	29.141	31.319
15	4.601	5.229	6.262	7.261	8.547	22.307	24.996	27.488	30.578	32.801
16	5.142	5.812	6.908	7.962	9.312	23.542	26.296	28.845	32.000	34.267
17	5.697	6.408	7.564	8.672	10.085	24.769	27.587	30.191	33.409	35.718
18	6.265	7.015	8.231	9.390	10.865	25.989	28.869	31.526	34.805	37.156
19	6.844	7.633	8.907	10.117	11.651	27.204	30.144	32.852	36.191	38.582
20	7.434	8.260	9.591	10.851	12.443	28.412	31.410	34.170	37.566	39.997
21	8.034	8.897	10.283	11.591	13.240	29.615	32.671	35.479	38.932	41.401
22	8.643	9.542	10.982	12.338	14.041	30.813	33.924	36.781	40.289	42.796
23	9.260	10.196	11.689	13.091	14.848	32.007	35.172	38.076	41.638	44.181
24	9.886	10.856	12.401	13.848	15.659	33.196	36.415	39.364	42.980	45.559
25	10.520	11.524	13.120	14.611	16.473	34.382	37.652	40.646	44.314	46.928
26	11.160	12.198	13.844	15.379	17.292	35.563	38.885	41.923	45.642	48.290
27	11.808	12.879	14.573	16.151	18.114	36.741	40.113	43.195	46.963	49.645
28	12.461	13.565	15.308	16.928	18.939	37.916	41.337	44.461	48.278	50.993
29	13.121	14.256	16.047	17.708	19.768	39.087	42.557	45.722	49.588	52.336
30	13.787	14.953	16.791	18.493	20.599	40.256	43.773	46.979	50.892	53.672
40	20.707	22.164	24.433	26.509	29.051	51.805	55.758	59.342	63.691	66.766
50	27.991	29.707	32.357	34.764	37.689	63.167	67.505	71.420	76.154	79.490
60	35.534	37.485	40.482	43.188	46.459	74.397	79.082	83.298	88.379	91.952
70	43.275	45.442	48.758	51.739	55.329	85.527	90.531	95.023	100.425	104.215
80	51.172	53.540	57.153	60.391	64.278	96.578	101.879	106.629	112.329	116.321
90	59.196	61.754	65.647	69.126	73.291	107.565	113.145	118.136	124.116	128.299
100	67.328	70.065	74.222	77.929	82.358	118.498	124.342	129.561	135.807	140.169

Table 5: Critical Values of the F Distribution with Alpha Level of 0.05

v_2 \ v_1	1	2	3	4	5	6	7	8	9	10	12	15	20	24	30	40	60	120	∞
1	161.4	199.5	215.7	224.6	230.2	234.0	236.8	238.9	240.5	241.9	243.9	245.9	248.0	249.1	250.1	251.1	252.2	253.3	254.3
2	18.51	19.00	19.16	19.25	19.30	19.33	19.35	19.37	19.38	19.40	19.41	19.43	19.45	19.45	19.46	19.47	19.48	19.49	19.50
3	10.13	9.55	9.28	9.12	9.01	8.94	8.89	8.85	8.81	8.79	8.74	8.70	8.66	8.64	8.62	8.59	8.57	8.55	8.53
4	7.71	6.94	6.59	6.39	6.26	6.16	6.09	6.04	6.00	5.96	5.91	5.86	5.80	5.77	5.75	5.72	5.69	5.66	5.63
5	6.61	5.79	5.41	5.19	5.05	4.95	4.88	4.82	4.77	4.74	4.68	4.62	4.56	4.53	4.50	4.46	4.43	4.40	4.36
6	5.99	5.14	4.76	4.53	4.39	4.28	4.21	4.15	4.10	4.06	4.00	3.94	3.87	3.84	3.81	3.77	3.74	3.70	3.67
7	5.59	4.74	4.35	4.12	3.97	3.87	3.79	3.73	3.68	3.64	3.57	3.51	3.44	3.41	3.38	3.34	3.30	3.27	3.23
8	5.32	4.46	4.07	3.84	3.69	3.58	3.50	3.44	3.39	3.35	3.28	3.22	3.15	3.12	3.08	3.04	3.01	2.97	2.93
9	5.12	4.26	3.86	3.63	3.48	3.37	3.29	3.23	3.18	3.14	3.07	3.01	2.94	2.90	2.86	2.83	2.79	2.75	2.71
10	4.96	4.10	3.71	3.48	3.33	3.22	3.14	3.07	3.02	2.98	2.91	2.85	2.77	2.74	2.70	2.66	2.62	2.58	2.54
11	4.84	3.98	3.59	3.36	3.20	3.09	3.01	2.95	2.90	2.85	2.79	2.72	2.65	2.61	2.57	2.53	2.49	2.45	2.40
12	4.75	3.89	3.49	3.26	3.11	3.00	2.91	2.85	2.80	2.75	2.69	2.62	2.54	2.51	2.47	2.43	2.38	2.34	2.30
13	4.67	3.81	3.41	3.18	3.03	2.92	2.83	2.77	2.71	2.67	2.60	2.53	2.46	2.42	2.38	2.34	2.30	2.25	2.21
14	4.60	3.74	3.34	3.11	2.96	2.85	2.76	2.70	2.65	2.60	2.53	2.46	2.39	2.35	2.31	2.27	2.22	2.18	2.13
15	4.54	3.68	3.29	3.06	2.90	2.79	2.71	2.64	2.59	2.54	2.48	2.40	2.33	2.29	2.25	2.20	2.16	2.11	2.07
16	4.49	3.63	3.24	3.01	2.85	2.74	2.66	2.59	2.54	2.49	2.42	2.35	2.28	2.24	2.19	2.15	2.11	2.06	2.01
17	4.45	3.59	3.20	2.96	2.81	2.70	2.61	2.55	2.49	2.45	2.38	2.31	2.23	2.19	2.15	2.10	2.06	2.01	1.96
18	4.41	3.55	3.16	2.93	2.77	2.66	2.58	2.51	2.46	2.41	2.34	2.27	2.19	2.15	2.11	2.06	2.02	1.97	1.92
19	4.38	3.52	3.13	2.90	2.74	2.63	2.54	2.48	2.42	2.38	2.31	2.23	2.16	2.11	2.07	2.03	1.98	1.93	1.88
20	4.35	3.49	3.10	2.87	2.71	2.60	2.51	2.45	2.39	2.35	2.28	2.20	2.12	2.08	2.04	1.99	1.95	1.90	1.84
21	4.32	3.47	3.07	2.84	2.68	2.57	2.49	2.42	2.37	2.32	2.25	2.18	2.10	2.05	2.01	1.96	1.92	1.87	1.81
22	4.30	3.44	3.05	2.82	2.66	2.55	2.46	2.40	2.34	2.30	2.23	2.15	2.07	2.03	1.98	1.94	1.89	1.84	1.78
23	4.28	3.42	3.03	2.80	2.64	2.53	2.44	2.37	2.32	2.27	2.20	2.13	2.05	2.01	1.96	1.91	1.86	1.81	1.76
24	4.26	3.40	3.01	2.78	2.62	2.51	2.42	2.36	2.30	2.25	2.18	2.11	2.03	1.98	1.94	1.89	1.84	1.79	1.73
25	4.24	3.39	2.99	2.76	2.60	2.49	2.40	2.34	2.28	2.24	2.16	2.09	2.01	1.96	1.92	1.87	1.82	1.77	1.71
26	4.23	3.37	2.98	2.74	2.59	2.47	2.39	2.32	2.27	2.22	2.15	2.07	1.99	1.95	1.90	1.85	1.80	1.75	1.69
27	4.21	3.35	2.96	2.73	2.57	2.46	2.37	2.31	2.25	2.20	2.13	2.06	1.97	1.93	1.88	1.84	1.79	1.73	1.67
28	4.20	3.34	2.95	2.71	2.56	2.45	2.36	2.29	2.24	2.19	2.12	2.04	1.96	1.91	1.87	1.82	1.77	1.71	1.65
29	4.18	3.33	2.93	2.70	2.55	2.43	2.35	2.28	2.22	2.18	2.10	2.03	1.94	1.90	1.85	1.81	1.75	1.70	1.64
30	4.17	3.32	2.92	2.69	2.53	2.42	2.33	2.27	2.21	2.16	2.09	2.01	1.93	1.89	1.84	1.79	1.74	1.68	1.62
40	4.08	3.23	2.84	2.61	2.45	2.34	2.25	2.18	2.12	2.08	2.00	1.92	1.84	1.79	1.74	1.69	1.64	1.58	1.51
60	4.00	3.15	2.76	2.53	2.37	2.25	2.17	2.10	2.04	1.99	1.92	1.84	1.75	1.70	1.65	1.59	1.53	1.47	1.39
120	3.92	3.07	2.68	2.45	2.29	2.17	2.09	2.02	1.96	1.91	1.83	1.75	1.66	1.61	1.55	1.50	1.43	1.35	1.25
∞	3.84	3.00	2.60	2.37	2.21	2.10	2.01	1.94	1.88	1.83	1.75	1.67	1.57	1.52	1.46	1.39	1.32	1.22	1.00

www.ingramcontent.com/pod-product-compliance
Lightning Source LLC
Chambersburg PA
CBHW062212220526
45471CB00009B/3174